Methods in Molecular Biology

Series Editor
John M. Walker
School of Life and Medical Sciences
University of Hertfordshire
Hatfield, Hertfordshire, UK

For further volumes:
http://www.springer.com/series/7651

For over 35 years, biological scientists have come to rely on the research protocols and methodologies in the critically acclaimed *Methods in Molecular Biology* series. The series was the first to introduce the step-by-step protocols approach that has become the standard in all biomedical protocol publishing. Each protocol is provided in readily-reproducible step-by step fashion, opening with an introductory overview, a list of the materials and reagents needed to complete the experiment, and followed by a detailed procedure that is supported with a helpful notes section offering tips and tricks of the trade as well as troubleshooting advice. These hallmark features were introduced by series editor Dr. John Walker and constitute the key ingredient in each and every volume of the *Methods in Molecular Biology* series. Tested and trusted, comprehensive and reliable, all protocols from the series are indexed in PubMed.

Transmembrane β-Barrel Proteins

Methods and Protocols

Edited by

Raffaele Ieva

Laboratoire de Microbiologie et Génétique Moléculaires, Centre de Biologie Intégrative, CNRS, Université de Toulouse, Toulouse, France

Humana Press

Editor
Raffaele Ieva
Laboratoire de Microbiologie et
Génétique Moléculaires
Centre de Biologie Intégrative, CNRS
Université de Toulouse
Toulouse, France

ISSN 1064-3745 ISSN 1940-6029 (electronic)
Methods in Molecular Biology
ISBN 978-1-0716-3733-3 ISBN 978-1-0716-3734-0 (eBook)
https://doi.org/10.1007/978-1-0716-3734-0

This Humana imprint is published by the registered company Springer Science+Business Media, LLC, part of Springer Nature.
The registered company address is: 1 New York Plaza, New York, NY 10004, U.S.A.

Paper in this product is recyclable.

Preface

Transmembrane proteins play crucial roles in cell physiology as they fulfill a variety of housekeeping or regulatory functions including compartment biogenesis, bioenergetics, signaling, and interactions with the cell exterior or with other cell compartments. These membrane-resident proteins integrate lipid bilayers mainly by means of two distinct structural motifs, transmembrane α-helices or transmembrane β-barrels. The presence of these two distinct fold signatures contributes to organize the pathways of transmembrane protein sorting in the cell as they undergo different assembly processes. For instance, proteins with an α-helical fold are typically inserted into the cytoplasmic membrane of bacteria or into the membranes of the endoplasmic reticulum and other organelles of eukaryotic cells. In contrast, transmembrane β-barrel proteins follow a different pathway. In diderm bacteria, transmembrane β-barrel proteins are transported unfolded across the inner membrane to be subsequently assembled into the outer membrane. Via a similarly selective sorting pathway, transmembrane β-barrel proteins of eukaryotic cells are specifically assembled into the outer membrane of cell organelles of endosymbiotic origin such as mitochondria and plastids.

Although representing the least abundant of the two groups, transmembrane β-barrels are nonetheless of primordial importance as they ensure the biogenesis and maintenance of the outer membranes surrounding bacteria, mitochondria, and plastids. Here, β-barrels also mediate multiple transactions with the surrounding environment, including the exchange of solutes, metabolites, nucleic acids, lipids, and proteins. During the past two decades, a great deal of research has tremendously advanced our understanding of transmembrane β-barrel proteins, elucidating their biogenesis and organization, and dissecting the large variety of their functions. Furthermore, therapy-discovery studies are exploiting specific properties of transmembrane β-barrel proteins to develop new inhibitors. Finally, given their properties as channels and transporters, some transmembrane β-barrels represent relevant biotechnological tools that are being optimized by computational design and protein engineering.

This book collects a broad range of experimental strategies and protocols useful to the expression, assembly, characterization, and exploitation of transmembrane β-barrel proteins. The book opens with an introductory review on β-barrel protein structures, functions, and industrial applications (Chapter 1). Then Chapters 2 and 3 illustrate methods of transmembrane β-barrel protein production in bacteria and cell sub-fractionation. Biochemical methodologies to study the assembly of β-barrel proteins (Chapters 4, 5, 6, and 7) are followed by protocols aiming at characterizing the landscape of transmembrane β-barrel protein interactions with other cellular factors (Chapters 8, 9, 10, and 11). The following chapters (Chapters 12, 13, and 14) describe methods to dissect the processes of protein transport in mitochondria and the dynamics of β-barrel proteins in the outer membrane of these organelles. The next part of the book (Chapters 15, 16, 17, and 18) focuses on different approaches for structural characterization and determination. These topics are complemented by a second review on recent developments on in silico structural modeling and molecular dynamics simulations of transmembrane β-barrel proteins and their assemblies

(Chapter 19), as well as a protocol example of molecular modeling and experimental validation (Chapter 20). The book closes with two chapters on industrial exploitation of β-barrel proteins, focusing on the rational design of nanopores as single-molecule analytical tools (Chapter 21) and on the screening of β-barrel protein inhibitors that can potentially function as novel antibiotics (Chapter 22).

Toulouse, France *Raffaele Ieva*

Contents

Contributors

ERIKO AOKI • *Frontier Science Research Center, University of Miyazaki, Kiyotake, Miyazaki, Japan*
THOMAS BECKER • *Institute for Biochemistry and Molecular Biology, Faculty of Medicine, University of Bonn, Bonn, Germany*
RITUPARNA BHOWMIK • *Institute for Biochemistry and Molecular Biology, Faculty of Medicine, University of Bonn, Bonn, Germany*
ISTVAN BOTOS • *Laboratory of Molecular Biology, National Institute of Diabetes & Digestive & Kidney Diseases, National Institutes of Health, Bethesda, MD, USA*
EMILY M. BOUZAN • *Chemistry Department, College of the Holy Cross, Worcester, MA, USA*
FABIAN DEN BRAVE • *Institute for Biochemistry and Molecular Biology, Faculty of Medicine, University of Bonn, Bonn, Germany*
SUSAN K. BUCHANAN • *Laboratory of Molecular Biology, National Institute of Diabetes & Digestive & Kidney Diseases, National Institutes of Health, Bethesda, MD, USA*
ANTONIO N. CALABRESE • *Astbury Centre for Structural Molecular Biology, School of Molecular and Cellular Biology, Faculty of Biological Sciences, University of Leeds, Leeds, UK*
MATTHIEU CHAVENT • *Laboratoire de Microbiologie et Génétique Moléculaires, Centre de Biologie Intégrative (CBI), Université de Toulouse, CNRS, UPS, Toulouse, France*
LAURENCE CLEENEWERK • *Section Molecular Microbiology, A-life Department and Amsterdam Institute of Molecular and Life Sciences, Vrije Universiteit, Amsterdam, The Netherlands*
IAN COLLINSON • *School of Biochemistry, University of Bristol, Bristol, UK*
REBECCA CONNERS • *Living Systems Institute, University of Exeter, Exeter, UK; Faculty of Health and Life Sciences, University of Exeter, Exeter, UK*
FEDERICO CORONA • *Genome Biology Unit, European Molecular Biology Laboratory, Heidelberg, Germany*
MICHAEL A. CURTIS • *Centre for Host-Microbiome Interactions, Faculty of Dentistry, Oral & Craniofacial Sciences, King's College London, London, UK*
MATTHEW THOMAS DOYLE • *Sydney Infectious Diseases Institute, The University of Sydney, Darlington, NSW, Australia; School of Medical Sciences, Faculty of Medicine and Health, The University of Sydney, Darlington, NSW, Australia*
ANNA L. DUNCAN • *Department of Chemistry, Aarhus University, Aarhus, Denmark*
RACHAEL DUODU • *Antimicrobial Resistance, Omics and Microbiota Group, Department of Biosciences, Nottingham Trent University, Nottingham, UK*
FRANCK DUONG VAN HOA • *Department of Biochemistry and Molecular Biology, Faculty of Medicine, Life Sciences Institute, University of British Columbia, Vancouver, BC, Canada*
HOLLY C. FORD • *School of Biochemistry, University of Bristol, Bristol, UK*
YA GAO • *School of Mathematics, Physics and Statistics, Shanghai University of Engineering Science, Shanghai, China; Department of Biological Sciences, Department of Chemistry, Department of Bioengineering, Lehigh University, Bethlehem, PA, USA*
JAMES A. GARNETT • *Centre for Host-Microbiome Interactions, Faculty of Dentistry, Oral & Craniofacial Sciences, King's College London, London, UK*

Edward Germany • *Frontier Science Research Center, University of Miyazaki, Kiyotake, Miyazaki, Japan*

Vicki A. M. Gold • *Living Systems Institute, University of Exeter, Exeter, UK; Faculty of Health and Life Sciences, University of Exeter, Exeter, UK*

Aathira Gopinath • *Institute of Biophysics, Department of Physics, Goethe University Frankfurt, Frankfurt/Main, Germany; Department of Physics, Freie Universität Berlin, Berlin, Germany*

Evert Haanappel • *Institut de Pharmacologie et Biologie Structurale, CNRS, Université de Toulouse, Toulouse, France*

Christine L. Hagan • *Chemistry Department, College of the Holy Cross, Worcester, MA, USA*

Scout Hayashi • *Laboratory of Molecular Biology, National Institute of Diabetes & Digestive & Kidney Diseases, National Institutes of Health, Bethesda, MD, USA*

Minh Sang Huynh • *Department of Biochemistry, University of Toronto, Toronto, ON, Canada; MaRS Center, West Tower, Toronto, ON, Canada*

Raffaele Ieva • *Laboratoire de Microbiologie et Génétique Moléculaires, Centre de Biologie Intégrative (CBI), Université de Toulouse, CNRS, UPS, Toulouse, France*

Wonpil Im • *Department of Biological Sciences, Department of Chemistry, Department of Bioengineering, Lehigh University, Bethlehem, PA, USA*

Benesh Joseph • *Department of Physics, Freie Universität Berlin, Berlin, Germany*

Sophie Ketter • *Institute of Biophysics, Department of Physics, Goethe University Frankfurt, Frankfurt/Main, Germany*

Xiangfeng Lai • *Department of Materials Science and Engineering, Faculty of Engineering, Monash University, Clayton, VIC, Australia*

Jack C. Leo • *Antimicrobial Resistance, Omics and Microbiota Group, Department of Biosciences, Nottingham Trent University, Nottingham, UK*

Dirk Linke • *Department of Biosciences, Section for Genetics and Evolutionary Biology, University of Oslo, Oslo, Norway; Department of Biosciences, University of Oslo, Oslo, Norway; Section for Evolution and Genetics, Department of Biosciences, University of Oslo, Oslo, Norway*

Christian D. Lorenz • *Biological Physics & Soft Matter Research Group, Department of Physics, King's College London, London, UK*

Joen Luirink • *Section Molecular Microbiology, A-life Department and Amsterdam Institute of Molecular and Life Sciences, Vrije Universiteit, Amsterdam, The Netherlands*

Mathew McLaren • *Living Systems Institute, University of Exeter, Exeter, UK; Faculty of Health and Life Sciences, University of Exeter, Exeter, UK*

Ina Meuskens • *Department of Biosciences, Section for Genetics and Evolutionary Biology, University of Oslo, Oslo, Norway; Institute of Molecular Biology & Biophysics, ETH Zurich, Zurich, Switzerland*

Trevor F. Moraes • *Department of Biochemistry, University of Toronto, Toronto, ON, Canada; MaRS Center, West Tower, Toronto, ON, Canada*

Violette Morales • *Laboratoire de Microbiologie et Génétique Moléculaires, Centre de Biologie Intégrative (CBI), Université de Toulouse, CNRS, UPS, Toulouse, France*

Stephan Nussberger • *Biophysics Department, Institute of Biomaterials and Biomolecular Systems, University of Stuttgart, Stuttgart, Germany*

Luis Orenday-Tapia • *Laboratoire de Microbiologie et Génétique Moléculaires, Centre de Biologie Intégrative (CBI), Université de Toulouse, CNRS, UPS, Toulouse, France; Department of Biochemistry, University of Oxford, Oxford, UK*

MARJORIE RUSSEL • *Rockefeller University, New York, NY, USA*
ATHANASIOS SARAGLIADIS • *Department of Biosciences, University of Oslo, Oslo, Norway*
HSIN-HUI SHEN • *Infection and Immunity Program, Monash Biomedicine Discovery Institute and Department of Microbiology, Monash University, Clayton, VIC, Australia*
TAKUYA SHIOTA • *Frontier Science Research Center, University of Miyazaki, Kiyotake, Miyazaki, Japan*
PETER VAN ULSEN • *Section Molecular Microbiology, A-life Department and Amsterdam Institute of Molecular and Life Sciences, Vrije Universiteit, Amsterdam, The Netherlands*
WALDEMAR VOLLMER • *Centre for Bacterial Cell Biology, Biosciences Institute, Newcastle University, Newcastle upon Tyne, UK; Institute for Molecular Bioscience, The University of Queensland, Brisbane, QLD, Australia*
ANASTASSIA A. VOROBIEVA • *Structural Biology Brussels, Vrije Universiteit Brussel and Center for Structural Biology, Brussels, Belgium; VIB-VUB Center for Structural Biology, VIB, Brussels, Belgium; VIB Center for AI and Computational Biology, VIB, Brussels, Belgium*
SHUO WANG • *Biophysics Department, Institute of Biomaterials and Biomolecular Systems, University of Stuttgart, Stuttgart, Germany*
JIAMING CAITLYN XU • *Department of Biochemistry, University of Toronto, Toronto, ON, Canada; MaRS Center, West Tower, Toronto, ON, Canada*
JOHN WILLIAM YOUNG • *Department of Biochemistry and Molecular Biology, Faculty of Medicine, Life Sciences Institute, University of British Columbia, Vancouver, BC, Canada; Department of Chemistry, Kavli Institute for Nanoscience Discovery, University of Oxford, Oxford, UK*
ZHIYU ZHAO • *Department of Biochemistry and Molecular Biology, Faculty of Medicine, Life Sciences Institute, University of British Columbia, Vancouver, BC, Canada*

Chapter 1

The Name Is Barrel, β-Barrel

Scout Hayashi, Susan K. Buchanan, and Istvan Botos

Abstract

β-barrels are a class of membrane proteins made up of a cylindrical, anti-parallel β-sheet with a hydrophobic exterior and a hydrophilic interior. The majority of proteins found in the outer membranes (OMs) of Gram-negative bacteria, mitochondria, and chloroplasts are β-barrel outer membrane proteins (OMPs). β-barrel OMPs have a diverse repertoire of functions, including nutrient transport, secretion, bacterial virulence, and enzymatic activity. Here, we discuss the broad functional classes of β-barrel OMPs, how they are folded into the membrane, and the future of β-barrel OMP research and its applications.

Key words Outer membrane β-barrels, BAM complex, SAM complex, TonB-dependent transporter

Abbreviations

BAM	Barrel-assembly machinery
cryo-EM	Cryo-electron microscopy
IM	Inner membrane
MOM	Mitochondrial outer membrane
OEM	Chloroplast outer envelope membrane
Oep80	Outer envelope protein 80
OM	Outer membrane
OMP	Outer membrane protein
POTRA domain	Polypeptide transport-associated domain
SAM	Sorting and assembly machinery
TBDT	TonB-dependent transporter
TM	Transmembrane
TOC	Translocase of the outer membrane of chloroplast
TOM	Translocase of the outer membrane
VDAC	Voltage-dependent anion-selective channel

Raffaele Ieva (ed.), *Transmembrane β-Barrel Proteins: Methods and Protocols*, Methods in Molecular Biology, vol. 2778, https://doi.org/10.1007/978-1-0716-3734-0_1,

1 Introduction to β-Barrel Outer Membrane Proteins

β-Barrel outer membrane proteins (OMPs) are found in Gram-negative bacteria and eukaryotic organelles: mitochondria and chloroplasts. Unlike Gram-positive bacteria, which have just a single thick peptidoglycan layer, Gram-negative bacteria are insulated from their environment by an inner membrane (IM), an outer membrane (OM), and a much thinner peptidoglycan layer [1]. Chloroplasts and mitochondria have an analogous double membrane structure, but no peptidoglycan.

The Gram-positive bacterial membrane and the IM of Gram-negative bacteria are very similar. Both are phospholipid bilayers and are the site of most cellular processes. Despite analogous proteins found in both membranes, almost none are β-barrels [2, 3].

The Gram-negative OM has a unique, asymmetrical composition. Lipopolysaccharides (LPS) are found in abundance in the outer leaflet and phospholipids in the inner leaflet [4]. Unlike the Gram-negative IM proteins, most Gram-negative OMPs are β-barrels, with a wide variety of functions [3, 5]. β-barrel OMPs are thought to account for 2–3% of encoded proteins in Gram-negative proteins. As of 2023, a database of Gram-negative OMPs lists published structures for approximately 200 β-barrel OMPs (https://blanco.biomol.uci.edu/mpstruc/).

2 β-Barrel Architecture

β-barrels are composed of a cylindrical antiparallel β-sheet consisting of between 8 and 36 β-strands. Notably absent are 20-stranded barrels. Closest to this missing type are the mitochondrial 19-stranded barrels, the only barrels with an odd number of strands. Genomic data analysis of Gram-negative bacteria reveals 16- or 22-stranded β-barrels to be the most abundant. Though strand length does not always correlate with function, β-barrels with 16 strands are mainly porins, and those with 22 strands are TonB-dependent transporters [6, 7] (Fig. 1).

The β-sheet structure sequesters polar backbone groups away from hydrophobic lipids in hydrogen bonds [8], so it is rare that β-strands terminate before reaching the end of the membrane. Hydrophobic residues tend to face the barrel exterior, while small polar residues line the interior. However, the interior surface of the barrel also contains some hydrophobic side chains, complicating efforts to identify β-barrels based on sequence alone. Snorkeling aromatic residues tend to sit closer to the rims of the barrel and interact with both polar headgroups and nonpolar tail groups of membrane lipids [9].

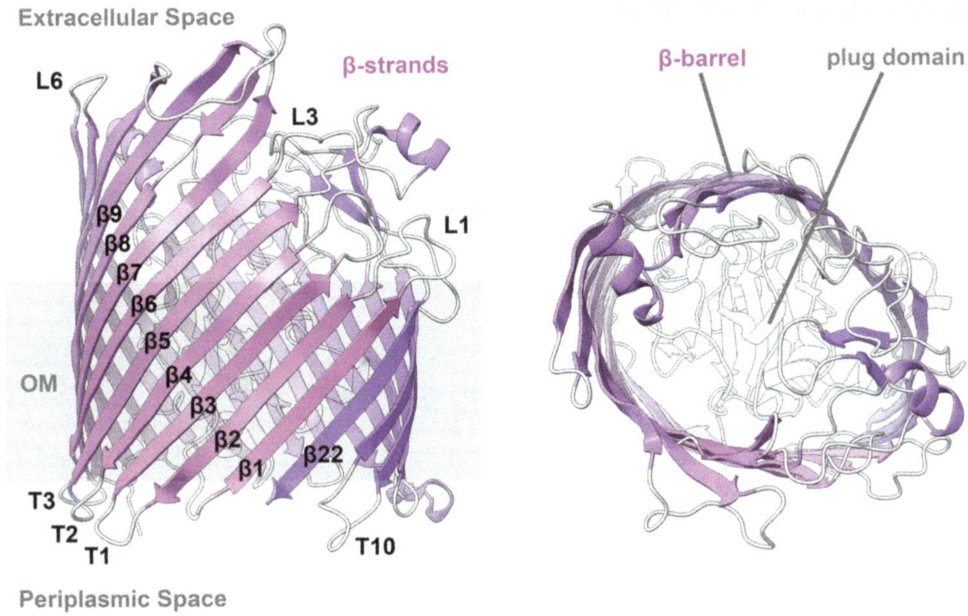

Fig. 1 Architecture of a typical β-barrel. Side and top view cartoon representations of the 22-stranded β-barrel FecA (PDB code 1PNZ [185]). The first β-strand (β1) starts from the periplasmic side and the strands run clockwise, ending with β22 on the periplasmic side. β1 hydrogen-bonds to the antiparallel β22 closing the barrel. There are 10 short periplasmic T loops and 11 long extracellular L loops. This structure also has an N-terminal plug domain that fills the barrel cavity

β-strands tend to be between 9 and 11 residues long and are tilted 20°–45° to the right, relative to the normal axis of the membrane [10]. A right-leaning tilt (Fig. 1) is universal as a left-leaning tilt is energetically unfavorable [9]. The extent of the tilt is sometimes denoted by the shear number. Though mathematically complex, the concept is simple. With two equal-length, parallel strands, each residue hydrogen-bonds to the adjacent residue from the other strand. When we tilt both strands to the right, the top residues from the right strand will stop hydrogen bonding to the left strand. The shear number indicates how many such residues exist in the barrel, with higher shear numbers corresponding to a more severe slant, and thus more unpaired residues at the barrel edges [11]. The shear number of a β-barrel is usually 2 greater than the number of strands in the barrel [9].

The strands are connected by extracellular and intracellular loops. In bacteria, the extracellular loops tend to be longer than the periplasmic loops, and many loops serve an additional purpose such as adhesion or transport specificity [12, 13].

3 Bacterial β-Barrel Functional Groups

Here, we describe the most common functional categories of β-barrel OMPs found in Gram-negative bacteria. The categories loosely align with those described in the OMPdb (www.ompdb.org), a database of β-barrel OMPs [14, 15]. Most structures discussed have been solved experimentally. However, a few structures predicted by AlphaFold2 [16, 17] and ESM-Fold [18] are also included (marked with *).

3.1 Nonspecific Diffusion Channels

The lipopolysaccharide-rich external leaflet of the bacterial OM is largely impermeable, so channels are required to transport nutrients and expel waste [19]. Most nonspecific channels have apertures close to 15 Å and allow passage of ions and small molecules up to 5000 Da [20, 21]. The overhanging loops of larger channels and porins often create extracellular funnels, constriction zones, or exit cones [20, 22].

The smallest category of bacterial β-barrel diffusion channels are the eight-stranded barrels, which include the well-studied *Escherichia coli* OmpW [23, 24] (Fig. 2a). OmpW forms small channels that allow small hydrophobic molecules to enter the bacteria [25]. OmpW has been studied as a drug target due to its high expression in the OM [23]. However, such eight-stranded β-barrels are rare. Few molecules are small enough to pass through the channel, so many eight-stranded β-barrels perform nontransport-related duties [26, 27].

16- to 18-stranded porins often appear as trimeric bundles of identical barrels and can be nonspecific or specific [22]. OmpC and OmpF in *E. coli* are trimers of 16-stranded barrels. Though they are considered nonspecific, negatively charged residues in the barrel lumen give both channels a slight preference for neutral and cationic molecules. Both barrels have homologs in other pathogenic bacteria, whose underexpression has been linked to antibiotic resistance [22, 28].

3.2 Specific Diffusion Channels

Substrate-specific channels allow passage of bulky molecules and have apertures closer to 6 Å [29]. The abundance of such channels could make them viable drug targets; however, certain strains such as *Pseudomonas aeruginosa* have reduced channel expression as an antibiotic resistance mechanism [30].

Most 12-stranded bacterial β-barrel OMPs are autotransporters. However, some passive diffusion proteins like *E. coli* NanC bear a high structural similarity, suggesting a common ancestor. NanC is a single β-barrel with no large overhanging loops that facilitates acidic oligosaccharide diffusion across the OM [13]. *E. coli* OmpG is a relatively stout 14-stranded barrel, with 7 large extracellular loops extending away from the membrane [31]

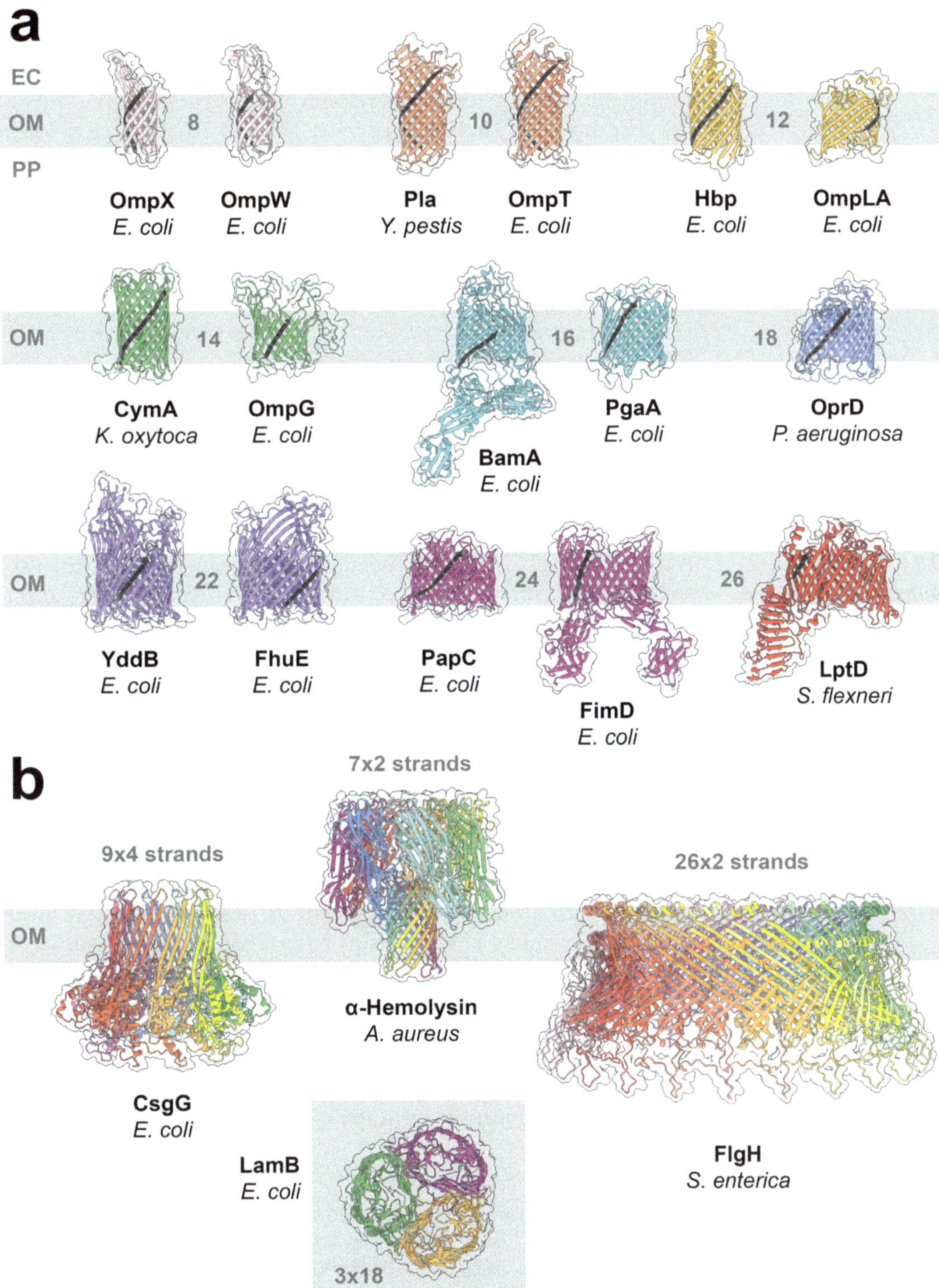

Fig. 2 Representative bacterial β-barrels. (**a**) Cartoon representation of monomeric β-barrel OMPs color coded by strand length. The number of strands is indicated next to the barrel, with β1 highlighted in black and the silhouette of the molecular surface outlined. OmpX (PDB: 1QJ8 [57]) and OmpW (PDB: 2F1V [23]) (pink) have 8 strands. Pla (PDB: 2X55 [77]) and OmpT (PDB: 1I78 [76]) (orange) have 10 strands. Hbp (PDB: 3AEH [51]) and

(Fig. 2a). This pH-gated channel passes large oligosaccharides into the bacterium [32] only under basic conditions [31]. *Klebsiella oxytoca* CymA is a taller 14-stranded barrel with small extracellular loops (Fig. 2a) that facilitate cyclodextrin diffusion [21].

The substrate specificity of most 16- and 18-stranded β-barrel porins prevents free diffusion of unnecessary or harmful molecules into the bacterium. The Opr class from *P. aeruginosa* contains many such channels, each with a unique substrate [33]. A representative member of the OprD family (Fig. 2a) is 18-stranded, with unstructured loops that curl into the lumen of the barrel, creating a constriction zone specific to basic amino acids. A row of basic residues next to the constriction contributes to specificity by binding the substrate carboxyl groups [34, 35].

There are also substrate-specific barrel trimers, such as LamB. This trimer of 18-stranded barrels allows maltodextrin through the outer membrane [36] (Fig. 2b). Like monomeric substrate-specific porins, unstructured loops form a constriction zone in the barrel lumen.

3.3 Transporters, Autotransporters, and Efflux Pumps

β-barrel transporters are generally larger than porins, with 20+ strand barrels being most common. Unlike porins, they often have additional domains and utilize conformational changes to drive transport [37].

One of the largest families of transporters are the TonB-dependent outer membrane receptors/transporters (TBDRs or TBDTs). The TonB complex, situated in the IM, contains an extension region that binds to TBDTs [38, 39]. TBDTs capitalize on the motion of the TonB complex, driven by the proton gradient, to move ions and small molecules across the OM [40, 41]. Most known TBDTs are 22-stranded and participate in iron transport [42]. *E. coli* TBDTs YddB and FhuE bind ferredoxin and iron in complex with a siderophore, respectively [43, 44] (Fig. 2a). Both contain N-terminal plug domains to prevent free ligand diffusion and extracellular loops that bind their substrate with high affinity. An active release mechanism resets the transporter. A C-terminal TonB box mediates TBDT interaction with the TonB complex [45]. Some TBDTs, like FecA, also act as regulators of transcription induction. Once bound to ferric citrate, FecA stimulates expression of ferric citrate transporters in the cytoplasm [46].

Fig. 2 (continued) OmpLA (PDB: 1QD6 [82]) (yellow) are 12-stranded. CymA (PDB: 4V3G [21]) and OmpG (PDB: 2JQY [31]) (green) have 14 strands. BamA (PDB: 6LYS [168]) and PgaA (PDB: 4Y25 [54]) (cyan) are 16-stranded, and OprD (PDB: 3SY7 [34]) (blue) is 18-stranded. YddB (PDB: 6OFR [44]) and FhuE (PDB: 6E4V [44]) (purple) have 22 strands, while PapC (PDB: 3FIP [64]) and FimD (PDB: 3RFZ [65]) (magenta) have 24 strands. LptD (PDB: 4Q35 [48]) (red) is 26-stranded. (**b**) Barrel multimers and assemblies, with subunits highlighted in different colors. CsgG (PDB: 7BRM [186]), α-hemolysin (PDB: 7AHL [67]), and FlgH (PDB: 7CBL [66]) are multimeric assemblies. LamB (PDB: 1MAL [36]) is a trimer of 18-stranded barrels. The number of subunits times the number of strands from a subunit are indicated (e.g., 3x18)

TonB-independent transporters do not rely on the TonB complex driven by the proton gradient and are more structurally diverse. The 14-stranded FadL transports fatty acids. Like TBDTs, FadL has a flexible plug domain but lacks high affinity for substrate binding. It is thought that transport occurs down the solute concentration gradient because of spontaneous conformational changes in the plug domain. The LptD-LptE complex (such as that of *Shigella flexneri* illustrated in Fig. 2a) is an extraordinarily tight complex of the LptD β-barrel with the LptE plug domain. Transport of LPS from the bacterial IM to the OM is carried out by a hydrophobic oligomeric LptA "chute" that connects the IM LptC to the N-terminal domain of LptD [47, 48]. LPS exits to the membrane through the LptD lateral gate [49]. This model is supported by a recent cryo-EM structure of LptDE with the lateral gate partially open [50].

Autotransporters are a unique class of protein transporters with two domains: a β-barrel and a passenger domain (often a virulence factor) that is threaded through the barrel and, in many cases, cleaved off [37]. In most cases, the barrel consists of 12 β-strands formed from a single polypeptide, but some autotransporters are trimeric—the 12-stranded barrel is formed from three copies of a polypeptide contributing 4 strands each. The passenger domain is secreted to the cell surface where it either remains attached (e.g., Hia) or is severed from the barrel by autocatalytic residues located within the barrel lumen (e.g., EspP, Hbp; Fig. 2a) [51].

Some β-barrels share structural similarities to autotransporters, but function as adhesins [52]. Intimin and invasin, found in most Gram-negative bacteria, are 12-stranded barrels with passenger domains. The barrel is approximately 8 Å wide and a short linker section connects it to C-terminal extracellular domains. These domains are able to bind to surface proteins on host cells [53].

Efflux pumps eject drugs, waste, and other small molecules from the bacterium. *E. coli* PgaA secretes a polymer that helps bacteria stick to one another forming a biofilm (Fig. 2a). The 16-stranded β-barrel is capped by long extracellular loops. The β-barrel and additional globular domains are responsible for polymer expulsion [54].

A final transporter, the *E. coli* curli complex, CsgG shares little with those above (Fig. 2b). The adhesion and biofilm formation protein is 36-stranded, with 9 subunits contributing four β-strands each to the barrel. In addition to a β-barrel, it contains a periplasmic funnel and a globular plug domain [55]. Examination of a single subunit gives little indication of belonging to a β-barrel OMP. Multimeric barrels such as CsgG are not assembled by the same machinery that assemble most β-barrel OMPs, instead rely on their own independent and less well understood insertion process involving small periplasmic proteins [56].

3.4 Adhesins/ Virulence Proteins

To successfully infect a host organism, bacteria must first attach to the host cell, and the β-barrel structure of adhesins makes them uniquely suited to the task. The transmembrane β-barrel acts as an anchor while a secondary domain binds to host cells. These proteins often have secondary evasive functions. *E. coli* OmpX, for example, both binds to host cell surface proteins and interferes with the complement system (Fig. 2a) [57]. A β-sheet domain attached to half of the barrel extends into extracellular space and may bind to complementary β-strands on host proteins [57]. OmpA, almost identical to OmpX, [58] is thought to be involved in *E. coli* adhesion and invasion of host cells and biofilm formation [26]. Its C-terminal domain is able to bind to bacterial peptidoglycans [59]. It may also function as a pore, but there is conflicting evidence on whether the channel is too narrow to allow ion or small molecule passage, or whether there is an "open-state" forming a larger pore not captured by existing structures [12, 60].

Usher pores participate in cell adherence through secretion and assembly of pili. The pili are transported as small globular subunits and polymerized into a fiber [61, 62]. Two different types of usher pores, both 24-stranded, are represented in Fig. 2a. The β-barrel PapC anchors a long string of globular PapA-G domains, which form P-type pili (Fig. 2a). Type 1 pili are formed by globular FimA-H domains, anchored by the β-barrel FimD [63]. The PapC lumen is kidney shaped and entirely blocked by a β-sandwich plug domain which prevents leakage of solutes when not actively building pili [64]. The FimD structure has an unoccluded pore with the plug domain dangling on the periplasmic side next to another C-terminal domain, demonstrating how the plug domain rotates into the periplasm during pili biogenesis [65] (Fig. 2a).

The *Salmonella enterica* FlgH complex is the largest known β-barrel assembly with a 52-strand circumference made of 26 subunits (Fig. 2b). Like the curli complex described above, FlgH is a large funnel facing the periplasmic space with double layered walls of β-barrel slanted in opposite directions. FlgH anchors the flagellar motor-hook complex that helps bacteria navigate their surroundings [66].

α-Hemolysin, a pore-forming toxin found in *Staphylococcus aureus*, is a 14-straded 7-mer (Fig. 2b). It spontaneously self-assembles in the membrane of the host and permits leakage of small molecules that damage and eventually kill the host cell. The tall transmembrane barrel is capped by a double layer of β-sheets forming a funnel on the extracellular face [67].

3.5 Assembly β-Barrel OMPs

The β-barrel OMPs that participate in β-barrel biogenesis, a process discussed in more detail in Subheading 6, are known as "assembly β-barrel OMPs." The Omp85 superfamily is involved in β-barrel OMP insertion and translocation of other proteins across the OM [68]. BamA is the most well studied of the Omp85 superfamily. In

E. coli, the main subunit of the BAM complex is BamA, a 16-stranded β-barrel, which associates with 4 lipoproteins BamB, BamC, BamD, and BamE to form the complex. N-terminal to the 16 stranded barrel, are five POlypeptide TRanslocation Associated (POTRA) domains which form an open ring at the base of the barrel. This ring of POTRA domains is stabilized by BamB-E [69–71]. POTRA 5 and BamD are known to associate, and both are essential for BAM complex function [70, 72, 73]. The extracellular opening is occluded by a dome of extracellular loops (most prominently L4, L6, and L7) and the periplasmic opening by the POTRA 5 domain, preventing free diffusion of metabolites [71].

The translocation and assembly module (TAM) complex is structurally similar to the BAM complex with two subunits TamA and TamB [73]. TamA, like BamA, has a barrel domain with a lateral gate and three POTRA domains [74]. TamB, built from a string of β-taco folds, associates with TamA via the POTRA domains and possibly the inner membrane. TamB may act as a "chute" that shuttles polypeptides across the periplasm [74]. TamA is less abundant than BamA, both being present in a subset of Gram-negative bacteria. Although they are thought to perform similar barrel biogenesis functions, TamA folds a much smaller group of OMPs. TamB is found in all Gram-negative bacteria [75].

3.6 Enzymes

Some smaller β-barrel OMPs have enzymatic activity. *Yersinia pestis* Pla and *E. coli* OmpT are 10-stranded proteases of highly similar structure (Fig. 2a). Both consist of a tall and semi-flattened cylinder with two β-sheets extending into the extracellular space, forming a deep groove where the active site lies. The barrel mouth aligns with the periplasmic surface of the membrane, while the rest of the barrel, including five extracellular loops, extends into extracellular space [76, 77].

OmpT, a serine protease, cleaves substrates between consecutive basic amino acids provided LPS is present [78, 79]. A negatively charged groove created by the two β-sheet flaps on the extracellular side of OmpT contains S99 and H212 that bind the substrate, while basic residues on the outside of the barrel bind LPS [76, 80].

Pla cleaves plasminogen into plasmin, which degrades fibrin and contributes to *Y. pestis* virulence [81]. Its putative active site contains D84, D86, D206, and H208, and like in OmpT, is situated between two β-sheets extending from the extracellular side of the barrel [77].

E. coli OmpLA is a phospholipase with a serine protease-like active site (Fig. 2a). The 12-stranded barrel sits comfortably in the membrane, widens slightly at the periplasmic face, and is not expected to function as a pore [82]. OmpLA is only active in a dimeric state, dissociating reversibly when inactive [83]. The dimer creates two active sites, containing S144, H142 and N156, that bind substrate only as a dimer. The pocket sits between barrels at the boundary between the membrane and extracellular space. Phospholipid substrates are cleaved into lysophospholipids which diffuse into the membrane [82].

4 Mitochondrial Outer Membrane β-Barrels

Like Gram-negative bacteria, mitochondria contain an IM and OM. Mitochondrial β-barrels are found exclusively in the OM, though this group has far less structural and functional variety than in bacteria. Only four distinct β-barrel OMPs have been identified (Fig. 3). A large majority are porins known as voltage-dependent anion channel (VDAC) [84]. Tom40 and Sam50, subunits of the translocase of outer membrane complex (TOM) and sorting and assembly machinery complex (SAM) respectively, are involved in β-barrel OMP import and assembly. Mitochondrial distribution and morphology protein 10 (Mdm10), found only in fungi, makes contact with the endoplasmic reticulum [85]. VDAC, Tom40, and Mdm10 each has 19 strands while Sam50 has 16 strands.

VDAC, Tom40, and Mdm10 belong to the "eukaryotic porins" superfamily, with no direct bacterial predecessor [86]. Strands β1 and β19 are parallel in their otherwise antiparallel 19-stranded β-barrels. They all have one or more N-terminal α-helices that span the interior of their barrels [87–89]. VDAC, the most abundant protein in the mitochondrial OM, is expressed as three isoforms VDAC1–3 and regulates metabolite transfer between the cytoplasm and mitochondria [90]. Tom40 is the mitochondrial entry gate for most preproteins encoded in the nucleus. The Tom40 barrel has a relatively large opening of ~30 Å, constricted by the α2 helix to ~13 Å. The barrel is tightly surrounded by the TM helices of Tom5–7, Tom22, and Tom40 to form the TOM core complex which can form dimers-tetramers [87, 91].

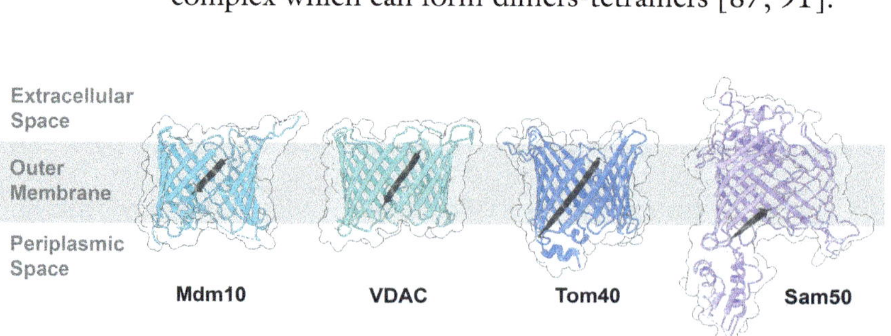

Fig. 3 Mitochondrial β-barrels. Cartoon representation of the four mitochondrial β-barrels. Mdm10 (PDB: 7BTX [89]), VDAC (PDB: 3EMN [88]), and Tom40 (PDB: 7E4H [187]) are 19-stranded barrels, while Sam50 (PDB: 6WUT [92]) has 16 strands. The first β-strand is highlighted in black in each barrel and the silhouette of the molecular surface is outlined

The main subunit of SAM is a 16-stranded β-barrel, Sam50, with much structural similarity to BamA, including a lateral gate with reduced hydrophobicity that thins the membrane. However, unlike BamA, it has only one POTRA domain oriented toward the intermembrane space. Additional subunits, Sam35 and Sam37, associate on the cytoplasmic side, opposite the POTRA domain [92]. Unlike BamA, the SAM complex may associate with a second barrel (Tom40, Mdm10, VDAC, or a second copy of Sam50), which sits next to Sam50 [89, 93–96]. Sam50 and Tom40 are conserved in higher eukaryotes, including humans. While the structures of human Sam50 and Tom40 are similar to their model organism counterparts cited above, their physiologies are not well understood. Sam50 appears to have some impact on cytotoxicity, mitophagy, and cristae morphology [97–99]. Tom40 may mitigate the effects of Parkinson's disease [100].

5 Chloroplast Outer Membrane β-Barrels

As recently reviewed by Roumia [84], chloroplasts house eight unique OM barrels: outer membrane proteins OEP21 [101], OEP23 [102], OEP24 [103], OEP37 [104], OEP40 [105]; trigalactosydiacylglyerol protein 4 TGD4 [106]; translocon of outer chloroplastic membrane TOC75 [107]; and OEP80 [108]. Because chloroplasts have fewer experimentally solved β-barrel OMP structures than bacteria and mitochondria, we discuss predicted structures generated by AlphaFold2 or ESM-Fold in addition to those found in the RCSB Protein Data Bank.

OEP21, OEP23, OEP24, OEP37, and OEP40 transport ions or small molecules and are thought to be regulated in part by voltage differences across the OM. OEP21 and OEP23 are ion transporters, both predicted to have eight β-strands (Fig. 4) [101, 102]. OEP24, another ion channel, is likely a homodimer of two seven-stranded barrels that is functionally related to mitochondrial VDAC [109]. The functions of OEP37 and TGD4 are still unknown, but are predicted to have 12 or 16 strands, and 20 strands, respectively [110]. If TGD4 is in fact a 20-stranded barrel, it would be the only one of this type. OEP40 is likely a 10-stranded glucose gate. Sister proteins TOC75 and OEP80 are involved in protein translocation and contain POTRA domains like those of BamA and Sam50 [111–113]. They are predicted to have 16 β-strands.

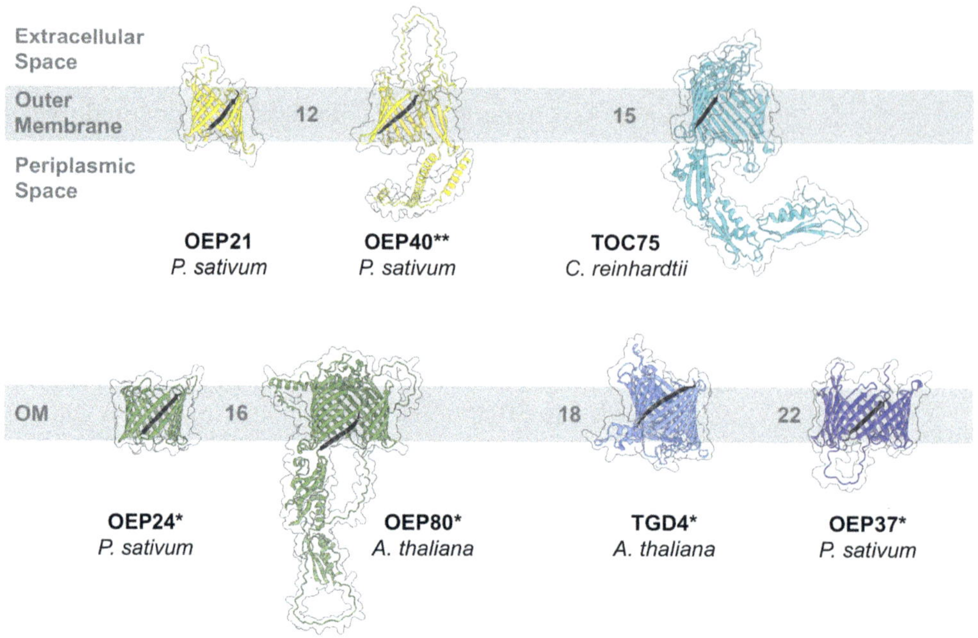

Fig. 4 Chloroplast β-barrels. Cartoon representation of the main chloroplast β-barrel structures and models. * indicates structures predicted by AlphaFold2 [16, 17] and ** by ESM-fold [18]. OEP21 (PDB: 7BHG [not published]) and OEP40** are 12-stranded, while OEP24* and OEP80* are 16-stranded. TOC75 (PDB: 7VCF [188]) has 15 strands, TGD4* 18 strands, and OEP37* 22 strands. The first β-strand in each barrel is highlighted in black and the silhouette of the molecular surface is outlined

6 β-Barrel Biogenesis

6.1 Bacterial Outer Membrane β-Barrel Biogenesis

β-Barrel OMP biogenesis in Gram-negative bacteria is uniquely complex; a newly synthesized polypeptide must cross both the IM and periplasmic space to reach its destination in the OM. First, the Sec translocon transports an unfolded precursor protein across the IM [114]. Chaperones such as SurA and Skp then shuttle the precursor protein to the OM, where it is folded and inserted into the OM by the β-barrel assembly machinery (BAM) complex [115, 116]. The BAM complex recognizes a short and highly conserved "β-signal" motif found in the last β-strand of every OMP. After binding the precursor, BAM folds it into a β-barrel and releases it into the membrane [117].

BamA opens laterally at the seam between the first strand and the short, kinked, 16th strand [69, 118, 119]. The two weakly associated strands can separate and reassociate, acting as a lateral gate that templates folding and eventually releases a fully formed new β-barrel into the lipid membrane [71]. Movement at the lateral gate is thought to be linked to the opening and closing of the pore by POTRA 5 via BamD, which makes contact both with the BamA barrel and with the POTRA domains [118, 120]. As a result, when POTRA 5 blocks the periplasmic face of the pore, BamA has an open lateral gate and when POTRA 5 points away from the pore, the lateral gate is closed [118].

6.2 Bacterial Assembly Models

Bacterial β-barrel assembly models can be largely grouped into "BamA-passive" models, where BamA merely facilitates conditions for spontaneous substrate folding, and "BamA-active" models, where BamA templates substrate folding. Each model illustrates the distinct steps taken in barrel folding, but they are not mutually exclusive. The pathway a barrel takes to fold correctly likely depends on many variables and a barrel might follow on a combination of the models to assemble.

6.2.1 BamA-Passive Models

BamA-passive models are more commonly known as "BamA-assisted" models [71, 121–123]. In this model, the primary role of BamA in β-barrel assembly is to thin the membrane such that spontaneous assembly is favorable [122, 124]. There is much data to support BamA-induced thinning of the membrane at the lateral gate in multiple species [71, 112, 122, 123, 125–127]. Hydrophilic residues in BamA near the lateral gate destabilize and thin the membrane, which may lower the energetic barrier to an assembling β-barrel precursor [122, 123].

6.2.2 BamA-Active Models

BamA-active models propose that BamA participates in folding. After the precursor localizes to BamA via its beta-signal, conformational changes in BamA facilitate the folding and release of the newly formed β-barrel [124, 128–130]. Different types of BamA-active models have emerged, including the barrel elongation model, the swing model, and the budding model, providing more clarity on the steps between β-signal binding and β-barrel release.

The barrel elongation model posits an amyloid-like self-assembly in which the barrel grows one hairpin at a time in a periplasmic folding funnel created by the POTRA domains and BamB-E. Experimental evidence for this model is scarce, but a rare BamA-LptDE intermediate complex was shown to interact with BamA and BamD in the periplasm [131].

The swing model also assumes assembly to begin in the periplasm with BamA flexing to create a hinge that can swing the mostly assembled substrate into the membrane [128].

The budding model, also known as the hybrid-barrel model, suggests that barrel assembly occurs in the membrane (Fig. 5). A structural study using BamA as both insertase and substrate showed that the C-terminal β-strand of the substrate hydrogen bonds with BamA-β1, while the other ends of each protein associate and curl inward with their exteriors touching forming a "B" shape if viewed from the top. This shape seals against membrane lipids while preventing free diffusion through the pore even as its circumference

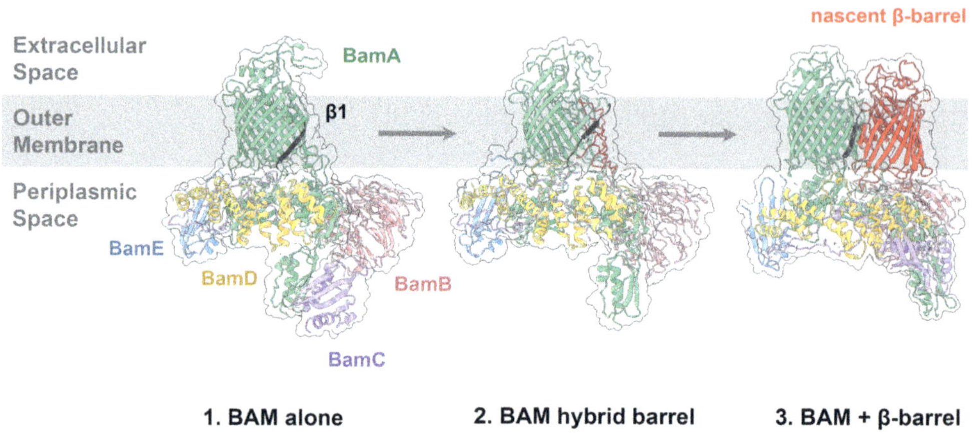

Fig. 5 Barrel biogenesis by the BAM machinery. Cartoon representation of three BAM complexes that illustrate the main steps of β-barrel biogenesis in bacteria by the hybrid barrel model. In the default stage (Step 1) the BAM complex has its β-barrel in a closed conformation, strand β1 (in black) hydrogen-bonding to β16 (PDB: 6LYS [168]). BamA-E are shown in green, pink, purple, yellow, and blue, respectively, In Step 2, the β-signal of a barrel precursor peptide (red) binds β1 and subsequent strands (red) form β-hairpins yielding a nascent barrel (PDB: 7RI4 [182]). Step 3 shows an almost completed nascent barrel in red (PDB: 6V05 [132]), which will dissociate from BamA into the OM

increases by one hairpin at a time. Once the N and C terminal β-strands of the substrate are sufficiently close, it is thought they begin to hydrogen-bond, displacing N-terminal β-strand associations between the substrate and BamA [132]. Additional experimental support for the hybrid-barrel model is shown in the folding of serine protease EspP that makes direct contact with the lateral seam of BamA [128]. The EspP appeared to grow by hairpins forming a "B"-shaped hybrid barrel [133].

6.2.3 Chaperone-Assisted Model

A third BamA assembly model suggests that the chaperone protein SurA binds the precursor before binding the BAM complex, while the chaperones Skp and DegP rescue OMPs not bound by SurA [134]. Upon binding, SurA facilitates a laterally closed gate, the "acceptor" state of BamA [135].

6.3 Mitochondrial Outer Membrane β-Barrel Biogenesis

Mitochondrial β-barrel OMP biogenesis follows a somewhat different mechanism than in bacteria. Substrates originate from outside the organelle but do not assemble at the boundary between mitochondrial OM and cytoplasm. Instead, they are imported through the OM and into intermembrane space by the translocase of the outer membrane (TOM) complex. Assembly occurs from the underside of the OM, a process inherited by mitochondria from their bacterial ancestors [136]. Chaperones known as translocase of the inner membrane (TIM) complexes escort the substrate through intermembrane space to the sorting and assembly machinery (SAM) complex, the mitochondrial homolog of BAM [137].

Due in part to the similarities between SAM substrates, our understanding of mitochondrial β-barrels is more complete. First, the substrate β-signal binds both Sam35 and Sam50 [138], and then the β1 strand of Sam50 associates with the β-signal of the substrate, which is located in the last β-strand of the new barrel just as in bacteria. The substrate is threaded through the lumen of Sam50 and assembled into a barrel one hairpin at a time, like the hybrid-barrel model proposed for the BAM complex [129]. Sam37 and the POTRA domain of Sam50 then release the substrate into the membrane [139].

In addition to β-barrel biogenesis, Sam50 is also involved in complex assembly, notably helping the TOM complex subunits associate with its main workhorse, Tom40 [95, 140]. The SAM complex may also function as a receptor for signals from other organelles and contribute to IM cristae and endoplasmic reticulum-mitochondrial structures [141–143].

7 β-Barrel Protein Evolution

The unanswered question of bacterial evolution is whether Gram-positive (monoderm) bacteria evolved from Gram-negative (diderm) bacteria or vice versa. The main argument for monoderm-first evolution is that natural antibiotic selection pressure gave rise to the OM for additional protection of the cell [144]. However, it is not clear how significantly antibiotics could have influenced the evolution of ancient bacteria. Another hypothesis suggests that the diderm to monoderm evolution occurred by loss of the OM by mutations in the peptidoglycan growth machinery [145]. The diderm-first hypothesis is supported by the discovery of independent diderm lineages in a typically monoderm phylum, suggesting a diderm ancestor from which the monoderm phenotype arose by loss of the OM. Gram-negative phenotypes dominate, pointing to an ancestral Gram-negative architecture from which the Gram-positive envelope evolved [146].

There is extensive evidence that most β-barrels in proteobacterial outer membranes have evolved from an ancestral protein or peptide (recently reviewed by Dhar & Slusky) [147]. β-barrels are formed by repeating β-hairpins (two strands connected by a loop), and a multi-prong bioinformatics approach identified a pool of β-hairpins as their common ancestor [148]. With β-hairpin basic building blocks, the smallest OM β-barrels still have 8 strands. Since bacterial β-barrels always have an even number of β-strands, it was suggested that larger barrels possibly evolved by gene duplication followed by fusion, insertion, deletion, circular permutation, shuffling, and mutation. This is also supported by experiments where the sequence of the known 8-stranded β-barrel OmpX was duplicated and the resulting β-barrel Omp2X folded into a larger β-barrel [149].

Large β-barrels did not evolve in a linear way, but rather via forking or alternate pathways. The multiple pathways for single-chain β-barrel evolution include domain duplications or loop to strand conversions. Barrels rarely transition to different sizes, instead they evolve maintaining the same strand number. The C-terminus of larger barrels retains conserved elements of an eight-strand sequence, in agreement with the importance of this region in folding [150]. While eight-strand β-barrel variants are impossible to design [151], new strands can be added to their N-terminus to engineer larger barrels. Certain classes of multi-chain β-barrel proteins, specifically the lysins and efflux pumps, evolved independently from the rest of the β-barrels, by possible evolutionary convergence into the barrel fold [150].

Most bacterial β-barrels can be traced to a common β-hairpin ancestor, and in mitochondria, the even-numbered barrel Sam50 can be traced to bacteria. The odd-numbered barrels (Tom40, VDAC, Mdm10) do not have a direct link to bacterial barrels. While these 19-strand barrels descend from bacteria, the fold formed independently in the proto-mitochondria [152]. They contain five repeats of a four-stranded motif with the N-terminal strand converted into an α-helical plug.

The conserved mechanism of biogenesis also suggests a common ancestry of β-barrel proteins [75]. In bacteria, β-barrels are all inserted into the outer membrane by the BAM complex [73], while in mitochondria, the corresponding machinery is the SAM complex [153], and in chloroplasts the TOC complex [154]. The core components of these complexes are evolutionarily conserved, presenting some differences in their folding signals and machineries [155]. The C-terminal beta signal of the barrel precursor peptide also shows a high degree of homology between bacteria and organelles [6].

Antibiotic use has put new evolutionary pressure on the outer membrane whose role is to protect the core cellular machineries; therefore, β-barrel proteins continue to evolve [3].

8 Methodology for Study of β-Barrels

8.1 X-Ray Crystallography and NMR

For the study of β-barrels, the major bottleneck remains the availability of these proteins. For the longest time, X-ray crystallography and NMR were the only methods available for the structural study of β-barrels. The initial difficulty to overexpress and purify homogenous membrane proteins in large quantities was further complicated by the almost insurmountable roadblock of obtaining diffraction quality crystals. Relatively recent developments in better detergents and lipidic cubic phase techniques somewhat lowered this barrier, but many membrane proteins could still not be crystallized and solved. NMR is also a great technique for studying dynamic systems in solution but was limited to the study of smaller β-barrel structures.

8.2 Cryo-EM

The recent Cryo-EM resolution revolution ushered in a dramatic increase in the number of solved membrane protein structures. Purified proteins that could not be crystallized despite prolonged attempts could readily be applied on grids and their structures solved by single-particle Cryo-EM reconstruction. Smaller membrane proteins could be purified complexed to larger proteins with already known structures and their structures solved. Monoclonal antibodies are becoming the tool of choice to immobilize small, flexible parts of a structure that otherwise would be too small for Cryo-EM. Further developments in detergent chemistry and lipid nanodiscs permitted capturing membrane proteins in patches of their native lipidic environments that are much closer to physiological conditions than detergent micelles. While X-ray crystallography is a relatively mature technique, Cryo-EM instrumentation and methods are still rapidly developing. Developments in grid preparation from very small amounts of sample make possible the structural study of much more exotic complexes, and though sample homogeneity is generally preferred, methodologies that can parse heterogenous protein samples are being developed for use in membrane protein structural studies [156]. The β-barrels can also be studied in their native environments with Cryo-EM tomography.

8.3 Machine Learning Algorithms

Algorithms that leverage known structures to predict structures of currently unsolved β-barrel OMP sequences may offer more efficient methods of structure generation [157, 158]. In recent years, we have witnessed major breakthroughs in machine learning techniques due to the development of massively parallel algorithms that can be run efficiently on commercial GPUs. Some algorithms, such as IsItABarrel, have become adept at classifying sequences as barrels or not-barrels with up to 95% accuracy [159]. However, the holy grail of structural biology, which is folding proteins from sequence, seems within reach by the development of protein prediction algorithms Alphafold2 by DeepMind [16, 17] and ESM-Fold by Meta [18]. These routines learn the general patterns of protein folding from existing databases and can predict a structure based on a protein sequence almost in real time, with a so far unmatched prediction accuracy. Previously, predicting β-barrels has been especially difficult for prediction routines; the alternating directions of β-strand residues tend to obfuscate patterns in the sequence. But Alphafold2 and ESM-Fold can model β-barrels from sequence (Fig. 4). Even if the resulting model is not fully correct, it could be used with success as a template for Cryo-EM particle picking and model building. Predicting proteins with unknown structure can help with developing strategies for the study of the respective protein: identifying domains or regions for truncation and locating key residues for mutagenesis. These algorithms are actively developed, and their newer versions already yield gradually better results.

9 β-Barrels in Medicine

Since β-barrels mediate essential bacterial functions, inhibiting their activity might be another way to fight Gram-negative pathogens. Most antibiotics are designed to enter the cell through porins, but surprisingly there are almost no antibiotics that target β-barrel porins themselves. Gram-negative bacteria have trimeric β-barrel porins that provide permeability to small molecules and antibiotics through the OM. Drug-resistant *E. coli* species, *Klebsiella pneumoniae*, *Pseudomonas aeruginosa*, and *Acinetobacter baumannii* express fewer porins and exhibit increased antibiotic resistance [160, 161]. This can be due to loss of the porin gene, lower expression levels, or porin mutations that render it ineffective. In *E coli*, two β-barrel proteins are essential: LptD involved in lipopolysaccahride biogenesis and BamA involved in OMP biogenesis. L27–11 is a compound that targets LptD specifically from *Pseudomonas* with nanomolar activity [162]. L27–11 is a macrocyclic peptidomimetic antibiotic related to murepavadin that blocks lipopolysaccharide transport by interacting with the periplasmic domain of LptD [163]. Another inhibitor, AOA-2, is a cyclic hexapeptide that inhibits adhesion of *E. coli*, *Pseudomonas*, and *Acinetobacter* bacteria to their host cells by binding to the extracellular loops of OmpA [164].

Other inhibitors target the β-barrel assembly machinery of Gram-negative pathogens and were recently reviewed by Steenhuis [165]. One such example is darobactin, a naturally occurring peptide that was identified by screening for *E. coli* antibacterial activity in extracts of *Photorhabdus* and *Xenorhabdus* [166]. Darobactin is a heptapeptide with a bicyclic core and targets BamA, the central component of the BAM complex, at the first β-strand of the lateral gate of BamA (Fig. 6, [167]), displacing the β-signal of the folding barrel [168]. MRL-494 is a peptide-like synthetic compound that does not need to cross the OM to act [169]. MRL-494 is moderately potent against Gram-negative bacteria by targeting BamA and is also active against Gram-positive bacteria by destabilizing their membrane. On its own, murepavadin targets LptD while polymyxin B binds lipid A. By combining the antibacterial activities of murepavadin and polymyxin B, new chimeric inhibitors were obtained [170]. One of the chimeric inhibitors, peptide 3, seems to target BamA and LptE from Gram-negative bacteria without the side-effects of murepavadin. JB-95 is similar to murepavadin, a β-hairpin peptidomimetic that inhibits LptD and BamA in Gram-negative bacteria and targets the membrane in Gram-positive bacteria [171].

The lectin-like bacteriocin LlpA, secreted by *Pseudomonas* strains, contains two lectin domains and a nonconserved C-terminal segment that target extracellular loop 6 of BamA [172]. Monoclonal antibodies can be developed against loop 4 of

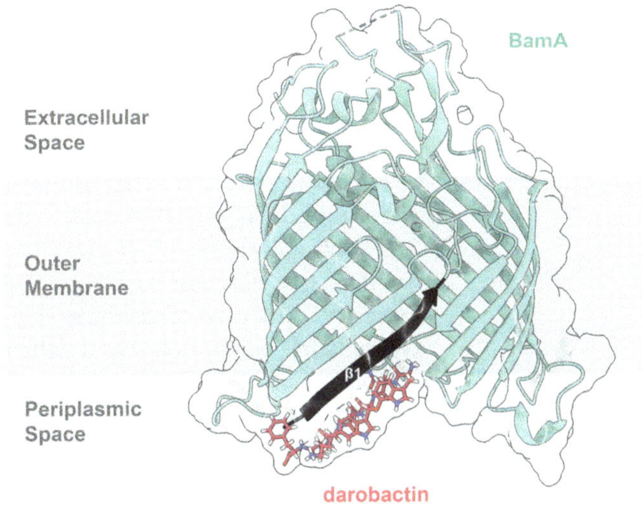

Fig. 6 BamA bound to darobactin. Cartoon representation of Bam in aqua with bound darobactin in red (PDB code 7NRF [167]). The silhouette of the molecular surface gives a better scale of the extent of the barrel. Darobactin is bound to the first β-strand of the barrel (black) blocking the binding of the β-signal and inhibiting novel barrel folding. All figures are generated with ChimeraX [189]

BamA, but their effectiveness is limited to *E coli* strains with truncated LPS chains [173]. In Gram-negative bacteria, long LPS chains occlude access of large molecules to the OMPs. β-barrels can also be used in vaccine development as scaffolds to present antigens to the immune system. Antigen display on bacteria can be mediated by autotransporters, but the size and complexity of the partner proteins can be a limiting factor. This can be overcome by ligating proteins to autotransporters already displayed on the surface of the cell or vesicle [174].

10 Building Novel Barrels

The β-barrel is a promising fold in biotechnology for engineering new types of pores, such as nanopores. Nanopores are large β-barrels of various biological origins that act as voltage sensors in single-molecule detection and nucleic acid sequencing [175]. However, the large diameter of proteins of interest necessitates the design of synthetic nanopores larger than those in the current library of β-barrels. These β-barrels may be designed via gene duplication of the whole barrel or by adding additional β-hairpins to existing barrels. The 8-stranded OmpX was duplicated to form the 16-stranded Omp2X barrel [149], and 22-strand FhuA was augmented to form a 24-strand barrel [176]. With the plug domain removed, the FhuA barrel can be extended all the way up to 34-strands [177]. The FhuA barrels can be transformed into highly conductive passive nanopores by removing the plug domain and trimming the extracellular loops [178]. The substrate specificity of pores can be changed by mutating the pore-lining residues, as in the case of the passive diffusion porin OmpG [170] or the ion-channel VDAC [179].

Barrel engineering is also useful for studying folding dynamics. More stable chimeric 8-stranded β-barrels can be engineered by combining different Omp barrels that sometimes create nanopores. One such chimera is of *E. coli* OmpX and *Y. pestis* Ail, built from two barrels with high structural, but low sequence and functional similarity. Analysis of folding dynamics found a central hairpin (typically β4–5) necessary and sufficient for correct β-barrel folding. The placement of the strand in the barrel did not appear to matter, and swapping central hairpins between barrels did not disrupt folding [180].

11 Conclusions and Future Perspectives

Mapping the structural diversity of the β-barrel OMP family remains critical to understanding the importance of the OM in bacteria and in its derivative eukaryotic organelles. In addition to advances in CryoEM and X-ray crystallography, the advent of AI-generated structures is a refreshing addition to β-barrel OMP research. As these algorithms mature, it may become more commonplace to see AI-generated structures alongside their experimental counterparts. Membrane proteins often demand unique purification conditions which slow the experimental structure determination process. Greater efficiency in structure determination could greatly expand the field of β-barrel OMP research. Focus may begin to shift toward improving the biological relevance of β-barrel OMP research methods. Recent advancements in cryo-electron tomography, molecular dynamics, and the development of native lipid nanodiscs have made it possible to study a wider variety of OMPs in their native membranes and more biologically relevant contexts [181–183].

The variety of β-barrel structures and functions is also an asset to be leveraged in biotechnology and medicine. The integral role played by β-barrel OMPs also offers the possibility of discovering new drug targets in the arms race of antibiotic resistance. Greater understanding of how β-barrel OMPs evolve and fold in the membrane may herald more efficient ways of genetically engineering new β-barrel OMPs with a variety of functions. For example, a β-barrel with aldolase activity was recently designed de novo through laboratory evolution [184]. Thus, β-barrel OMPs remain a promising subject of study for structural biologists and biomedical scientists alike.

Acknowledgments

S.H., I.B., and S.K.B. were supported by the Intramural Research Program of the NIH, National Institute of Diabetes and Digestive and Kidney Diseases (NIDDK).

References

1. Silhavy TJ, Ruiz N, Kahne D (2006) Advances in understanding bacterial outer-membrane biogenesis. Nat Rev Microbiol 4(1):57–66. https://doi.org/10.1038/nrmicro1322
2. Navarre WW, Schneewind O (1999) Surface proteins of Gram-positive bacteria and mechanisms of their targeting to the cell wall envelope. Microbiol Mol Biol Rev 63(1): 174–229. https://doi.org/10.1128/MMBR.63.1.174-229.1999
3. Hermansen S, Linke D, Leo JC (2022) Transmembrane β-barrel proteins of bacteria: from structure to function. Adv Protein Chem Struct Biol 128:113–161. https://doi.org/10.1016/BS.APCSB.2021.07.002
4. Nikaido H (2003) Molecular basis of bacterial outer membrane permeability revisited. Microbiol Mol Biol Rev 67(4):593–656. https://doi.org/10.1128/MMBR.67.4.593-656.2003
5. Zhang XC, Han L (2016) How does a β-barrel integral membrane protein insert into the membrane? Protein Cell 7(7): 471–477. https://doi.org/10.1007/s13238-016-0273-6
6. Paramasivam N, Habeck M, Linke D (2012) Is the C-terminal insertional signal in Gram-negative bacterial outer membrane proteins species-specific or not? BMC Genomics 13: 510. https://doi.org/10.1186/1471-2164-13-510
7. Udho E, Jakes KS, Buchanan SK, James KJ, Jiang X, Klebba PE et al (2009) Reconstitution of bacterial outer membrane TonB-dependent transporters in planar lipid bilayer membranes. Proc Natl Acad Sci U S A 106(51):21990–21995. https://doi.org/10.1073/pnas.0910023106
8. Bishop CM, Walkenhorst WF, Wimley WC (2001) Folding of β-sheets in membranes: specificity and promiscuity in peptide model systems. J Mol Biol 309(4):975–988. https://doi.org/10.1006/jmbi.2001.4715
9. Schulz GE (2000) β-Barrel membrane proteins. Curr Opin Struct Biol 10(4):443–447. https://doi.org/10.1016/S0959-440X(00)00120-2
10. Wimley WC (2003) The versatile β-barrel membrane protein. Curr Opin Struct Biol 13(4):404–411. https://doi.org/10.1016/S0959-440X(03)00099-X
11. Liu W-M (1998) Shear numbers of protein β-barrels: definition refinements and statistics. J Mol Biol 275(4):541–545. https://doi.org/10.1006/jmbi.1997.1501
12. Sugawara E, Nikaido H (1992) Pore-forming activity of OmpA protein of Escherichia coli. J Biol Chem 267(4):2507–2511
13. Wirth C, Condemine G, Boiteux C, Bernèche S, Schirmer T, Peneff CM (2009) NanC crystal structure, a model for outer-membrane channels of the acidic sugar-specific KdgM porin family. J Mol Biol 394(4):718–731. https://doi.org/10.1016/j.jmb.2009.09.054
14. Tsirigos KD, Bagos PG, Hamodrakas SJ (2011) OMPdb: a database of {beta}-barrel outer membrane proteins from Gram-negative bacteria. Nucleic Acids Res 39(Database issue):D324–D331. https://doi.org/10.1093/nar/gkq863
15. Roumia AF, Tsirigos KD, Theodoropoulou MC, Tamposis IA, Hamodrakas SJ, Bagos PG (2021) OMPdb: a global hub of beta-barrel outer membrane proteins. Front Bioinform 1:646581. https://doi.org/10.3389/fbinf.2021.646581
16. Jumper J, Evans R, Pritzel A, Green T, Figurnov M, Ronneberger O et al (2021) Highly accurate protein structure prediction with AlphaFold. Nature 596(7873): 583–589. https://doi.org/10.1038/s41586-021-03819-2
17. Varadi M, Anyango S, Deshpande M, Nair S, Natassia C, Yordanova G et al (2021) AlphaFold Protein Structure Database: massively expanding the structural coverage of protein-sequence space with high-accuracy models. Nucleic Acids Res 50(D1):D439–DD44. https://doi.org/10.1093/nar/gkab1061
18. Lin Z, Akin H, Rao R, Hie B, Zhu Z, Lu W et al (2022) Evolutionary-scale prediction of atomic level protein structure with a language model. bioRxiv:2022.07.20.500902.

https://doi.org/10.1101/2022.07.20.500902

19. Nikaido H, Vaara M (1985) Molecular basis of bacterial outer membrane permeability. Microbiol Rev 49(1):1–32. https://doi.org/10.1128/mr.49.1.1-32.1985
20. Jap BK, Walian PJ (1996) Structure and functional mechanism of porins. Physiol Rev 76(4):1073–1088. https://doi.org/10.1152/physrev.1996.76.4.1073
21. van den Berg B, Prathyusha Bhamidimarri S, Dahyabhai Prajapati J, Kleinekathöfer U, Winterhalter M (2015) Outer-membrane translocation of bulky small molecules by passive diffusion. Proc Natl Acad Sci U S A 112(23):E2991–E2999. https://doi.org/10.1073/pnas.1424835112
22. Sharma A, Yadav SP, Sarma D, Mukhopadhaya A (2022) Chapter two – modulation of host cellular responses by gram-negative bacterial porins. In: Donev R (ed) Advances in protein chemistry and structural biology. Academic Press, London, pp 35–77
23. Hong H, Patel DR, Tamm LK, van den Berg B (2006) The outer membrane protein OmpW forms an eight-stranded beta-barrel with a hydrophobic channel. J Biol Chem 281(11):7568–7577. https://doi.org/10.1074/jbc.M512365200
24. Shahryari S, Mohammadnejad P, Noghabi KA (2021) Screening of anti-Acinetobacter baumannii phytochemicals, based on the potential inhibitory effect on OmpA and OmpW functions. R Soc Open Sci 8(8):201652. https://doi.org/10.1098/rsos.201652
25. Catel-Ferreira M, Marti S, Guillon L, Jara L, Coadou G, Molle V et al (2016) The outer membrane porin OmpW of Acinetobacter baumannii is involved in iron uptake and colistin binding. FEBS Lett 590(2):224–231. https://doi.org/10.1002/1873-3468.12050
26. Smith SGJ, Mahon V, Lambert MA, Fagan RP (2007) A molecular Swiss army knife: OmpA structure, function and expression. FEMS Microbiol Lett 273(1):1–11. https://doi.org/10.1111/j.1574-6968.2007.00778.x
27. Moraes TF, Yu RH, Strynadka NC, Schryvers AB (2009) Insights into the bacterial transferrin receptor: the structure of transferrin-binding protein B from Actinobacillus pleuropneumoniae. Mol Cell 35(4):523–533. https://doi.org/10.1016/j.molcel.2009.06.029
28. Masi M, Vergalli J, Ghai I, Barba-Bon A, Schembri T, Nau WM et al (2022) Cephalosporin translocation across enterobacterial OmpF and OmpC channels, a filter across the outer membrane. Commun Biol 5(1):1059. https://doi.org/10.1038/s42003-022-04035-y
29. Galdiero S, Falanga A, Cantisani M, Tarallo R, Elena Della Pepa M et al (2012) Microbe-host interactions: structure and role of gram-negative bacterial porins. Curr Protein Pept Sci 13(8):843–854. https://doi.org/10.2174/138920312804871120
30. Garnacho-Montero J, Amaya-Villar R (2022) The problem of multi-resistance in gram-negative bacilli in intensive care units: treatment and prevention strategies. Med Intensiva (Engl Ed) 46(6):326–335. https://doi.org/10.1016/j.medine.2022.04.006
31. Liang B, Tamm LK (2007) Structure of outer membrane protein G by solution NMR spectroscopy. Proc Natl Acad Sci U S A 104(41):16140–16145. https://doi.org/10.1073/pnas.0705466104
32. Fajardo DA, Cheung J, Ito C, Sugawara E, Nikaido H, Misra R (1998) Biochemistry and regulation of a novel Escherichia coli K-12 porin protein, OmpG, which produces unusually large channels. J Bacteriol 180(17):4452–4459. https://doi.org/10.1128/jb.180.17.4452-4459.1998
33. Tamber S, Ochs MM, Hancock RE (2006) Role of the novel OprD family of porins in nutrient uptake in Pseudomonas aeruginosa. J Bacteriol 188(1):45–54. https://doi.org/10.1128/jb.188.1.45-54.2006
34. Eren E, Vijayaraghavan J, Liu J, Cheneke BR, Touw DS, Lepore BW et al (2012) Substrate specificity within a family of outer membrane carboxylate channels. PLoS Biol 10(1):e1001242. https://doi.org/10.1371/journal.pbio.1001242
35. Biswas S, Mohammad MM, Movileanu L, van den Berg B (2008) Crystal structure of the outer membrane protein OpdK from Pseudomonas aeruginosa. Structure 16(7):1027–1035. https://doi.org/10.1016/j.str.2008.04.009
36. Schirmer T, Keller TA, Wang YF, Rosenbusch JP (1995) Structural basis for sugar translocation through maltoporin channels at 3.1 A resolution. Science 267(5197):512–514. https://doi.org/10.1126/science.7824948
37. Meuskens I, Saragliadis A, Leo JC, Linke D (2019) Type V secretion systems: an overview of passenger domain functions. Front Microbiol 10:1163
38. Ratliff AC, Buchanan SK, Celia H (2022) The ton motor. Front Microbiol 13:852955.

https://doi.org/10.3389/fmicb.2022.852955
39. Celia H, Noinaj N, Buchanan SK (2020) Structure and stoichiometry of the ton molecular motor. Int J Mol Sci 21(2). https://doi.org/10.3390/ijms21020375
40. Fujita M, Mori K, Hara H, Hishiyama S, Kamimura N, Masai E (2019) A TonB-dependent receptor constitutes the outer membrane transport system for a lignin-derived aromatic compound. Commun Biol 2(1):1–10. https://doi.org/10.1038/s42003-019-0676-z
41. Guérin J, Buchanan SK (2021) Protein import and export across the bacterial outer membrane. Curr Opin Struct Biol 69:55–62. https://doi.org/10.1016/j.sbi.2021.03.007
42. Noinaj N, Guillier M, Barnard TJ, Buchanan SK (2010) TonB-dependent transporters: regulation, structure, and function. Annu Rev Microbiol 64:43–60. https://doi.org/10.1146/annurev.micro.112408.134247
43. Grinter R, Leung PM, Wijeyewickrema LC, Littler D, Beckham S, Pike RN et al (2019) Protease-associated import systems are widespread in Gram-negative bacteria. PLoS Genet 15(10):e1008435. https://doi.org/10.1371/journal.pgen.1008435
44. Grinter R, Lithgow T (2019) Determination of the molecular basis for coprogen import by Gram-negative bacteria. IUCrJ 6(Pt 3):401–411. https://doi.org/10.1107/s2052252519002926
45. Oeemig JS, Ollila OHS, Iwaï H (2018) NMR structure of the C-terminal domain of TonB protein from Pseudomonas aeruginosa. PeerJ 6:e5412. https://doi.org/10.7717/peerj.5412
46. Härle C, Kim I, Angerer A, Braun V (1995) Signal transfer through three compartments: transcription initiation of the Escherichia coli ferric citrate transport system from the cell surface. EMBO J 14(7):1430–1438. https://doi.org/10.1002/j.1460-2075.1995.tb07129.x
47. Dong H, Xiang Q, Gu Y, Wang Z, Paterson NG, Stansfeld PJ et al (2014) Structural basis for outer membrane lipopolysaccharide insertion. Nature 511(7507):52–56. https://doi.org/10.1038/nature13464
48. Qiao S, Luo Q, Zhao Y, Zhang XC, Huang Y (2014) Structural basis for lipopolysaccharide insertion in the bacterial outer membrane. Nature 511(7507):108–111. https://doi.org/10.1038/nature13484
49. Botos I, Majdalani N, Mayclin SJ, McCarthy JG, Lundquist K, Wojtowicz D et al (2016) Structural and functional characterization of the LPS transporter LptDE from Gram-negative pathogens. Structure 24(6):965–976. https://doi.org/10.1016/j.str.2016.03.026
50. Botte M, Ni D, Schenck S, Zimmermann I, Chami M, Bocquet N et al (2022) Cryo-EM structures of a LptDE transporter in complex with Pro-macrobodies offer insight into lipopolysaccharide translocation. Nat Commun 13(1):1826. https://doi.org/10.1038/s41467-022-29459-2
51. Tajima N, Kawai F, Park SY, Tame JR (2010) A novel intein-like autoproteolytic mechanism in autotransporter proteins. J Mol Biol 402(4):645–656. https://doi.org/10.1016/j.jmb.2010.06.068
52. Newman CL, Stathopoulos C (2004) Autotransporter and two-partner secretion: delivery of large-size virulence factors by gram-negative bacterial pathogens. Crit Rev Microbiol 30(4):275–286. https://doi.org/10.1080/10408410490499872
53. Fairman JW, Noinaj N, Buchanan SK (2011) The structural biology of β-barrel membrane proteins: a summary of recent reports. Curr Opin Struct Biol 21(4):523. https://doi.org/10.1016/J.SBI.2011.05.005
54. Wang Y, Andole Pannuri A, Ni D, Zhou H, Cao X, Lu X et al (2016) Structural basis for translocation of a biofilm-supporting exopolysaccharide across the bacterial outer membrane. J Biol Chem 291(19):10046–10057. https://doi.org/10.1074/jbc.M115.711762
55. Liu C, Ma J, Wang J, Wang H, Zhang L (2020) Cryo-EM structure of a bacterial lipid transporter YebT. J Mol Biol 432(4):1008–1019. https://doi.org/10.1016/J.JMB.2019.12.008
56. Dunstan RA, Hay ID, Wilksch JJ, Schittenhelm RB, Purcell AW, Clark J et al (2015) Assembly of the secretion pores GspD, Wza and CsgG into bacterial outer membranes does not require the Omp85 proteins BamA or TamA. Mol Microbiol 97(4):616–629. https://doi.org/10.1111/mmi.13055
57. Vogt J, Schulz GE (1999) The structure of the outer membrane protein OmpX from Escherichia coli reveals possible mechanisms of virulence. Structure 7(10):1301–1309. https://doi.org/10.1016/S0969-2126(00)80063-5
58. Sugawara E, Steiert M, Rouhani S, Nikaido H (1996) Secondary structure of the outer membrane proteins OmpA of Escherichia coli and OprF of Pseudomonas aeruginosa. J

Bacteriol 178(20):6067–6069. https://doi.org/10.1128/jb.178.20.6067-6069.1996

59. Park JS, Lee WC, Yeo KJ, Ryu KS, Kumarasiri M, Hesek D et al (2012) Mechanism of anchoring of OmpA protein to the cell wall peptidoglycan of the gram-negative bacterial outer membrane. FASEB J 26(1): 219–228. https://doi.org/10.1096/fj.11-188425
60. Pautsch A, Schulz GE (2000) High-resolution structure of the OmpA membrane domain. J Mol Biol 298(2):273–282. https://doi.org/10.1006/jmbi.2000.3671
61. Capitani G, Eidam O, Glockshuber R, Grütter MG (2006) Structural and functional insights into the assembly of type 1 pili from Escherichia coli. Microbes Infect 8(8): 2284–2290. https://doi.org/10.1016/j.micinf.2006.03.013
62. Sauer FG, Remaut H, Hultgren SJ, Waksman G (2004) Fiber assembly by the chaperone-usher pathway. Biochim Biophys Acta 1694(1–3):259–267. https://doi.org/10.1016/j.bbamcr.2004.02.010
63. Busch A, Waksman G (2012) Chaperone-usher pathways: diversity and pilus assembly mechanism. Philos Trans R Soc Lond Ser B Biol Sci 367(1592):1112–1122. https://doi.org/10.1098/rstb.2011.0206
64. Huang Y, Smith BS, Chen LX, Baxter RH, Deisenhofer J (2009) Insights into pilus assembly and secretion from the structure and functional characterization of usher PapC. Proc Natl Acad Sci U S A 106(18): 7403–7407. https://doi.org/10.1073/pnas.0902789106
65. Phan G, Remaut H, Wang T, Allen WJ, Pirker KF, Lebedev A et al (2011) Crystal structure of the FimD usher bound to its cognate FimC-FimH substrate. Nature 474(7349): 49–53. https://doi.org/10.1038/nature10109
66. Tan J, Zhang X, Wang X, Xu C, Chang S, Wu H et al (2021) Structural basis of assembly and torque transmission of the bacterial flagellar motor. Cell 184(10):2665–79.e19. https://doi.org/10.1016/j.cell.2021.03.057
67. Song L, Hobaugh MR, Shustak C, Cheley S, Bayley H, Gouaux JE (1996) Structure of staphylococcal alpha-hemolysin, a heptameric transmembrane pore. Science (New York, NY) 274(5294):1859–1866. https://doi.org/10.1126/science.274.5294.1859
68. Doyle MT, Bernstein HD (2022) Function of the Omp85 superfamily of outer membrane protein assembly factors and polypeptide transporters. Annu Rev Microbiol 76:259–279. https://doi.org/10.1146/annurev-micro-033021-023719
69. Gu Y, Li H, Dong H, Zeng Y, Zhang Z, Paterson NG et al (2016) Structural basis of outer membrane protein insertion by the BAM complex. Nature 531(7592):64–69. https://doi.org/10.1038/nature17199
70. Han L, Zheng J, Wang Y, Yang X, Liu Y, Sun C et al (2016) Structure of the BAM complex and its implications for biogenesis of outer-membrane proteins. Nat Struct Mol Biol 23(3):192–196. https://doi.org/10.1038/nsmb.3181
71. Noinaj N, Kuszak AJ, Gumbart JC, Lukacik P, Chang H, Easley NC et al (2013) Structural insight into the biogenesis of β-barrel membrane proteins. Nature 501(7467):385–390. https://doi.org/10.1038/nature12521
72. Malinverni JC, Werner J, Kim S, Sklar JG, Kahne D, Misra R et al (2006) YfiO stabilizes the YaeT complex and is essential for outer membrane protein assembly in Escherichia coli. Mol Microbiol 61(1):151–164. https://doi.org/10.1111/j.1365-2958.2006.05211.x
73. Wu T, Malinverni J, Ruiz N, Kim S, Silhavy TJ, Kahne D (2005) Identification of a multicomponent complex required for outer membrane biogenesis in Escherichia coli. Cell 121(2):235–245. https://doi.org/10.1016/j.cell.2005.02.015
74. Selkrig J, Mosbahi K, Webb CT, Belousoff MJ, Perry AJ, Wells TJ et al (2012) Discovery of an archetypal protein transport system in bacterial outer membranes. Nat Struct Mol Biol 19(5):506–510. https://doi.org/10.1038/nsmb.2261
75. Webb CT, Heinz E, Lithgow T (2012) Evolution of the β-barrel assembly machinery. Trends Microbiol 20(12):612–620. https://doi.org/10.1016/j.tim.2012.08.006
76. Vandeputte-Rutten L, Kramer RA, Kroon J, Dekker N, Egmond MR, Gros P (2001) Crystal structure of the outer membrane protease OmpT from Escherichia coli suggests a novel catalytic site. EMBO J 20(18):5033–5039. https://doi.org/10.1093/emboj/20.18.5033
77. Eren E, Murphy M, Goguen J, van den Berg B (2010) An active site water network in the plasminogen activator pla from Yersinia pestis. Structure 18(7):809–818. https://doi.org/10.1016/j.str.2010.03.013
78. Dekker N, Cox RC, Kramer RA, Egmond MR (2001) Substrate specificity of the integral membrane protease OmpT determined by

spatially addressed peptide libraries. Biochemistry 40(6):1694–1701. https://doi.org/10.1021/bi0014195

79. Kramer RA, Zandwijken D, Egmond MR, Dekker N (2000) In vitro folding, purification and characterization of Escherichia coli outer membrane protease ompT. Eur J Biochem 267(3):885–893. https://doi.org/10.1046/j.1432-1327.2000.01073.x
80. Kramer RA, Dekker N, Egmond MR (2000) Identification of active site serine and histidine residues in Escherichia coli outer membrane protease OmpT. FEBS Lett 468(2–3):220–224. https://doi.org/10.1016/S0014-5793(00)01231-X
81. Sodeinde OA, Subrahmanyam Y, Stark K, Quan T, Bao Y, Goguen JD (1992) A surface protease and the invasive character of plague. Science 258(5084):1004–1007
82. Snijder HJ, Ubarretxena-Belandia I, Blaauw M, Kalk KH, Verheij HM, Egmond MR et al (1999) Structural evidence for dimerization-regulated activation of an integral membrane phospholipase. Nature 401(6754):717–721. https://doi.org/10.1038/44890
83. Dekker N, Tommassen J, Lustig A, Rosenbusch J, Verheij H (1997) Dimerization regulates the enzymatic activity of Escherichia coli outer membrane phospholipase A. J Biol Chem 272:3179–3184
84. Roumia AF, Theodoropoulou MC, Tsirigos KD, Nielsen H, Bagos PG (2020) Landscape of eukaryotic transmembrane beta barrel proteins. J Proteome Res 19(3):1209–1221. https://doi.org/10.1021/acs.jproteome.9b00740
85. Eaglesfield R, Tokatlidis K (2021) Targeting and insertion of membrane proteins in mitochondria. Front Cell Dev Biol 9:803205
86. Bay DC, Hafez M, Young MJ, Court DA (2012) Phylogenetic and coevolutionary analysis of the β-barrel protein family comprised of mitochondrial porin (VDAC) and Tom40. Biochim Biophys Acta 1818(6):1502–1519. https://doi.org/10.1016/j.bbamem.2011.11.027
87. Tucker K, Park E (2019) Cryo-EM structure of the mitochondrial protein-import channel TOM complex at near-atomic resolution. Nat Struct Mol Biol 26(12):1158–1166. https://doi.org/10.1038/s41594-019-0339-2
88. Ujwal R, Cascio D, Colletier J-P, Faham S, Zhang J, Toro L et al (2008) The crystal structure of mouse VDAC1 at 2.3 Å resolution reveals mechanistic insights into metabolite gating. Proc Natl Acad Sci 105(46):17742–17747. https://doi.org/10.1073/pnas.0809634105
89. Takeda H, Tsutsumi A, Nishizawa T, Lindau C, Busto JV, Wenz LS et al (2021) Mitochondrial sorting and assembly machinery operates by β-barrel switching. Nature 590(7844):163–169. https://doi.org/10.1038/s41586-020-03113-7
90. Colombini M (2004) VDAC: the channel at the interface between mitochondria and the cytosol. Mol Cell Biochem 256–257(1–2):107–115. https://doi.org/10.1023/b:mcbi.0000009862.17396.8d
91. Bausewein T, Mills DJ, Langer JD, Nitschke B, Nussberger S, Kühlbrandt W (2017) Cryo-EM structure of the TOM core complex from Neurospora crassa. Cell 170(4):693–700.e7. https://doi.org/10.1016/j.cell.2017.07.012
92. Diederichs KA, Ni X, Rollauer SE, Botos I, Tan X, King MS et al (2020) Structural insight into mitochondrial β-barrel outer membrane protein biogenesis. Nat Commun 11(1):1–13. https://doi.org/10.1038/s41467-020-17144-1
93. Chacinska A, Koehler CM, Milenkovic D, Lithgow T, Pfanner N (2009) Importing mitochondrial proteins: machineries and mechanisms. Cell 138(4):628–644. https://doi.org/10.1016/j.cell.2009.08.005
94. Meisinger C, Pfannschmidt S, Rissler M, Milenkovic D, Becker T, Stojanovski D et al (2007) The morphology proteins Mdm12/Mmm1 function in the major β-barrel assembly pathway of mitochondria. EMBO J 26(9):2229–2239. https://doi.org/10.1038/sj.emboj.7601673
95. Meisinger C, Rissler M, Chacinska A, Szklarz LKS, Milenkovic D, Kozjak V et al (2004) The mitochondrial morphology protein Mdm10 functions in assembly of the preprotein translocase of the outer membrane. Dev Cell 7(1):61–71. https://doi.org/10.1016/j.devcel.2004.06.003
96. Qiu J, Wenz LS, Zerbes RM, Oeljeklaus S, Bohnert M, Stroud DA et al (2013) Coupling of mitochondrial import and export translocases by receptor-mediated supercomplex formation. Cell 154(3):596–608. https://doi.org/10.1016/j.cell.2013.06.033
97. Lionello S, Marzaro G, Martinvalet D (2020) SAM50, a side door to the mitochondria: the case of cytotoxic proteases. Pharmacol Res 160:105196. https://doi.org/10.1016/j.phrs.2020.105196
98. Jian F, Chen D, Chen L, Yan C, Lu B, Zhu Y et al (2018) Sam50 regulates PINK1-parkin-

mediated mitophagy by controlling PINK1 stability and mitochondrial morphology. Cell Rep 23(10):2989–3005. https://doi.org/10.1016/j.celrep.2018.05.015

99. Tang J, Zhang K, Dong J, Yan C, Hu C, Ji H et al (2020) Sam50–Mic19–Mic60 axis determines mitochondrial cristae architecture by mediating mitochondrial outer and inner membrane contact. Cell Death Differ 27(1):146–160. https://doi.org/10.1038/s41418-019-0345-2

100. Bender A, Desplats P, Spencer B, Rockenstein E, Adame A, Elstner M et al (2013) TOM40 mediates mitochondrial dysfunction induced by α-synuclein accumulation in Parkinson's disease. PLoS One 8(4):e62277. https://doi.org/10.1371/journal.pone.0062277

101. Hemmler R, Becker T, Schleiff E, Bölter B, Stahl T, Soll J et al (2006) Molecular properties of Oep21, an ATP-regulated anion-selective solute channel from the outer chloroplast membrane. J Biol Chem 281(17):12020–12029. https://doi.org/10.1074/jbc.M513586200

102. Goetze TA, Patil M, Jeshen I, Bölter B, Grahl S, Soll J (2015) Oep23 forms an ion channel in the chloroplast outer envelope. BMC Plant Biol 15:47. https://doi.org/10.1186/s12870-015-0445-1

103. Pohlmeyer K, Soll J, Grimm R, Hill K, Wagner R (1998) A high-conductance solute channel in the chloroplastic outer envelope from Pea. Plant Cell 10(7):1207–1216. https://doi.org/10.1105/tpc.10.7.1207

104. Goetze TA, Philippar K, Ilkavets I, Soll J, Wagner R (2006) OEP37 is a new member of the chloroplast outer membrane ion channels. J Biol Chem 281(26):17989–17998. https://doi.org/10.1074/jbc.M600700200

105. Harsman A, Schock A, Hemmis B, Wahl V, Jeshen I, Bartsch P et al (2016) OEP40, a regulated glucose-permeable β-barrel solute channel in the chloroplast outer envelope membrane. J Biol Chem 291(34):17848–17860. https://doi.org/10.1074/jbc.M115.712398

106. Wang Z, Xu C, Benning C (2012) TGD4 involved in endoplasmic reticulum-to-chloroplast lipid trafficking is a phosphatidic acid binding protein. Plant J 70(4):614–623. https://doi.org/10.1111/j.1365-313X.2012.04900.x

107. Hinnah SC, Wagner R, Sveshnikova N, Harrer R, Soll J (2002) The chloroplast protein import channel Toc75: pore properties and interaction with transit peptides. Biophys J 83(2):899–911. https://doi.org/10.1016/s0006-3495(02)75216-8

108. Hsu SC, Nafati M, Inoue K (2012) OEP80, an essential protein paralogous to the chloroplast protein translocation channel Toc75, exists as a 70-kD protein in the Arabidopsis thaliana chloroplast outer envelope. Plant Mol Biol 78(1–2):147–158. https://doi.org/10.1007/s11103-011-9853-2

109. Röhl T, Motzkus M, Soll J (1999) The outer envelope protein OEP24 from pea chloroplasts can functionally replace the mitochondrial VDAC in yeast. FEBS Lett 460(3):491–494. https://doi.org/10.1016/s0014-5793(99)01399-x

110. Schleiff E, Eichacker LA, Eckart K, Becker T, Mirus O, Stahl T et al (2003) Prediction of the plant beta-barrel proteome: a case study of the chloroplast outer envelope. Protein Sci 12(4):748–759. https://doi.org/10.1110/ps.0237503

111. Simmerman RF, Dave AM, Bruce BD (2014) Structure and function of POTRA domains of Omp85/TPS superfamily. Int Rev Cell Mol Biol 308:1–34. https://doi.org/10.1016/b978-0-12-800097-7.00001-4

112. Gessmann D, Chung YH, Danoff EJ, Plummer AM, Sandlin CW, Zaccai NR et al (2014) Outer membrane β-barrel protein folding is physically controlled by periplasmic lipid head groups and BamA. Proc Natl Acad Sci U S A 111(16):5878–5883. https://doi.org/10.1073/PNAS.1322473111/SUPPL_FILE/SAPP.PDF

113. Baldwin A, Wardle A, Patel R, Dudley P, Park SK, Twell D et al (2005) A molecular-genetic study of the Arabidopsis Toc75 gene family. Plant Physiol 138(2):715–733. https://doi.org/10.1104/pp.105.063289

114. Arkowitz RA, Joly JC, Wickner W (1993) Translocation can drive the unfolding of a preprotein domain. EMBO J 12(1):243–253. https://doi.org/10.1002/j.1460-2075.1993.tb05650.x

115. Kumamoto CA, Beckwith J (1983) Mutations in a new gene, secB, cause defective protein localization in Escherichia coli. J Bacteriol 154(1):253–260. https://doi.org/10.1128/jb.154.1.253-260.1983

116. Weiss JB, Ray PH, Bassford PJ (1988) Purified secB protein of Escherichia coli retards folding and promotes membrane translocation of the maltose-binding protein in vitro. Proc Natl Acad Sci U S A 85(23):8978–8982. https://doi.org/10.1073/pnas.85.23.8978

117. Walther DM, Rapaport D (2009) Biogenesis of mitochondrial outer membrane proteins.

Biochim Biophys Acta 1793(1):42–51. https://doi.org/10.1016/J.BBAMCR.2008.04.013
118. Iadanza MG, Higgins AJ, Schiffrin B, Calabrese AN, Brockwell DJ, Ashcroft AE et al (2016) Lateral opening in the intact β-barrel assembly machinery captured by cryo-EM. Nat Commun 7:12865. https://doi.org/10.1038/ncomms12865
119. Noinaj N, Gumbart JC, Buchanan SK (2017) The β-barrel assembly machinery in motion. Nat Rev Microbiol 15(4):197–204. https://doi.org/10.1038/nrmicro.2016.191
120. Bakelar J, Buchanan SK, Noinaj N (2016) The structure of the β-barrel assembly machinery complex. Science (New York, NY) 351(6269):180–186. https://doi.org/10.1126/science.aad3460
121. Iadanza MG, Schiffrin B, White P, Watson MA, Horne JE, Higgins AJ et al (2020) Distortion of the bilayer and dynamics of the BAM complex in lipid nanodiscs. Commun Biol 3:766. https://doi.org/10.1038/s42003-020-01419-w
122. Liu J, Gumbart JC (2020) Membrane thinning and lateral gating are consistent features of BamA across multiple species. PLoS Comput Biol 16(10):e1008355. https://doi.org/10.1371/journal.pcbi.1008355
123. Schiffrin B, Calabrese AN, Higgins AJ, Humes JR, Ashcroft AE, Kalli AC et al (2017) Effects of periplasmic chaperones and membrane thickness on BamA-catalyzed outer-membrane protein folding. J Mol Biol 429(23):3776–3792. https://doi.org/10.1016/j.jmb.2017.09.008
124. Noinaj N, Rollauer SE, Buchanan SK (2015) The β-barrel membrane protein insertase machinery from Gram-negative bacteria. Curr Opin Struct Biol 31:35–42. https://doi.org/10.1016/j.sbi.2015.02.012
125. Doerner PA, Sousa MC (2017) Extreme dynamics in the BamA β-barrel seam. Biochemistry 56(24):3142–3149. https://doi.org/10.1021/acs.biochem.7b00281
126. Patel GJ, Kleinschmidt JH (2013) The lipid bilayer-inserted membrane protein BamA of Escherichia coli facilitates insertion and folding of outer membrane protein a from its complex with Skp. Biochemistry 52(23):3974–3986. https://doi.org/10.1021/bi400103t
127. Schiffrin B, Brockwell DJ, Radford SE (2017) Outer membrane protein folding from an energy landscape perspective. BMC Biol 15(1):123. https://doi.org/10.1186/s12915-017-0464-5
128. Doyle MT, Bernstein HD (2019) Bacterial outer membrane proteins assemble via asymmetric interactions with the BamA β-barrel. Nat Commun 10(1):3358. https://doi.org/10.1038/s41467-019-11230-9
129. Höhr AIC, Lindau C, Wirth C, Qiu J, Stroud DA, Kutik S et al (2018) Membrane protein insertion through a mitochondrial β-barrel gate. Science 359(6373):eaah6834. https://doi.org/10.1126/science.aah6834
130. Lee J, Tomasek D, Santos TM, May MD, Meuskens I, Kahne D (2019) Formation of a β-barrel membrane protein is catalyzed by the interior surface of the assembly machine protein BamA. elife 8:e49787. https://doi.org/10.7554/eLife.49787
131. Lee J, Xue M, Wzorek JS, Wu T, Grabowicz M, Gronenberg LS et al (2016) Characterization of a stalled complex on the β-barrel assembly machine. Proc Natl Acad Sci U S A 113(31):8717–8722. https://doi.org/10.1073/pnas.1604100113
132. Tomasek D, Rawson S, Lee J, Wzorek JS, Harrison SC, Li Z et al (2020) Structure of a nascent membrane protein as it folds on the Bam complex. Nature 583(7816):473–478. https://doi.org/10.1038/s41586-020-2370-1
133. Kuo KM, Ryoo D, Lundquist K, Gumbart JC (2022) Modeling intermediates of BamA folding an outer membrane protein. Biophys J 121(17):3242–3252. https://doi.org/10.1016/j.bpj.2022.07.027
134. Sklar JG, Wu T, Kahne D, Silhavy TJ (2007) Defining the roles of the periplasmic chaperones SurA, Skp, and DegP in Escherichia coli. Genes Dev 21(19):2473–2484. https://doi.org/10.1101/gad.1581007
135. Schiffrin B, Machin JM, Karamanos TK, Zhuravleva A, Brockwell DJ, Radford SE et al (2022) Dynamic interplay between the periplasmic chaperone SurA and the BAM complex in outer membrane protein folding. Commun Biol 5:560. https://doi.org/10.1038/s42003-022-03502-w
136. Schmidt O, Pfanner N, Meisinger C (2010) Mitochondrial protein import: from proteomics to functional mechanisms. Nat Rev Mol Cell Biol 11(9):655–667. https://doi.org/10.1038/nrm2959
137. Jiang J-H, Tong J, Tan KS, Gabriel K (2012) From evolution to pathogenesis: the link between β-barrel assembly machineries in the outer membrane of mitochondria and Gram-negative bacteria. Int J Mol Sci 13(7):8038–8050. https://doi.org/10.3390/ijms13078038

138. Kutik S, Stojanovski D, Becker L, Becker T, Meinecke M, Krüger V et al (2008) Dissecting membrane insertion of mitochondrial β-barrel proteins. Cell 132(6):1011–1024. https://doi.org/10.1016/j.cell.2008.01.028
139. Chan NC, Lithgow T (2008) The peripheral membrane subunits of the SAM complex function codependently in mitochondrial outer membrane biogenesis. Mol Biol Cell 19(1):126–136. https://doi.org/10.1091/mbc.E07-08-0796
140. Thornton N, Stroud DA, Milenkovic D, Guiard B, Pfanner N, Becker T (2010) Two modular forms of the mitochondrial sorting and assembly machinery are involved in biogenesis of alpha-helical outer membrane proteins. J Mol Biol 396(3):540–549. https://doi.org/10.1016/j.jmb.2009.12.026
141. Wiedemann N, Meisinger C, Pfanner N (2009) Cell biology. Connecting organelles. Science 325(5939):403–404. https://doi.org/10.1126/science.1178016
142. Kornmann B, Currie E, Collins SR, Schuldiner M, Nunnari J, Weissman JS et al (2009) An ER-mitochondria tethering complex revealed by a synthetic biology screen. Science (New York, NY) 325(5939):477–481. https://doi.org/10.1126/science.1175088
143. Ott C, Ross K, Straub S, Thiede B, Götz M, Goosmann C et al (2012) Sam50 functions in mitochondrial intermembrane space bridging and biogenesis of respiratory complexes. Mol Cell Biol 32(6):1173–1188. https://doi.org/10.1128/MCB.06388-11
144. Gupta RS (2011) Origin of diderm (Gram-negative) bacteria: antibiotic selection pressure rather than endosymbiosis likely led to the evolution of bacterial cells with two membranes. Antonie Van Leeuwenhoek 100(2):171–182. https://doi.org/10.1007/s10482-011-9616-8
145. Cavalier-Smith T (2006) Rooting the tree of life by transition analyses. Biol Direct 1:19. https://doi.org/10.1186/1745-6150-1-19
146. Megrian D, Taib N, Witwinowski J, Beloin C, Gribaldo S (2020) One or two membranes? Diderm Firmicutes challenge the Gram-positive/Gram-negative divide. Mol Microbiol 113(3):659–671. https://doi.org/10.1111/mmi.14469
147. Dhar R, Slusky JS (2021) Outer membrane protein evolution. Curr Opin Struct Biol 68:122–128. https://doi.org/10.1016/j.sbi.2021.01.002
148. Remmert M, Biegert A, Linke D, Lupas AN, Söding J (2010) Evolution of outer membrane beta-barrels from an ancestral beta beta hairpin. Mol Biol Evol 27(6):1348–1358. https://doi.org/10.1093/molbev/msq017
149. Arnold T, Poynor M, Nussberger S, Lupas AN, Linke D (2007) Gene duplication of the eight-stranded beta-barrel OmpX produces a functional pore: a scenario for the evolution of transmembrane beta-barrels. J Mol Biol 366(4):1174–1184. https://doi.org/10.1016/j.jmb.2006.12.029
150. Franklin MW, Nepomnyachiy S, Feehan R, Ben-Tal N, Kolodny R, Slusky JSG (2018) Efflux pumps represent possible evolutionary convergence onto the β-barrel fold. Structure 26(9):1266–74.e2. https://doi.org/10.1016/j.str.2018.06.007
151. Stapleton JA, Whitehead TA, Nanda V (2015) Computational redesign of the lipid-facing surface of the outer membrane protein OmpA. Proc Natl Acad Sci U S A 112(31):9632–9637. https://doi.org/10.1073/pnas.1501836112
152. Pereira J, Lupas AN (2018) The origin of mitochondria-specific outer membrane β-barrels from an ancestral bacterial fragment. Genome Biol Evol 10(10):2759–2765. https://doi.org/10.1093/gbe/evy216
153. Wiedemann N, Kozjak V, Chacinska A, Schönfisch B, Rospert S, Ryan MT et al (2003) Machinery for protein sorting and assembly in the mitochondrial outer membrane. Nature 424(6948):565–571. https://doi.org/10.1038/nature01753
154. Töpel M, Ling Q, Jarvis P (2012) Neofunctionalization within the Omp85 protein superfamily during chloroplast evolution. Plant Signal Behav 7(2):161–164. https://doi.org/10.4161/psb.18677
155. Walther DM, Rapaport D, Tommassen J (2009) Biogenesis of β-barrel membrane proteins in bacteria and eukaryotes: evolutionary conservation and divergence. Cell Mol Life Sci 66(17):2789–2804. https://doi.org/10.1007/s00018-009-0029-z
156. Su CC, Lyu M, Morgan CE, Bolla JR, Robinson CV, Yu EW (2021) A 'Build and Retrieve' methodology to simultaneously solve cryo-EM structures of membrane proteins. Nat Methods 18(1):69–75. https://doi.org/10.1038/s41592-020-01021-2
157. Bernhofer M, Rost B (2022) TMbed: transmembrane proteins predicted through language model embeddings. BMC

Bioinformatics 23(1):326. https://doi.org/10.1186/s12859-022-04873-x

158. Bigelow HR, Petrey DS, Liu J, Przybylski D, Rost B (2004) Predicting transmembrane beta-barrels in proteomes. Nucleic Acids Res 32(8):2566–2577. https://doi.org/10.1093/nar/gkh580
159. Montezano D, Bernstein R, Copeland MM, Slusky JSG (2022) Is it a barrel? General features of transmembrane beta barrels from a large database. bioRxiv:2022.04.13.488089. https://doi.org/10.1101/2022.04.13.488089
160. Vergalli J, Bodrenko IV, Masi M, Moynié L, Acosta-Gutiérrez S, Naismith JH et al (2020) Porins and small-molecule translocation across the outer membrane of Gram-negative bacteria. Nat Rev Microbiol 18(3):164–176. https://doi.org/10.1038/s41579-019-0294-2
161. Zgurskaya HI, López CA, Gnanakaran S (2015) Permeability barrier of Gram-negative cell envelopes and approaches to bypass it. ACS Infect Dis 1(11):512–522. https://doi.org/10.1021/acsinfecdis.5b00097
162. Srinivas N, Jetter P, Ueberbacher BJ, Werneburg M, Zerbe K, Steinmann J et al (2010) Peptidomimetic antibiotics target outer-membrane biogenesis in Pseudomonas aeruginosa. Science 327(5968):1010–1013. https://doi.org/10.1126/science.1182749
163. Andolina G, Bencze LC, Zerbe K, Müller M, Steinmann J, Kocherla H et al (2018) A peptidomimetic antibiotic interacts with the periplasmic domain of LptD from Pseudomonas aeruginosa. ACS Chem Biol 13(3):666–675. https://doi.org/10.1021/acschembio.7b00822
164. Vila-Farrés X, Parra-Millán R, Sánchez-Encinales V, Varese M, Ayerbe-Algaba R, Bayó N et al (2017) Combating virulence of Gram-negative bacilli by OmpA inhibition. Sci Rep 7(1):14683. https://doi.org/10.1038/s41598-017-14972-y
165. Steenhuis M, van Ulsen P, Martin NI, Luirink J (2021) A ban on BAM: an update on inhibitors of the β-barrel assembly machinery. FEMS Microbiol Lett 368(11). https://doi.org/10.1093/femsle/fnab059
166. Imai Y, Meyer KJ, Iinishi A, Favre-Godal Q, Green R, Manuse S et al (2019) A new antibiotic selectively kills Gram-negative pathogens. Nature 576(7787):459–464. https://doi.org/10.1038/s41586-019-1791-1
167. Kaur H, Jakob RP, Marzinek JK, Green R, Imai Y, Bolla JR et al (2021) The antibiotic darobactin mimics a β-strand to inhibit outer membrane insertase. Nature 593(7857):125–129. https://doi.org/10.1038/s41586-021-03455-w
168. Xiao L, Han L, Li B, Zhang M, Zhou H, Luo Q et al (2021) Structures of the β-barrel assembly machine recognizing outer membrane protein substrates. FASEB J 35(1):e21207. https://doi.org/10.1096/fj.202001443RR
169. Hart EM, Mitchell AM, Konovalova A, Grabowicz M, Sheng J, Han X et al (2019) A small-molecule inhibitor of BamA impervious to efflux and the outer membrane permeability barrier. Proc Natl Acad Sci U S A 116(43):21748–21757. https://doi.org/10.1073/pnas.1912345116
170. Luther A, Urfer M, Zahn M, Müller M, Wang SY, Mondal M et al (2019) Chimeric peptidomimetic antibiotics against Gram-negative bacteria. Nature 576(7787):452–458. https://doi.org/10.1038/s41586-019-1665-6
171. Urfer M, Bogdanovic J, Lo Monte F, Moehle K, Zerbe K, Omasits U et al (2016) A peptidomimetic antibiotic targets outer membrane proteins and disrupts selectively the outer membrane in Escherichia coli. J Biol Chem 291(4):1921–1932. https://doi.org/10.1074/jbc.M115.691725
172. Ghequire MGK, Swings T, Michiels J, Buchanan SK, De Mot R (2018) Hitting with a BAM: selective killing by lectin-like bacteriocins. MBio 9(2). https://doi.org/10.1128/mBio.02138-17
173. Storek KM, Auerbach MR, Shi H, Garcia NK, Sun D, Nickerson NN et al (2018) Monoclonal antibody targeting the β-barrel assembly machine of *Escherichia coli* is bactericidal. Proc Natl Acad Sci 115(14):3692–3697. https://doi.org/10.1073/pnas.1800043115
174. van Ulsen P, Zinner KM, Jong WSP, Luirink J (2018) On display: autotransporter secretion and application. FEMS Microbiol Lett 365(18). https://doi.org/10.1093/femsle/fny165
175. Larimi MG, Ha JH, Loh SN, Movileanu L (2021) Insertion state of modular protein nanopores into a membrane. Biochim Biophys Acta Biomembr 1863(5):183570. https://doi.org/10.1016/j.bbamem.2021.183570
176. Krewinkel M, Dworeck T, Fioroni M (2011) Engineering of an E. coli outer membrane protein FhuA with increased channel diameter. J Nanobiotechnology 9:33. https://doi.org/10.1186/1477-3155-9-33

177. Liu Z, Ghai I, Winterhalter M, Schwaneberg U (2017) Engineering enhanced pore sizes using FhuA Δ1-160 from E. coli outer membrane as template. ACS Sens 2(11): 1619–1626. https://doi.org/10.1021/acssensors.7b00481
178. Mohammad MM, Howard KR, Movileanu L (2011) Redesign of a plugged beta-barrel membrane protein. J Biol Chem 286(10): 8000–8013. https://doi.org/10.1074/jbc.M110.197723
179. Choudhary OP, Ujwal R, Kowallis W, Coalson R, Abramson J, Grabe M (2010) The electrostatics of VDAC: implications for selectivity and gating. J Mol Biol 396(3): 580–592. https://doi.org/10.1016/j.jmb.2009.12.006
180. Chaturvedi D, Mahalakshmi R (2018) Folding determinants of transmembrane β-barrels using engineered OMP chimeras. Biochemistry 57(13):1987–1996. https://doi.org/10.1021/acs.biochem.8b00012
181. Gibson K, Chu JK, Zhu S, Nguyen D, Mrázek J, Liu J et al (2022) A tripartite efflux system affects flagellum stability in Helicobacter pylori. Int J Mol Sci 23(19). https://doi.org/10.3390/ijms231911609
182. Wu R, Bakelar JW, Lundquist K, Zhang Z, Kuo KM, Ryoo D et al (2021) Plasticity within the barrel domain of BamA mediates a hybrid-barrel mechanism by BAM. Nat Commun 12(1):7131. https://doi.org/10.1038/s41467-021-27449-4
183. Fleming PJ, Patel DS, Wu EL, Qi Y, Yeom MS, Sousa MC et al (2016) BamA POTRA domain interacts with a native lipid membrane surface. Biophys J 110(12): 2698–2709. https://doi.org/10.1016/j.bpj.2016.05.010
184. Kipnis Y, Chaib AO, Vorobieva AA, Cai G, Reggiano G, Basanta B et al (2022) Design and optimization of enzymatic activity in a de novo β-barrel scaffold. Protein Sci 31(11): e4405. https://doi.org/10.1002/pro.4405
185. Yue WW, Grizot S, Buchanan SK (2003) Structural evidence for iron-free citrate and ferric citrate binding to the TonB-dependent outer membrane transporter FecA. J Mol Biol 332(2):353–368. https://doi.org/10.1016/s0022-2836(03)00855-6
186. Zhang M, Shi H, Zhang X, Zhang X, Huang Y (2020) Cryo-EM structure of the nonameric CsgG-CsgF complex and its implications for controlling curli biogenesis in Enterobacteriaceae. PLoS Biol 18(6): e3000748. https://doi.org/10.1371/journal.pbio.3000748
187. Wang Q, Guan Z, Qi L, Zhuang J, Wang C, Hong S et al (2021) Structural insight into the SAM-mediated assembly of the mitochondrial TOM core complex. Science 0704 (August):eabh0704. https://doi.org/10.1126/science.abh0704
188. Jin Z, Wan L, Zhang Y, Li X, Cao Y, Liu H et al (2022) Structure of a TOC-TIC supercomplex spanning two chloroplast envelope membranes. Cell 185(25):4788–800.e13. https://doi.org/10.1016/j.cell.2022.10.030
189. Pettersen EF, Goddard TD, Huang CC, Meng EC, Couch GS, Croll TI et al (2021) UCSF ChimeraX: structure visualization for researchers, educators, and developers. Protein Sci 30(1):70–82. https://doi.org/10.1002/pro.3943

Chapter 2

Recombinant Expression and Overproduction of Transmembrane β-Barrel Proteins

Ina Meuskens, Jack C. Leo, and Dirk Linke

Abstract

Transmembrane β-barrel proteins reside in the outer membrane of Gram-negative bacteria and are thus in direct contact with the environment. Because of that, they are involved in many key processes stretching from cellular survival to virulence. Hence, they are an attractive target for the development of novel antimicrobials, in addition to being of fundamental biological interest. To study this class of proteins, they are often required to be expressed in *Escherichia coli*. Recombinant expression of β-barrel proteins can be achieved using two fundamentally different strategies. The first alternative uses a complete coding sequence that includes a signal peptide for targeting the protein to its native cellular location, the bacterial outer membrane. The second alternative omits the signal peptide in the gene, leading to mislocalization and aggregation of the protein in the bacterial cytoplasm. These aggregates, called inclusion bodies, can be solubilized and the protein can be folded into its native form in vitro. In this chapter, we present example protocols for both strategies and discuss their advantages and disadvantages.

Key words Outer membrane protein, Protein expression, Membrane protein folding, Inclusion bodies, Detergents, Membrane insertion

1 Introduction

β-Barrel outer membrane proteins only occur in the outer membrane of Gram-negative bacteria and the outer membranes of eukaryotic organelles that originate from bacteria (mitochondria and plastids) [1–3]. They are fundamentally different in structure from other membrane proteins, as they do not contain hydrophobic transmembrane α-helices, but are made from amphipathic, antiparallel β-strands that fold into a β-sheet, where the first and last strands then interact to form the resulting barrel shape [3, 4].

There are different reasons for expressing outer membrane β-barrel proteins (OMPs) recombinantly. It might be that the protein of interest is expressed at extremely low levels natively or that it originates from an organism that is not easily grown in the lab. For structural biology or substrate binding experiments, it

Raffaele Ieva (ed.), *Transmembrane β-Barrel Proteins: Methods and Protocols*, Methods in Molecular Biology, vol. 2778,
https://doi.org/10.1007/978-1-0716-3734-0_2,

might be desirable to have large quantities of purified protein available. Last but not least, labeling with natural or non-natural amino acids for photo-crosslinking [5], labeling with recombinant tags for antibody-based detection or for purification [6], spin labeling [7], or labeling with radioactive [8] or non-radioactive isotopes [9] will typically require recombinant approaches.

For the purpose of this chapter, we will only consider recombinant expression in a Gram-negative bacterial host (typically *Escherichia coli*). In Gram-negative bacteria in vivo, OMPs take a complex path from the point of synthesis, the ribosome, to their final destination, the outer membrane [10]. For the purpose of recombinant expression, many of the factors and checkpoints in this transport and folding pathway constitute potential bottlenecks. The presence or absence of an N-terminal signal peptide determines whether the nascent polypeptide leaving the ribosome is secreted through the inner membrane general secretory (SEC) machinery to the periplasmic space [11], and not all signal peptides are equally efficient in guiding the recombinant protein of interest via the SEC machinery [12]. Periplasmic chaperones and proteases are involved in quality control and in keeping the polypeptide in an unfolded state that is competent for outer membrane insertion [10]. In *E. coli*, these include, for example, the chaperones Skp and SurA [13] and the periplasmic protease DegP that can degrade misfolded β-barrel proteins in transit [14] and that has an alternative chaperone function at low temperatures [15]. In addition, periplasmic prolyl-*cis-trans* isomerases and enzymes that catalyze the formation of disulfide bonds can directly affect recombinant protein expression in the outer membrane [16–18]. Recombinant transmembrane β-barrel proteins, after passing all these steps, will then be recognized by the β-barrel assembly machinery (BAM) through a C-terminal sequence motif and inserted into the outer membrane [19, 20]. This motif is to some extent species-specific, which might affect heterologous expression in *E. coli* [21, 22].

Transmembrane β-barrel proteins can be recombinantly expressed without an N-terminal signal peptide. This circumvents the complex machineries for export and membrane insertion mentioned above but does not yield properly folded, membrane-inserted protein. Instead, so-called "inclusion bodies" are formed in the *E.coli* cytosol. Inclusion bodies are protein aggregates that typically contain extremely high quantities of the recombinant protein, but in an unfolded or misfolded state [23]. Additionally, they might contain co-precipitated cellular chaperones, RNA, and other cellular components [24]. It is worth noting that in some cases, also expression of constructs that include signal peptides can yield inclusion bodies—these are then typically found in the periplasm [11]. Inclusion bodies of transmembrane β-barrel proteins have been described as amyloid-like structures, because the amphipathic β-strands can easily form β-sheet interactions intermolecularly (between neighboring polypeptides in the aggregates) [25, 26]. These strong interactions make it relatively easy to separate inclusion bodies from other cellular components [27]. To yield properly folded β-barrel proteins, the interactions then need to be broken by the use of chaotropic reagents, and the resulting solubilized, unfolded polypeptides then need to be refolded using detergents [28, 29] or lipids [30] to mimic the natural outer membrane environment [31, 32].

Independent of whether expression was performed in the outer membrane or in inclusion bodies with subsequent refolding, each individual protein has to be checked for proper folding and, if appropriate, for membrane insertion [33]. This chapter can only give a glimpse of the two fundamentally different expression strategies and provides sample protocols for both cases (*see* **Note 1**). The protocols are restricted to expression, as downstream protocols for purification (or refolding if applicable) can be highly variable for different recombinant transmembrane β-barrel proteins (Fig. 1).

2 Materials

All buffers and solutions must be prepared using ultrapure water (>18 MΩcm at 25 °C) and chemicals should conform to reagent grade.

2.1 Equipment

1. Floor centrifuge with rotor for 500 mL or 1 L bottles.
2. Benchtop centrifuge with rotors for 50 mL Falcon tubes and for 1.5 mL reaction tubes.
3. Temperature-controlled shaker for small culture tubes and for 2 L Erlenmeyer flasks.
4. French Press (or an alternative device for rupturing bacterial cells).
5. Chromatography system including columns for protein purification (affinity purification columns, size exclusion columns, and ion-exchange columns). This item is recommended, but gravity-flow columns can be used for most applications as a cheaper alternative.

2.2 Expression in the Native Membrane

1. *E. coli* BL21 (DE3) Gold ΔABCF [34]. This strain lacks the genes encoding for the most abundant outer membrane proteins OmpA, OmpC, OmpF, and LamB (maltoporin) (*see* **Note 2**).
2. pET3b_OmpX_HA [34]. pET vectors are IPTG-inducible and require expression strains that harbor the DE3 lysogen to express the protein of interest. The plasmid used for this protocol harbors the gene encoding for OmpX from *E. coli*, which

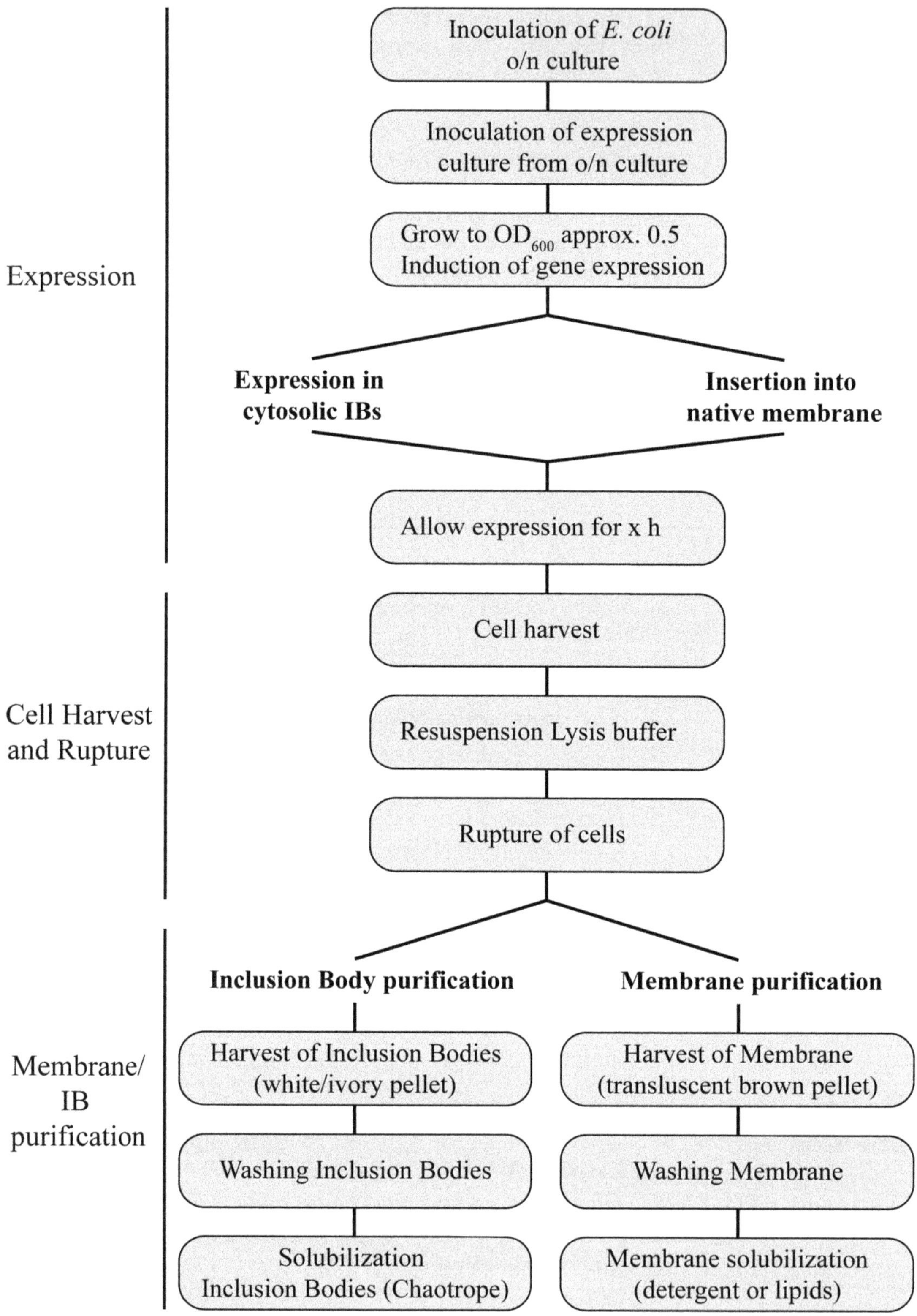

Fig. 1 Flowchart showing the alternative routes for expression and purification of OMPs. o/n, overnight; IB, inclusion bodies

was additionally double-tagged with an HA-tag for better detection, e.g., in Western blots.

3. Lysogeny broth (LB—Lennox variant): 10 g/L tryptone, 5 g/L yeast extract, 5 g/L NaCl. Sterilize using an autoclave at standard settings.
4. Ampicillin (Amp). Supplemented to autoclaved and cooled-down LB media to a final concentration of 100 mg/L.
5. Erlenmeyer flasks (250 mL).
6. Baffled Erlenmeyer flasks (2 L).
7. LB-Amp agar plates. Add 1.5 g/L agar to LB medium (not containing antibiotics). Autoclave to dissolve the agar. Shake well and let cool before adding 100 μg/mL ampicillin. Pour plates immediately (ca. 25 mL medium per 10 cm Petri dish). Let the agar fully solidify before use.
8. Lysis buffer: 50 mL of buffer containing 50 mM Tris–HCl pH 8.0, 100 mM NaCl, 1 mM $MgCl_2$. Sterile-filter. Directly before use, add a pinch of lyophilized DNAse I and 8 μg/mL lysozyme and shake well (*see* **Note 3**).
9. Wash buffer: 50 mM Tris–HCl pH 8.0.
10. Inner membrane solubilization buffer: 50 mM Tris–HCl pH 8.0, 1% *N*-laurylsarcosine. *N*-laurylsarcosine selectively solubilized the inner membrane and ensures less contamination of the outer membrane protein of interest [35].
11. Outer membrane solubilization buffer: 50 mM Tris–HCl pH 8.0, 100 mM NaCl, 1 mM EDTA, 1% (w/v) sulfobetaine SB-12 (*see* **Note 4**).

2.3 Expression in Inclusion Bodies

1. *E. coli* BL21 C41 (DE3) [36]. This strain not only carries mutations for enhanced expression of membrane proteins but also generates inclusion bodies with high yield in our hands.
2. pET3b_OmpX [37, 38].
3. LB—Lennox variant.
4. LB—Amp agar plates.
5. Small culture tube for a 5 mL preculture.
6. Baffled Erlenmeyer flask (2 L) for expression.
7. Induction stock solution: 1 M IPTG (isopropyl-β-D-thiogalactopyranoside). Store 1 mL aliquots in closed reaction tubes at −20 °C. Use at 1:1000 dilution.
8. Lysis buffer: 50 mM Tris–HCl pH 8.0, 50 mM NaCl, 1 mM $MgCl_2$. Sterile-filter. Directly before use, add a pinch of lyophilized DNAse I and shake well (*see* **Note 3**).
9. Wash buffer: 1% (v/v) Triton X-100, 50 mM Tris–HCl pH 8.0.

10. Solubilization buffer: 8 M urea, 50 mM Tris–HCl pH 8, 1 mM DTT.
11. Refolding buffer: 1% (w/v) sulfobetaine SB-12, 50 mM Tris–HCl pH 8.0 (*see* **Note 4**).

3 Methods

3.1 Expression in the Native Membrane

This protocol is based on published protocols for the expression of different OMPs in the native outer membrane of *E. coli* [34, 39].

1. Transform *E. coli* BL21 (DE3) ΔABCF with pET3-b_OmpX_HA using $CaCl_2$-competent cells and standard protocols [40]. Plate the cells on LB agar plates containing a final concentration of 100 mg/L ampicillin.
2. Select a single colony from the plate after overnight growth at 30 °C and inoculate in 50 mL LB medium in a 250 mL Erlenmeyer flask. Incubate this preculture in a shaking incubator at 200 rpm, 30 °C overnight.
3. The next morning, inoculate 5 mL of the preculture in a new subculture of 500 mL LB-Amp medium and incubate at 30 °C shaking at 100–150 rpm. Grow the culture to an OD_{600} of 0.5 and induce by supplementing IPTG to a final concentration of 1 mM. Allow expression for 4 h at 30 °C in a shaking incubator set to 150 rpm.
4. Harvest the bacterial cells by centrifugation at 4500 × g using suitable centrifuge bottles (e.g., 500 mL bottles) for 10 min at 4 °C. Cell pellets can be frozen and stored at −20 °C for later use, or kept at 4 °C and directly subjected to the protein purification process.
5. Completely resuspend the cell pellet in lysis buffer (approx. 25 mL/4 g of cell pellet).
6. Lyse the cells by passing the smooth suspension through a French press at 1000 psi three to five times until the lysate changes from slightly milky to clear (with a brown tint). Alternatively, a French Press comparable device can be used.
7. Remove unlysed cells and cell debris by centrifugation at 15,000 × g for 10 min at 4 °C.
8. Remove the supernatant and transfer into suitable ultracentrifugation tubes. Centrifuge at 100,000 × g for 60 min to pellet the membranes.
9. Resuspend the membrane pellet in "inner membrane solubilization buffer" to selectively solubilize the bacterial inner membrane. Incubate the solution at 4 °C for 1 h on a shaking incubator (*see* **Note 5**).

10. Centrifuge the sample again to pellet the (non-solubilized) outer membrane at 100,000 × g for 60 min at 4 °C.
11. Wash the outer membrane pellet three times with wash buffer to remove residual *N*-laurylsarcosine from the preparation. Do this by resuspending the pellet carefully in wash buffer and centrifuge at 100,000 rpm for 60 min at 4 °C after each step. If necessary, resuspended membranes in wash buffer may be snap-frozen in liquid nitrogen at this stage and stored at −80 °C for later use.
12. Finally, resuspend the pure outer membrane pellet in "outer membrane solubilization buffer" using a Hamilton syringe. Incubate a rocking incubator at 4 °C for 2 h (*see* **Note 6**).

3.2 Expression in Inclusion Bodies

This protocol largely follows published procedures for OmpX expression in inclusion bodies [37, 38, 41].

1. Transform *E. coli* BL21 C41 (DE3) with the plasmid pET3-b_OmpX using your preferred transformation protocol (we use chemically competent cells). Plate the bacteria on an LB-Amp plate and incubate at 37 °C overnight.
2. Select a single colony from the LB-Amp plate. Inoculate 5 mL of LB-Amp medium in a suitable culture tube and grow overnight at 37 °C in an orbital shaker at 200 rpm.
3. Use the preculture to inoculate 500 mL of LB-Amp medium in a 2 L baffled Erlenmeyer flask. Shake at 37 °C, 200 rpm in an orbital shaker until the OD_{600} reaches 0.5.
4. Induce gene expression by adding IPTG to a final concentration of 0.5 mM.
5. Continue incubation at 37 °C in an orbital shaker set to 200 rpm for 4 h.
6. Harvest the bacteria by centrifugation at 5000 × g, 4 °C for 15 min using suitable centrifuge bottles (e.g., 2× 500 mL bottles). Keep all samples on ice from here on.
7. Drain the bacterial pellets by pouring the LB-Amp supernatant into an appropriate waste container. Dispose of the spent medium in accordance with local guidelines (*see* **Note 7**).
8. Resuspend the bacterial pellet/pellets in 20–30 mL of lysis buffer.
9. Lyse the bacterial cells using a French Press at 1000 psi or comparable device (*see* **Note 8**).
10. Centrifuge the cell lysate at low speed (4000 × g for 10 min) in a 50 mL Falcon tube. Discard the supernatant. The supernatant should still be turbid at this point and contain ruptured cell membranes.

11. The pellet should contain inclusion bodies that appear white to ivory in color. Wash the pellet with wash buffer three times (4000 × g for 10 min) to remove membrane debris and other loosely attached cellular material.
12. Wash an additional three times with pure water to remove the detergent-containing wash buffer.
13. For storage, inclusion bodies can be suspended in a small amount of pure water and frozen in suitable aliquots. Inclusion bodies in water have an almost unlimited shelf life if stored at −20 °C.
14. To refold the protein, first dissolve a small amount of inclusion bodies in solubilization buffer at room temperature under shaking for 1 h in a 1.5 mL reaction tube. The resulting solution should be centrifuged in a benchtop centrifuge at maximum speed for 10–15 min to remove any insoluble debris.
15. Determine the protein concentration of the solution using, e.g., a spectrophotometer at 280 nm wavelength (OmpX has an extinction coefficient of 34,840 M^{-1} cm^{-1}) (*see* **Note 9**).
16. Dilute the protein to a concentration of 1 mg/mL using solubilization buffer.
17. Refold the protein by adding the solution dropwise to the refolding buffer under shaking or stirring. Use enough refolding buffer to achieve a 1:20 dilution of the original protein solution. Keep the refolded solution on ice for 1 h (*see* **Note 10**).
18. The protein solution at this point is already very pure and contains only minimal amounts of other proteins, but it can be further purified using chromatographic methods, if, for example, the residual urea needs to be removed.

4 Notes

1. This chapter can only give example protocols for expression of outer membrane proteins. Exact volumes for cultures, OD_{600} at induction, or concentrations of buffer components can vary in every protocol and should be optimized based on the protein of interest. Each protein will behave differently, and special care needs to be taken with recombinant proteins that originate from heterologous sources (i.e., not *E. coli* or related enterobacterial species). In these cases, inclusion body expression can avoid some of the potential problems with membrane insertion that result from heterologous expression. Many other variables can be modified to enable or enhance OMP expression in *E. coli*, all of them strongly depending on the exact protein target: the choice of expression vector and expression strain,

the type of media used, and growth parameters such as temperature and oxygen supply that will have significant impact on expression yields.

2. Growth of bacterial strains for recombinant gene expression should be performed in laboratories approved for BSL-1 GMO work (in accordance with national guidelines). The *E. coli* BL21 (DE3) ΔABCF strain is sensitive to higher temperatures (>30 °C) and high salt concentrations (e.g., >5 g/L NaCl). The single, double, and triple knockout strains are less temperature-sensitive and more tolerant toward high salt concentrations than the quadruple knockout strain [34]. Alternatively, strains with single, double, and triple knockouts of these key outer membrane proteins are available in all combinations [34].

3. By a "pinch," we mean dipping a pipette tip in the DNAse I powder and transferring the adherent powder to the lysis buffer. This works well in our hands. If you are uncomfortable with this measurement, prepare a stock solution of 1 mg/mL DNAse I and add this in a ratio of 1:1000 to the buffer. The procedure is compatible with most buffers at pH 7.5–8.5, including phosphate buffers or HEPES buffer.

4. After successful expression, membrane proteins require additional buffer additives compared to soluble proteins. For OMPs, the choice of detergent either for solubilization of native membranes or for refolding from inclusion bodies largely depends on the target protein, and detergent screening might be necessary to find optimal solubilization or refolding conditions [42]. For OmpX, SB-12 works well for both solubilization and refolding. Buffer and detergent choices might be further limited by downstream applications—obvious examples include the incompatibility of EDTA or DTT as buffer components in Ni-NTA-based affinity chromatography.

5. A Hamilton syringe or douncer can be used to fully solubilize the pellet.

6. The detergent in this buffer will depend on the downstream application. 1% SDS can be used if the downstream application is SDS-PAGE or Western blotting, but should be avoided if the protein needs to remain in a native state for purification. In the latter case, use a non-denaturing detergent at suitable concentration, such as 1% sulfobetaine SB-12 for OmpX. *See* **Note 3** about optimizing detergents for other OMPs.

7. Antibiotics-containing media should not be disposed with normal wastewater.

8. Alternative lysis protocols with lysozyme and/or sonication are also acceptable but usually give lower yields.

9. Other protein quantification methods can also be used, but note that the solution contains high concentrations of urea that disturb some of the available colorimetric protein quantitation assays.

10. *See* **Note 4** about optimizing the detergent for refolding other OMPs. In any case, we highly recommend using functional assays to monitor the activity of the target protein during all steps of the protocol. If a simple functional assay is not available, as is the case for many of the OMP diffusion pores, we recommend using spectroscopic methods such as CD or fluorescence spectroscopy to determine the folding state of the expression target during and after purification. Alternatively, the heat modifiability assay can inform on proper OMP folding in many but not all cases. This assay makes use of the fact that many OMPs are not easily denatured by SDS sample buffer if not excessively heated, leading to a shift in apparent molecular weight in SDS-PAGE [28, 33].

Acknowledgments

D.L. receives support from the Research Council of Norway, grants 294605, 302723, and 331752.

References

1. Ulrich T, Rapaport D (2015) Biogenesis of beta-barrel proteins in evolutionary context. Int J Med Microbiol 305:259–264
2. Schulz GE (2003) Transmembrane beta-barrel proteins. Adv Protein Chem 63:47–70
3. Hermansen S, Linke D, Leo JC (2022) Transmembrane beta-barrel proteins of bacteria: from structure to function. Adv Protein Chem Struct Biol 128:113–161
4. Schulz GE (2002) The structure of bacterial outer membrane proteins. Biochim Biophys Acta 1565:308–317
5. Wzorek JS, Lee J, Tomasek D et al (2017) Membrane integration of an essential beta-barrel protein prerequires burial of an extracellular loop. Proc Natl Acad Sci U S A 114: 2598–2603
6. Malhotra A (2009) Tagging for protein expression. Methods Enzymol 463:239–258
7. Fanucci GE, Cadieux N, Piedmont CA et al (2002) Structure and dynamics of the beta-barrel of the membrane transporter BtuB by site-directed spin labeling. Biochemistry 41: 11543–11551
8. Koebnik R (1996) In vivo membrane assembly of split variants of the E.coli outer membrane protein OmpA. EMBO J 15:3529–3537
9. Mahalakshmi R, Franzin CM, Choi J et al (2007) NMR structural studies of the bacterial outer membrane protein OmpX in oriented lipid bilayer membranes. Biochim Biophys Acta 1768:3216–3224
10. Mogensen JE, Otzen DE (2005) Interactions between folding factors and bacterial outer membrane proteins. Mol Microbiol 57:326–346
11. Georgiou G, Segatori L (2005) Preparative expression of secreted proteins in bacteria: status report and future prospects. Curr Opin Biotechnol 16:538–545
12. Low KO, Muhammad Mahadi N, Md Illias R (2013) Optimisation of signal peptide for recombinant protein secretion in bacterial hosts. Appl Microbiol Biotechnol 97:3811–3826
13. Mas G, Thoma J, Hiller S (2019) The periplasmic chaperones Skp and SurA. Subcell Biochem 92:169–186

14. Grosskinsky U, Schutz M, Fritz M et al (2007) A conserved glycine residue of trimeric autotransporter domains plays a key role in Yersinia adhesin A autotransport. J Bacteriol 189: 9011–9019
15. Subrini O, Betton JM (2009) Assemblies of DegP underlie its dual chaperone and protease function. FEMS Microbiol Lett 296:143–148
16. Stull F, Betton JM, Bardwell JCA (2018) Periplasmic chaperones and prolyl isomerases. EcoSal Plus 8. https://doi.org/10.1128/ecosalplus.ESP-0005-2018
17. Duguay AR, Silhavy TJ (2004) Quality control in the bacterial periplasm. Biochim Biophys Acta 1694:121–134
18. Merdanovic M, Clausen T, Kaiser M et al (2011) Protein quality control in the bacterial periplasm. Annu Rev Microbiol 65:149–168
19. Ricci DP, Silhavy TJ (2019) Outer membrane protein insertion by the beta-barrel assembly machine. EcoSal Plus 8. https://doi.org/10.1128/ecosalplus.ESP-0035-2018
20. Robert V, Volokhina EB, Senf F et al (2006) Assembly factor Omp85 recognizes its outer membrane protein substrates by a species-specific C-terminal motif. PLoS Biol 4:e377
21. Paramasivam N, Habeck M, Linke D (2012) Is the C-terminal insertional signal in Gram-negative bacterial outer membrane proteins species-specific or not? BMC Genomics 13:510
22. Paramasivam N, Linke D (2015) Strategies for the analysis of bam recognition motifs in outer membrane proteins. Methods Mol Biol 1329: 271–277
23. Singhvi P, Saneja A, Srichandan S et al (2020) Bacterial inclusion bodies: a treasure trove of bioactive proteins. Trends Biotechnol 38:474–486
24. Maachupallireddy J, Kelley BD, Clark ED (1997) Effect of inclusion body contaminants on the oxidative renaturation of hen egg white lysozyme. Biotechnol Prog 13:144–150
25. Danoff EJ, Fleming KG (2015) Aqueous, unfolded OmpA forms amyloid-like fibrils upon self-association. PLoS One 10:e0132301
26. De Marco A, Ferrer-Miralles N, Garcia-Fruitos E et al (2019) Bacterial inclusion bodies are industrially exploitable amyloids. FEMS Microbiol Rev 43:53–72
27. Georgiou G, Valax P (1999) Isolating inclusion bodies from bacteria. Methods Enzymol 309: 48–58
28. Orwick-Rydmark M, Arnold T, Linke D (2016) The use of detergents to purify membrane proteins. Curr Protoc Protein Sci 84: 4.8.1–4.8.35
29. Linke D (2009) Detergents: an overview. Methods Enzymol 463:603–617
30. Zardeneta G, Horowitz PM (1994) Detergent, liposome, and micelle-assisted protein refolding. Anal Biochem 223:1–6
31. Tamm LK, Hong H, Liang B (2004) Folding and assembly of beta-barrel membrane proteins. Biochim Biophys Acta 1666:250–263
32. Vallejo LF, Rinas U (2004) Strategies for the recovery of active proteins through refolding of bacterial inclusion body proteins. Microb Cell Factories 3:11
33. Leo JC, Oberhettinger P, Linke D (2015) Assessing the outer membrane insertion and folding of multimeric transmembrane beta-barrel proteins. Methods Mol Biol 1329:157–167
34. Meuskens I, Michalik M, Chauhan N et al (2017) A new strain collection for improved expression of outer membrane proteins. Front Cell Infect Microbiol 7:464
35. Thein M, Sauer G, Paramasivam N et al (2010) Efficient subfractionation of gram-negative bacteria for proteomics studies. J Proteome Res 9:6135–6147
36. Miroux B, Walker JE (1996) Over-production of proteins in Escherichia coli: mutant hosts that allow synthesis of some membrane proteins and globular proteins at high levels. J Mol Biol 260:289–298
37. Pautsch A, Vogt J, Model K et al (1999) Strategy for membrane protein crystallization exemplified with OmpA and OmpX. Proteins 34: 167–172
38. Arnold T, Poynor M, Nussberger S et al (2007) Gene duplication of the eight-stranded beta-barrel OmpX produces a functional pore: a scenario for the evolution of transmembrane beta-barrels. J Mol Biol 366:1174–1184
39. Wollmann P, Zeth K, Lupas AN et al (2006) Purification of the YadA membrane anchor for secondary structure analysis and crystallization. Int J Biol Macromol 39:3–9
40. Swords WE (2003) Chemical transformation of E. coli. In: Casali N, Preston A (eds) E. coli plasmid vectors: methods and applications. Humana Press, Totowa, pp 49–53
41. Hermansen S, Ryoo D, Orwick-Rydmark M et al (2022) The role of extracellular loops in the folding of outer membrane protein X (OmpX) of Escherichia coli. Front Mol Biosci 9:918480
42. Linke D (2014) Explanatory chapter: choosing the right detergent. Methods Enzymol 541: 141–148

Chapter 3

Bacterial Envelope Fractionation

Athanasios Saragliadis and Dirk Linke

Abstract

Numerous bioinformatics tools allow predicting the localization of membrane proteins in the outer or inner membrane of *Escherichia coli* with high precision. Nevertheless, it might be desirable to experimentally verify such predictions or to assay the correct localization of recombinant or mutated variants of membrane proteins. Here we describe two methods (preferential detergent solubilization and sucrose-gradient fractionation) that allow to fractionate Gram-negative bacterial membranes and subsequently to enrich inner or outer membrane proteins.

Key words Inner membrane, Outer membrane, Fractionation, Cell envelope, Detergent, Sucrose gradient

1 Introduction

Gram-negative bacteria are enveloped by a dual-membrane system: the inner or plasma membrane and an outer membrane. The space between the two membranes, called the periplasmic space, contains soluble and membrane-attached proteins of various function [1] and a peptidoglycan layer that confers structural integrity to the cell. The inner membrane is a permeability barrier of around 8 nm thickness with no particular rigidity and is composed of a phospholipid bilayer and of transmembrane proteins, with functions ranging from energy production to lipid biosynthesis and protein secretion and transport [2]. The outer membrane is of an asymmetric nature, composed of phopsholipids on the inner leaflet and lipopolysaccharides on the outer leaflet. As a lipid bilayer, it has similar dimensions as the inner membrane [3, 4], but in addition, its outer layer is extended by the carbohydrate component of the lipopolysaccharides that extends its thickness depending on the species and strain, ranging up to roughly 40 nm for some *E. coli* strains [5]. This makes the outer membrane a very selective permeability barrier [6, 7]. In addition, it contains different transmembrane proteins forming pores, as well as lipoproteins that can be covalently linked

Raffaele Ieva (ed.), *Transmembrane β-Barrel Proteins: Methods and Protocols*, Methods in Molecular Biology, vol. 2778, https://doi.org/10.1007/978-1-0716-3734-0_3,

to the peptidoglycan layer [8]. The lipopolysaccharide layer of the outer membrane usually forms ionic bonds with cations, which adds to the overall rigidity and by consequence influences the shape and integrity of the cell.

The number of membrane-associated proteins is predicted to be roughly 20–30% of the total cell proteins [9, 10]. For example, in *E. coli*, for which 4289 genes have been annotated, at least 900 of these encode membrane proteins, although this number may vary depending on the method of membrane prediction analysis used [11]. According to more recent publications, the predicted number of inner membrane proteins is 875 (~20% of the total) and outer membrane–associated proteins is 98 [12]. This is in line with earlier estimates that outer membrane proteins (OMPs) comprise only a small fraction (approximately 2%) of the total Gram-negative bacterial proteome [13, 14].

Due to their hydrophobic and amphiphilic attributes, membrane proteins are challenging substrates to isolate and analyze [15]. In bacteria, they are a very interesting target for drug development [16], especially concerning antimicrobial substances, where membrane proteins are key players both as targets and as resistance factors. Antimicrobial drug resistance is a major global concern as described by the World Health Organization (WHO) [17]. In addition, outer membrane proteins can be utilized as recombinant surface display tools [18] and to elicit immune responses [19, 20] rendering them potential vaccine candidates.

There are several approaches that can help with recombinant production of membrane proteins, ranging from *E. coli* strains with improved production traits, cell-free systems, strains alternative to *E. coli*, or overexpression of different and easier-to-produce protein variants than the wild type [21, 22]. Some protocol examples for protein production methods can be found in Chap. 2.

The specific cellular localization of proteins can be predicted with the help of bioinformatics [23]. But sometimes, it can be desirable to validate such data experimentally, for example, by isolating various cellular compartments or membranes. Membrane isolation can also be a useful step in the purification of membrane proteins. Several methods have been evaluated in the past for their efficiency in bacterial membrane fractionation [24]. Traditionally, separation of different membrane fractions based on isopycnic sucrose gradients or preferential detergent solubilization has been used [25–27]. From a practical application perspective, each method has its own advantages and disadvantages. With methods based on isopycnic separation through a sucrose gradient, relatively high enrichment levels can be achieved. However, for carrying out this specific protocol, dedicated, sophisticated, and expensive equipment (ultracentrifuge) and consumables (suitable compatible containers) are required with cost that in 2020 were estimated to exceed the $300,000 mark [28]. On the other hand, the preferential detergent solubilization method requires no sophisticated equipment, which is a definite advantage. Nevertheless, the membrane fractions produced following the detergent solubilization protocol usually appear more cross-contaminated since there is no absolute solubilization occurring that works for either the inner or the outer fraction [27].

Independent of whichever protocol is used, methods that allow for enrichment, isolation, and eventually characterization of a specific category of membrane proteins represent a very useful toolkit. In this chapter, we provide two isolation and enrichment protocols for the bacterial outer membrane, based on literature descriptions [27, 29, 30] but refined for use with the model organism, *E. coli.* Considering that membrane protein expression is heavily dependent on strain and growth conditions [31] and that membranes can have different composition depending on the Gram-negative species used, optimization of these protocols will be necessary to achieve optimal results with different bacterial model organisms.

2 Materials

All solutions are prepared in double-distilled water (ddH_2O) sterilized either by filtration or by autoclaving. All reagents are of analytical grade. Store buffers and dilute solutions at 4 °C. If solutions contain detergents, warm them before use to dissolve any possible precipitation.

2.1 Bacterial Storage, Growth, and Cultivation

1. Lysogeny broth (or LB medium, Miller): dissolve 10 g tryptone, 5 g yeast extract, and 10 g NaCl in 800–900 mL of ddH_2O with gentle stirring. Adjust the pH to 7.0 ± 0.2 with 3 N NaOH and 3 N HCl and adjust final volume to 1 L (*see* **Note 1**). Sterilize by autoclaving (121 °C for 20 min).
2. LB agar: 15 g of Bacto agar in 1.0 L of LB medium. Sterilize by autoclaving (121 °C for 20 min).
3. Glycerol stock: 20–25% (v/v) glycerol in LB (*see* **Note 2**).
4. Spectrophotometer.
5. Erlenmeyer flasks and glass tubes.
6. Beckman Avanti J-26-XP centrifuge and Beckman JA-14 rotor.
7. Polycarbonate bottles of 250 mL with screw-on caps (for Beckman JA-14 rotor).
8. Beckman Allegra X-30R centrifuge and Beckman SX4400 rotor.
9. Buckets (for Beckman SX4400 rotor) to hold Falcon-type polypropylene tubes.

2.2 Cell Lysis

1. Lysis buffer: 200 mM Tris–HCl, pH 7, 0.5 M sucrose, supplemented with 60 μg/mL lysozyme and 0.5 mM EDTA.
2. Sonication device: SONICS with CV33 needle at 40% amplitude.
3. Beckman Avanti J-26-XP centrifuge and Beckman JA-25.50 rotor.
4. Beckman polycarbonate bottles with screw-on caps for Beckman JA-25.50 rotor.

2.3 Total Membrane Harvesting, Solubilization, and Sucrose-Based Enrichment

1. Tabletop microcentrifuge.
2. Beckman Optima L-80 Ultracentrifuge with Beckman SW 32 Ti rotor.
3. Open-top thickwall polycarbonate tubes for Beckman SW 32 Ti rotor.
4. Beckman Optima L-80 Ultracentrifuge with Beckman SW 41 Ti rotor.
5. Open-top thinwall Ultra-Clear tubes for Beckman SW 41 Ti rotor.
6. Thermoshaker for microtubes.
7. Detergents: Triton X-100 or N-Lauroylsarcosine sodium salt (sarkosyl).
8. Sucrose gradient: 30–55% (w/v) sucrose in 3 mM EDTA, pH 7.6.
9. Bicinchoninic acid or similar reagents for protein quantification.

3 Methods

3.1 Bacterial Growth and Cultivation

1. Streak bacterial strain of interest from −80 °C glycerol stock on LB agar plates containing appropriate antibiotics depending on individual strain needs and incubate overnight at 37 °C (*see* **Note 3**).
2. On the following day, pick one single colony using sterile loop from the plate and inoculate a starting culture of 5 mL medium in an appropriate vessel (shake flask or culture tube). Grow the strain overnight at 37 °C with (orbital) shaking at 200 rpm (*see* **Note 4**).
3. On the following day, inoculate 200 mL of fresh medium in a 1 L Erlenmeyer flask, using 2 mL of the overnight culture. Grow overnight at 37 °C with (orbital) shaking at 200 rpm.
4. Check the optical density at 600 nm (OD600) of the culture using a spectrophotometer that has been blanked using the same medium.

5. Harvest the bacterial cells by centrifugation at 10,000 × *g* for 20 min at 12 °C using JA14 rotor and the appropriate sample bottles.
6. Discard the supernatant and wash the pellet with 35 mL of 200 mM Tris-HCl pH 7 and transfer to 50 mL Falcon-type tubes.
7. Repeat centrifugation using SW4400 swinging-bucket rotor at 4200 × g for 20 min at 12 °C. Discard supernatant and proceed with the cell lysis.

3.2 Cell Lysis

The lysis method is adapted from [29].

1. Resuspend the cell pellet in lysis buffer (*see* **Note 5**).
2. After vortexing, incubate at 4 °C for 30 min.
3. Add slowly the suspension to the same volume of ice-cold water with continuous shaking.
4. Incubate at 4 °C for 30 min.
5. Sonicate the solution on ice for a minimum of 3 min using 10 s bursts ON cycle followed by 10 s OFF cycle (*see* **Note 6**).
6. Centrifuge cell debris at 12,000 × g for 20 min in a JA25.50 rotor at 12 °C.

3.3 Total Membrane Harvest

1. Transfer the supernatant from the previous step into ultracentrifuge tubes compatible with the SW32 Ti rotor, and centrifuge for 60 min at 125,000 × *g* at 18 °C to harvest the total membrane protein fraction.
2. Wash the pellet once using 25 mL of 200 mM Tris-HCl pH 7 using the same centrifugation conditions.

   The total membrane protein pellet can be further fractionated using either the selective detergent solubilization method or the sucrose-gradient fractionation method (Fig. 1).

3.4 Selective Detergent Solubilization

1. Suspend total membranes from Subheading 3.3, **step 2** in 1 mL of 50 mM Tris–HCl pH 7 and supplement with 0.1% (w/v) Triton X-100.
2. Incubate with shaking at 800 rpm for 30 min at room temperature (RT).
3. Transfer in 1.5 mL tubes and centrifuge for 30 min at 17,000 × g at 12 °C.
4. Collect the supernatant as inner-membrane enriched fraction.
5. Resuspend pellet in 1 mL of 50 mM Tris–HCl pH 7 and supplement with 0.2% (w/v) sarkosyl.
6. Incubate with shaking at 800 rpm for 30 min at room temperature (RT).

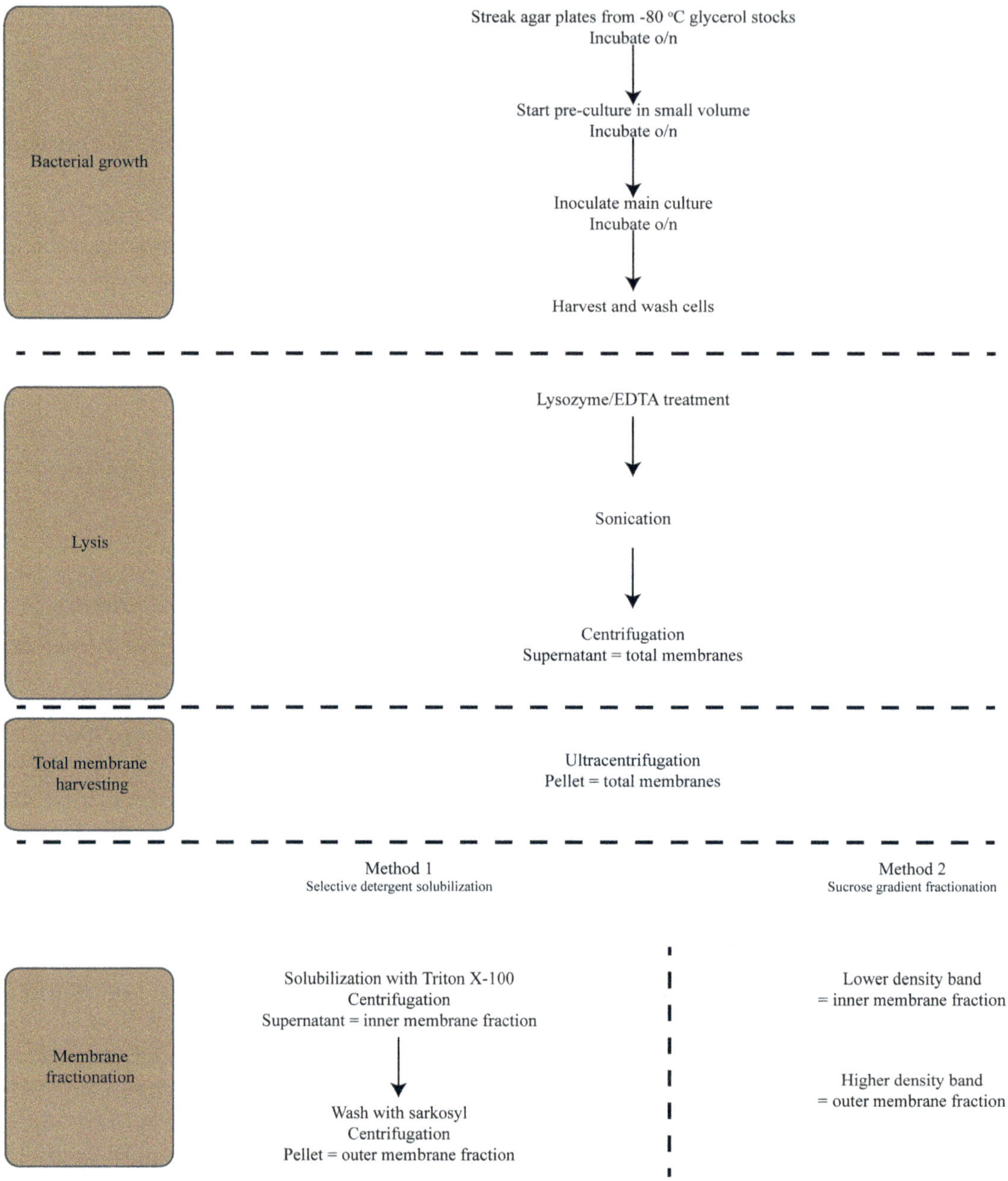

Fig. 1 Flowchart summarizing the steps that lead from cell growth to the two bacterial envelope fractionation methods

7. Centrifuge for 30 min at 17,000 × g at 12 °C.
8. Remove the supernatant as wash fraction.
9. Resuspend pellet in 1 mL of 50 mM Tris–HCl, pH 7, and incubate with shaking for 30 min at room temperature (RT). This will be the outer-membrane enriched fraction (*see* **Note 7**).

3.5 Sucrose-Gradient-Based Fractionation

The sucrose gradient method is adapted from the literature as described here [29].

1. Suspend total membranes from Subheading 3.3, **step 2** in 1 mL of 50 mM Tris–HCl, pH 7.
2. Layer 500 μL onto the sucrose gradient (*see* **Note 8**).
3. Centrifuge samples for 16 h at 173,000 × *g* at 4 °C using a SW 41 Ti rotor.
4. At the end of the centrifugation round, collect 500 μL samples from the top of the tube.

3.6 Analysis and Characterization of Membrane Fractions

Total protein can be quantified using bicinchoninic acid [32] or any other equivalent method (*see* **Note 9**), and samples can be further analyzed via SDS-polyacrylamide gel electrophoresis [33] and subsequently immunoblotted and probed using antibodies or antisera against protein(s) of interest [12].

4 Notes

1. When working with acids and bases, use appropriate personal protective equipment and carry out the handling inside chemical fume hood.
2. For easier handling, first prepare a 50% (v/v) glycerol solution by diluting 100% glycerol in distilled water and then filter (0.2 μm) sterilize it.
3. When inoculating bacterial cultures, make sure to maintain sterile conditions with aseptic techniques in order to limiting contamination opportunities.
4. Having a preculture avoids the growth lag that some strains might exhibit if they are inoculated directly from −80 °C cryostock into the main culture volume. Additionally, there is flexibility of inoculating multiple main cultures of different end volume simultaneously depending on the scale of the experiment and the analysis required.
5. Depending on the actual optical density of the harvested cell culture, adjust the volume accordingly.
6. Inspect the optical density of the cell lysate and evaluate the sonication procedure. If cells appear to not have lysed within the proposed time, repeat the procedure. Alternatively, one can use other methods as well if they are available and depending on the scale. For instance, for relatively big samples, high-pressure homogenizers such as French press can be used or, conversely, for smaller samples, bead mills such as "bead beaters" can be utilized. For an overview of different lysis methods, please refer to ref. [34].

7. Regarding detergent wash method, there are several publications that describe the use of detergents as membrane solubilization agents for bacterial and mammalian membranes [27, 30, 35, 36]. We observed that we could obtain relatively pure outer membrane fractions if we introduce a 0.1% Triton X-100 in 50 mM Tris–HCl pH 7 wash, followed by 0.2% sarkosyl in 50 mM Tris–HCl pH 7 wash of the membrane fraction. Additionally, if there is access to ultracentrifuge instrumentation, then all the steps described to be carried out at 17,000 × *g* could be alternatively performed at higher centrifugation speeds for shorter times.
8. The order of the sucrose gradient is 0.8 mL 30%, 1.4 mL 35%, followed by 2.3 mL for the 40%, 45%, 50% and final cushion of 1.4 mL of 55% in 3 mM EDTA.
9. Depending on the protein quantification method used and potential compatibility issues that may arise, protein samples could be diluted, dialyzed, or precipitated in order to remove or reduce the concentration of interfering components such as detergents or sucrose or simply to increase the concentration of the protein samples.

Acknowledgments

This work was supported by the Research Council of Norway, Grant 294605 (Center for Digital Life).

References

1. Miller SI, Salama NR (2018) The gram-negative bacterial periplasm: size matters. PLoS Biol 16:e2004935. https://doi.org/10.1371/journal.pbio.2004935
2. Silhavy TJ, Kahne D, Walker S (2010) The bacterial cell envelope. Cold Spring Harb Perspect Biol 2:a000414. https://doi.org/10.1101/cshperspect.a000414
3. DiRienzo JM, Nakamura K, Inouye M (1978) The outer membrane proteins of Gram-negative bacteria: biosynthesis, assembly, and functions. Annu Rev Biochem 47:481–532. https://doi.org/10.1146/annurev.bi.47.070178.002405
4. Osborn MJ, Gander JE, Parisi E, Carson J (1972) Mechanism of assembly of the outer membrane of Salmonella typhimurium. Isolation and characterization of cytoplasmic and outer membrane. J Biol Chem 247:3962–3972
5. Strauss J, Burnham NA, Camesano TA (2009) Atomic force microscopy study of the role of LPS O-antigen on adhesion of E. coli. J Mol Recognit 22:347–355. https://doi.org/10.1002/jmr.955
6. Zgurskaya HI, Lopez CA, Gnanakaran S (2015) Permeability barrier of gram-negative cell envelopes and approaches to bypass it. ACS Infect Dis 1:512–522. https://doi.org/10.1021/acsinfecdis.5b00097
7. Delcour AH (2009) Outer membrane permeability and antibiotic resistance. Biochim Biophys Acta 1794:808–816. https://doi.org/10.1016/j.bbapap.2008.11.005
8. Braun V (1975) Covalent lipoprotein from the outer membrane of Escherichia coli. Biochim Biophys Acta 415:335–377. https://doi.org/10.1016/0304-4157(75)90013-1
9. Almen MS, Nordstrom KJ, Fredriksson R, Schioth HB (2009) Mapping the human membrane proteome: a majority of the human membrane proteins can be classified according to function and evolutionary origin. BMC Biol

7:50. https://doi.org/10.1186/1741-7007-7-50
10. Wallin E, von Heijne G (1998) Genome-wide analysis of integral membrane proteins from eubacterial, archaean, and eukaryotic organisms. Protein Sci 7:1029–1038. https://doi.org/10.1002/pro.5560070420
11. Krogh A, Larsson B, von Heijne G, Sonnhammer EL (2001) Predicting transmembrane protein topology with a hidden Markov model: application to complete genomes. J Mol Biol 305:567–580. https://doi.org/10.1006/jmbi.2000.4315
12. Sueki A, Stein F, Savitski MM, Selkrig J, Typas A (2020) Systematic localization of Escherichia coli membrane proteins. mSystems 5. https://doi.org/10.1128/mSystems.00808-19
13. Wimley WC (2003) The versatile beta-barrel membrane protein. Curr Opin Struct Biol 13:404–411. https://doi.org/10.1016/s0959-440x(03)00099-x
14. Wimley WC (2002) Toward genomic identification of beta-barrel membrane proteins: composition and architecture of known structures. Protein Sci 11:301–312. https://doi.org/10.1110/ps.29402
15. von Heijne G (1999) Recent advances in the understanding of membrane protein assembly and structure. Q Rev Biophys 32:285–307. https://doi.org/10.1017/s0033583500003541
16. Rosas NC, Lithgow T (2022) Targeting bacterial outer-membrane remodelling to impact antimicrobial drug resistance. Trends Microbiol 30:544–552. https://doi.org/10.1016/j.tim.2021.11.002
17. WHO (2021) Antimicrobial resistance. https://www.who.int/news-room/fact-sheets/detail/antimicrobial-resistance
18. Lang H (2000) Outer membrane proteins as surface display systems. Int J Med Microbiol 290:579–585. https://doi.org/10.1016/S1438-4221(00)80004-1
19. Massari P, Ram S, Macleod H, Wetzler LM (2003) The role of porins in neisserial pathogenesis and immunity. Trends Microbiol 11:87–93. https://doi.org/10.1016/s0966-842x(02)00037-9
20. Liu C, Chen Z, Tan C, Liu W, Xu Z, Zhou R, Chen H (2012) Immunogenic characterization of outer membrane porins OmpC and OmpF of porcine extraintestinal pathogenic Escherichia coli. FEMS Microbiol Lett 337:104–111. https://doi.org/10.1111/1574-6968.12013
21. Schlegel S, Hjelm A, Baumgarten T, Vikstrom D, de Gier JW (2014) Bacterial-based membrane protein production. Biochim Biophys Acta 1843:1739–1749. https://doi.org/10.1016/j.bbamcr.2013.10.023
22. Meuskens I, Michalik M, Chauhan N, Linke D, Leo JC (2017) A new strain collection for improved expression of outer membrane proteins. Front Cell Infect Microbiol 7:464. https://doi.org/10.3389/fcimb.2017.00464
23. Orfanoudaki G, Economou A (2014) Proteome-wide subcellular topologies of E. coli polypeptides database (STEPdb). Mol Cell Proteomics 13:3674–3687. https://doi.org/10.1074/mcp.O114.041137
24. Thein M, Sauer G, Paramasivam N, Grin I, Linke D (2010) Efficient subfractionation of gram-negative bacteria for proteomics studies. J Proteome Res 9:6135–6147. https://doi.org/10.1021/pr1002438
25. Miura T, Mizushima S (1968) Separation by density gradient centrifugation of two types of membranes from spheroplast membrane of Escherichia coli K12. Biochim Biophys Acta 150:159–161. https://doi.org/10.1016/0005-2736(68)90020-5
26. Schnaitman CA (1971) Solubilization of the cytoplasmic membrane of Escherichia coli by Triton X-100. J Bacteriol 108:545–552. https://doi.org/10.1128/jb.108.1.545-552.1971
27. Filip C, Fletcher G, Wulff JL, Earhart CF (1973) Solubilization of the cytoplasmic membrane of Escherichia coli by the ionic detergent sodium-lauryl sarcosinate. J Bacteriol 115:717–722. https://doi.org/10.1128/jb.115.3.717-722.1973
28. Edwards GB, Muthurajan UM, Bowerman S, Luger K (2020) Analytical ultracentrifugation (AUC): an overview of the application of fluorescence and absorbance AUC to the study of biological macromolecules. Curr Protoc Mol Biol 133:e131. https://doi.org/10.1002/cpmb.131
29. Laskowska E, Bohdanowicz J, Kuczynska-Wisnik D, Matuszewska E, Kedzierska S, Taylor A (2004) Aggregation of heat-shock-denatured, endogenous proteins and distribution of the IbpA/B and Fda marker-proteins in Escherichia coli WT and grpE280 cells. Microbiology (Reading) 150:247–259. https://doi.org/10.1099/mic.0.26470-0
30. Arachea BT, Sun Z, Potente N, Malik R, Isailovic D, Viola RE (2012) Detergent selection for enhanced extraction of membrane proteins. Protein Expr Purif 86:12–20. https://doi.org/10.1016/j.pep.2012.08.016
31. Mathieu K, Javed W, Vallet S, Lesterlin C, Candusso MP, Ding F, Xu XN, Ebel C, Jault JM, Orelle C (2019) Functionality of membrane

proteins overexpressed and purified from E. coli is highly dependent upon the strain. Sci Rep 9:2654. https://doi.org/10.1038/s41598-019-39382-0

32. Smith PK, Krohn RI, Hermanson GT, Mallia AK, Gartner FH, Provenzano MD, Fujimoto EK, Goeke NM, Olson BJ, Klenk DC (1985) Measurement of protein using bicinchoninic acid. Anal Biochem 150:76–85. https://doi.org/10.1016/0003-2697(85)90442-7

33. Schagger H (2006) Tricine-SDS-PAGE. Nat Protoc 1:16–22. https://doi.org/10.1038/nprot.2006.4

34. Islam MS, Aryasomayajula A, Selvaganapathy PR (2017) A review on macroscale and microscale cell lysis methods. Micromachines 8:83. https://doi.org/10.3390/mi8030083

35. Beis K, Whitfield C, Booth I, Naismith JH (2006) Two-step purification of outer membrane proteins. Int J Biol Macromol 39:10–14. https://doi.org/10.1016/j.ijbiomac.2005.12.008

36. Rampado R, Giordano F, Moracci L, Crotti S, Caliceti P, Agostini M, Taraballi F (2022) Optimization of a detergent-based protocol for membrane proteins purification from mammalian cells. J Pharm Biomed Anal 219:114926. https://doi.org/10.1016/j.jpba.2022.114926

Chapter 4

Fluorescent Labeling of Outer Membrane Proteins Using the SpyCatcher-SpyTag System

Rachael Duodu, Dirk Linke, and Jack C. Leo

Abstract

The SpyCatcher-SpyTag system has become a popular and versatile tool for protein ligation. It is based on a small globular protein (SpyCatcher) that binds to a 13-residue peptide (SpyTag), which subsequently leads to the formation of a covalent isopeptide bond. Thus, the reaction is essentially irreversible. Here, we describe how the SpyCatcher-SpyTag system can be used to label surface-exposed bacterial outer membrane proteins, e.g., for topology mapping or fluorescent time-course experiments. We cover using fluorescence measurements and microscopy to measure labeling efficiency using SpyCatcher fused with superfolder GFP in this chapter.

Key words Outer membrane protein, Fluorescent labeling, Protein ligation, Topology mapping, SpyCatcher, SpyTag

1 Introduction

The fluorescent labeling of specific proteins is key to many state-of-the-art detection systems, including super-resolution microscopy applications or cell sorting. For surface-exposed proteins in Gram-negative bacteria, fluorescent labeling has been used for determining the localization [1], interactions [2], dynamics [3], and topology [4] of proteins residing in the outer membrane. Topology mapping, i.e., determining which part(s) of the protein are on the outside of the outer membrane versus located in the periplasm, is still important, even with the availability of good structure prediction programs and homology models. For example, investigating mutations that impede surface exposure or circular permutations of transmembrane β-barrel proteins still requires experimental validation of the resulting topology [5, 6]. Proper surface exposure is also an important parameter in verifying

Raffaele Ieva (ed.), *Transmembrane β-Barrel Proteins: Methods and Protocols*, Methods in Molecular Biology, vol. 2778,
https://doi.org/10.1007/978-1-0716-3734-0_4,

successful heterologous protein expression, e.g., of surface-localized virulence factors for functional studies or vaccine development [7–10].

Antibodies have traditionally been used for detecting exposed structures on the surfaces of bacteria, either by using monoclonal antibodies recognizing known epitopes on the proteins of interest [11, 12] or by genetically introducing epitope tags such as FLAG, HA, or Strep tags that can be recognized by specific antibodies [6, 8]. Permeabilizing or solubilizing the outer membrane allows detection of periplasmically located epitopes and full topology mapping [8]. Antibodies are still the most popular and often the most convenient tools for labeling cells and topology mapping, but they do have some limitations. In addition to nonspecific interactions that may cause background noise, the antibody-epitope interaction is not a covalent one and, therefore, antibodies may detach from their targets during long time-course experiments. Furthermore, secondary antibodies conjugated to fluorophores are often required. Though this may have the advantage of boosting the fluorescent signal, additional incubation and wash steps will be needed, and the use of additional antibodies increases the likelihood of background fluorescence.

An alternative to antibodies would be covalent labeling of proteins of interest. Numerous methods have been developed where small molecules are bound to the protein covalently. This generally requires the proteins to either have an endogenous residue that can be modified, usually a free cysteine, or a residue can be introduced through mutagenesis. Surface-exposed cysteines can be modified by thiol-reactive compounds unable to cross the outer membrane such as fluorescein-maleimide [11, 13]. As free cysteines are rare in surface proteins, this strategy can be quite specific to introduced cysteines on target proteins [13].

A relatively recent development is the SpyCatcher-SpyTag system [14]. This is based on a pilin protein, where an isopeptide bond is formed spontaneously within the protein, between the side chains of a lysine and an aspartate residue, once the protein folds. This protein can be split into two components, a globular protein (SpyCatcher) containing the reactive lysine, and a 13-amino-acid peptide (SpyTag) with the aspartate, which still retain the ability to form the isopeptide bond once they come into contact [15]. The isopeptide bond is covalent and therefore for all intents and purposes irreversible (Fig. 1a). For control experiments, isopeptide bond formation can be prevented by using a variant of SpyCatcher where a catalytic glutamate is changed to a glutamine (SpyCatcherEQ) [15]. Both SpyCatcher and SpyTag can be genetically fused to a variety of proteins, allowing a multitude of applications, from stabilizing enzymes through cyclisation, immobilizing proteins to a scaffold, or labeling proteins on live cells [14, 16]. SpyTag in particular can be fused to protein termini but, like epitope tags, is

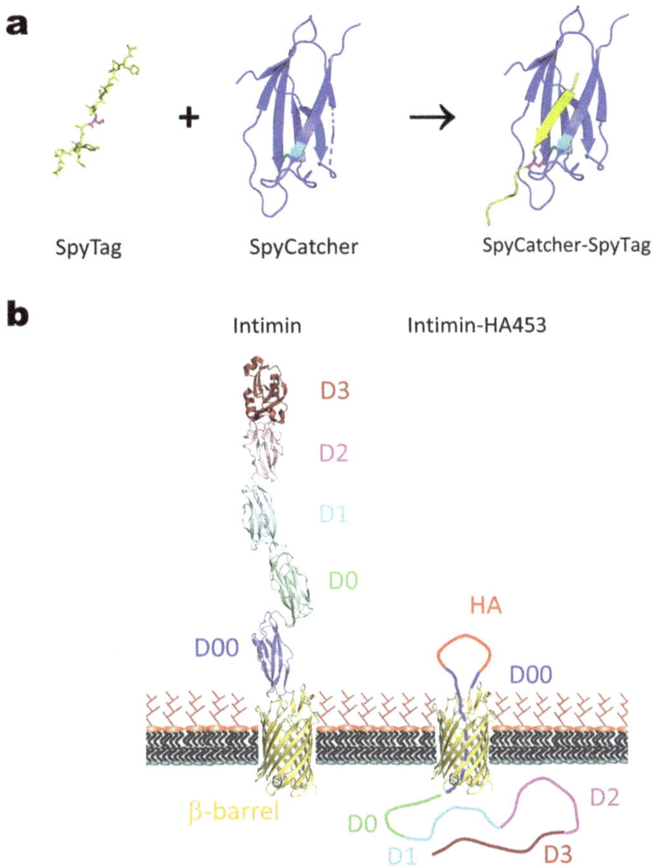

Fig. 1 Structure of proteins used in this chapter. (**a**) The mechanism of the SpyCatcher-SpyTag reaction. The SpyTag peptide (yellow) is recognized by the SpyCatcher protein (blue). Upon binding, an aspartate in SpyTag (magenta) reacts with a lysine in SpyCatcher (cyan) to form a covalent isopeptide bond. The images are based on the SpyCatcher-SpyTag crystal structure (PDB ID: 4MLI). (**b**) Structure of intimin and intiminHA453. Intimin is anchored in the outer membrane by the N-terminal β-barrel domain (yellow; PDB ID: 4E1S). The extracellular passenger consists of four immunoglobulin-like domains, D00 (blue), D0 (light green), D1 (cyan), and D2 (pink), and the lectin-like D3 domain (dark red) at the C-terminus. The passenger was constructed using the D00-D0 (PDB ID: 6TQD) and D1-D3 (PDB ID: 1F00) crystal structures. In the HA453 variant, a double hemagglutinin tag (red) inserted into D00 disrupts the folding of the domain and consequently prevents secretion of the rest of the passenger through the β-barrel. Thus, the C-terminus is located in the periplasm. The periplasmic LysM domain is omitted for clarity. The structure figures were prepared using PyMol (Schroedinger)

often also well tolerated at internal sites [15]. We have previously exploited the SpyCatcher-SpyTag system to probe the topology of outer membrane proteins on bacterial cells [4, 17]. To this end, we created a genetic fusion of SpyCatcher and superfolder green fluorescent protein (sfGFP). SpyCatcher-sfGFP is too large to cross the outer membrane; therefore, it can only bind to surface-exposed SpyTags in intact cells, but it can bind to periplasmic SpyTags if the integrity of the outer membrane is compromised.

In this chapter, we describe using the SpyCatcher-SpyTag to label bacterial outer membrane proteins and map their topology. As an example protein for topology mapping, we use the outer membrane-anchored adhesin intimin, which is involved in the pathogenesis of enterohaemorrhagic and enteropathogenic *E. coli* [18]. Intimin is an inverse autotransporter protein, the C-terminus of which is exported via the N-terminal β-barrel domain to the outside of the cell [19]. The extracellular region of intimin (the "passenger") contains four immunoglobulin-like domains capped at the C-terminus by a lectin-like domain (Fig. 1b). Epitope tags such as SpyTag can be inserted at the very C-terminus of intimin without affecting its secretion or function [6, 20]. We previously found that disrupting one of the immunoglobulin-like domains (D00) resulted in a stalled secretion intermediate of intimin, where the C-terminus was stuck in the periplasm [6] (Fig. 1b). We present two methods to compare labeling efficiency of the two intimin variants: fluorescence measurements using a plate reader (*see* Subheading 3.2) and observation of labeled bacteria using fluorescence microscopy (*see* Subheading 3.3). Additionally, we provide a protocol for the production and purification of SpyCatcher-sfGFP (*see* Subheading 3.1). Further methods, not covered in this chapter, include detecting the binding of SpyCatcher through Western blotting or imaging using in-gel fluorescence [4].

2 Materials

2.1 SpyCatcher-sfGFP Production and Purification

1. *Escherichia coli* BL21Gold (DE3) transformed with either pIBA3-SpyCatcher sfGFP (Addgene #107420) or pIBA3-SpyCatcherEQ-sfGFP (Addgene #107421).
2. Lysogeny broth (LB) medium: 10 g tryptone, 5 g yeast extract, 5 g sodium chloride, add water to 1 L. Autoclave before use.
3. ZYP medium [21]: 10 g tryptone, 5 g yeast extract, 2 mM magnesium sulfate, 25 mM ammonium sulfate, 50 mM monopotassium phosphate, 50 mM disodium phosphate, add distilled water to 1 L. Autoclave before use.
4. Ampicillin stock: 100 mg/mL ampicillin in 50% ethanol. Store at −20 °C.
5. Anhydrotetracycline stock: 1 mg/mL in ethanol. Store at −20 °C (*see* **Note 1**).
6. Bioreactor and shaking incubator.

7. Antifoam A concentrate.
8. Spectrophotometer and cuvettes.
9. Large centrifuge.
10. Sonicator.
11. 0.22 μm syringe filters.
12. Fast protein liquid chromatography (FPLC) system.
13. Ni-nitrilotriacetic acid (NTA) column.
14. Phosphate-buffered saline (PBS): 10 mM sodium phosphate, 150 mM sodium chloride, pH 7.4.
15. Elution buffer: PBS supplemented with 0.5 M imidazole.
16. Dialysis membrane and clips.
17. Storage buffer: PBS supplemented with 10% glycerol.
18. SDS-PAGE equipment and buffers.
19. Bicinchoninic acid (BCA) assay kit.
20. Liquid nitrogen.

2.2 Fluorescence Measurements

1. *Escherichia coli* BL21Gold (DE3) transformed with either pIBA2-Int-SpyTag (Addgene #198038) or pIBA2-IntHA453-SpyTag (Addgene #198039).
2. LB medium (as in Subheading 2.1).
3. Ampicillin stock (as in Subheading 2.1).
4. Spectrophotometer and cuvettes.
5. Anhydrotetracycline stock (as in Subheading 2.1).
6. Centrifuge tubes.
7. Benchtop centrifuge.
8. PBS (as in Subheading 2.1).
9. Rotary mixer.
10. SpyCatcher-sfGFP and SpyCatcherEQ-sfGFP (produced as described in Subheading 3.1).
11. Microcentrifuge.
12. Microcentrifuge tubes.
13. Fluorescence plate reader.
14. Black 96-well plates.

2.3 Fluorescence Microscopy

1. Same requirements as described in Subheading 2.2, **points 1–12**.
2. Microscope slides and coverslips.
3. Molten 1% agarose in PBS (*see* **Note 2**).
4. Fluorescence microscope.

3 Methods

3.1 Production and Purification of SpyCatcher-sfGFP

For fluorescent labeling of SpyTagged proteins, we fused sfGFP to the C-terminus of SpyCatcher [4]. In addition, we included an N-terminal His tag on SpyCatcher for purification and a Strep II tag at the C-terminus of sfGFP for an alternative detection method. This construct was cloned into the pASK-IBA3 plasmid (from IBA Lifesciences). Expression from this plasmid is induced by the addition of anhydrotetracycline. After lysis of the bacteria and clarification of the supernatant, the SpyCatcher-sfGFP protein can be purified using nickel affinity chromatography.

1. Inoculate BL21Gold(DE3) strains containing pIBA3-SpyCatcher-sfGFP and pIBA3-SpyCatcherEQ-sfGFP separately in 5 mL of LB medium supplemented with 100 μg/mL of ampicillin. Grow the bacteria overnight at 37 °C with shaking at 180 rpm.
2. The following morning, dilute the bacterial cultures (1:200) separately into 1 L of ZYP medium supplemented with 100 μg/mL of ampicillin.
3. Place the bacterial cultures in a bioreactor set to 37 °C and add antifoam (*see* **Note 3**).
4. Grow until mid-log phase (until the optical density at 600 nm [OD_{600}] reaches ~0.5), normally approximately 3 h.
5. Induce protein expression by adding 200 ng/mL of anhydrotetracycline, i.e., a 1:5000 dilution of the stock.
6. Grow the bacteria for a further 2 h at 37 °C, then collect the cells by centrifuging for 15 min at 5000× *g*.
7. Remove the supernatant and resuspend the pellets in a total of ~30 mL of PBS. The bacteria can be stored at −80 °C for later processing.
8. Use a sonicator to lyse the bacteria (*see* **Note 4**).
9. Take a 40 μL of the sample of the lysate, move it to a microcentrifuge tube with 10 μL of 5× SDS-PAGE sample buffer, and store at −20 °C for later analysis by SDS-PAGE.
10. Clarify the lysate by centrifuging for 30 min at 20000× *g*.
11. Pass the supernatant through a 0.22 μm filter and take a 40 μL sample for SDS-PAGE as in **point 9** above.
12. Apply the supernatant to a Ni-NTA column attached to a FPLC unit and elute the attached SpyCatcher(EQ)-sfGFP protein using an imidazole gradient (*see* **Note 5**). Take 40 μL samples of the flow-through, wash and elution fractions for SDS-PAGE as described in **point 9** above.

13. Run SDS-PAGE to determine which fractions are pure and pool them (*see* **Note 6**).
14. Dialyze the pooled fractions against storage buffer either overnight at 4 °C or at room temperature for 3–4 h.
15. Determine the concentration of the samples using a BCA kit (*see* **Note 7**).
16. Make 100–500 μL aliquots of the sample, flash freeze in liquid nitrogen, and store at −80 °C. Once thawed, do not refreeze an aliquot. Short-term storage at 4 °C is also possible.

3.2 Quantifying Surface Exposure of Intimin by Fluorescence Using SpyCatcher-sfGFP

Quantification of bound SpyCatcher-sfGFP can be easily done using a plate reader with fluorescence measurement capabilities [4, 17]. Bacteria expressing the SpyTagged protein of interest are first incubated with SpyCatcher-sfGFP, washed, and then moved to a suitable 96-well plate for the measurements. As a control, the SpyCatcherEQ-sfGFP variant is used, which fails to form the covalent isopeptide bond between SpyTag and SpyCatcher and can consequently be easily washed away from the bacteria. The resulting fluorescence measurements can then be plotted; an example is given in Fig. 2a.

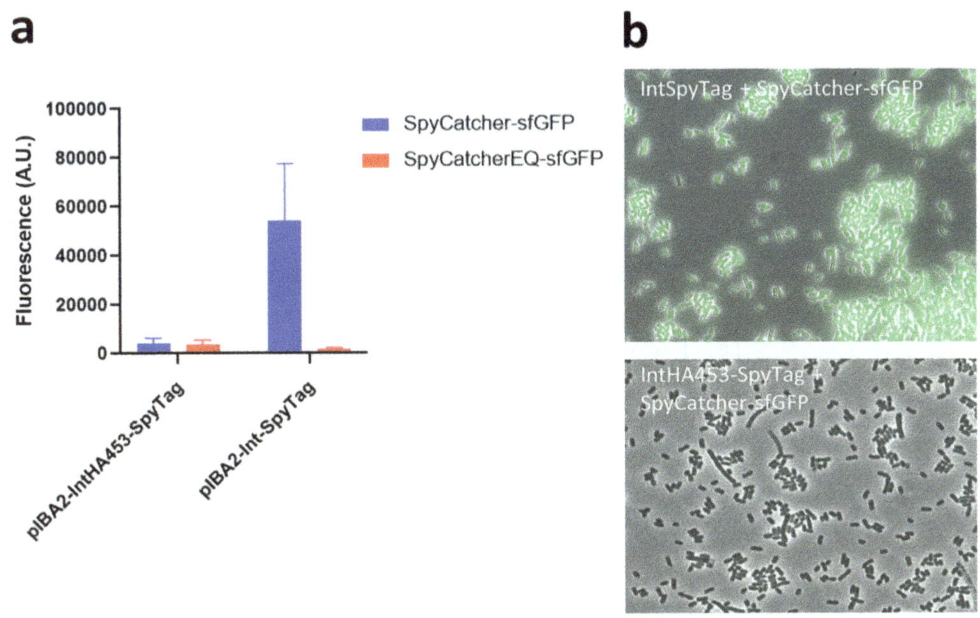

Fig. 2 Example results for labeling with SpyCatcher-sfGFP. (**a**) Fluorescence measurement of cultures labeled with SpyCatcher-sfGFP (blue) or SpyCatcherEQ-sfGFP (red). Because the SpyTag in the pIBA2-IntHA453-SpyTag construct is protected in the periplasm, only background levels of fluorescence are seen for this construct. Error bars denote standard deviations. (**b**) Fluorescence microscopy of cultures labeled with SpyCatcher-sfGFP. Merged images of fluorescent and phase contrast images are shown. The SpyCatcherEQ-sfGFP control images are omitted

1. Inoculate 5 mL LB supplemented with 100 μg/mL of ampicillin with BL21Gold(DE3) strains containing pIBA2-Int-SpyTag and pIBA2-IntHA453-SpyTag separately, each culture in three biological replicates. Grow overnight at 37 °C with shaking at 180 rpm.
2. The following morning, dilute the cultures 1:50 in 20 mL fresh LB + ampicillin 100 μg/mL.
3. Grow the cultures in flasks in a shaking incubator (180 rpm) at 37 °C, until they reach mid-log phase (OD_{600} ~ 0.5, usually around 2 h).
4. Induce intimin production by adding 50 ng/mL of anhydrotetracycline (1:20,000 dilution of stock).
5. Grow the cultures for another 2 h at 37 °C, and then measure the OD_{600} of the cultures again.
6. For each culture, calculate the volume corresponding to 10 mL at an OD_{600} value of 1.0 and transfer this amount to a centrifuge tube (*see* **Note 8**).
7. Pellet the cells by centrifuging 10 min at 5000× *g*.
8. Resuspend each pellet in 1 mL PBS, and then split the sample into two 500 μL aliquots in microcentrifuge tubes.
9. To one of the aliquots, add 0.2 mg/mL of SpyCatcher-sfGFP; to the other, add same concentration of SpyCatcherEQ-sfGFP.
10. Incubate the samples for 1 h at room temperature in a rotary mixer.
11. Centrifuge the cells for 2 min at 6000× *g* in a microcentrifuge.
12. Resuspend the pellets in 500 μL PBS.
13. Repeat **steps 11–12** twice.
14. Centrifuge as in **step 12**, and then resuspend the pellets in 200 μL of PBS.
15. Transfer the suspensions to a black 96-well plate.
16. Measure the fluorescence of the samples using a fluorescence plate reader with excitation at 485 nm and emission at 510 nm (*see* **Note 9**).
17. Calculate the average and standard deviation for the triplicate samples and plot.

3.3 Observing Surface Exposure of Intimin by Fluorescence Microscopy Using SpyCatcher-sfGFP

A complimentary method to fluorescence measurements is observation of the labeled bacteria using a fluorescent microscope. In addition to qualitatively demonstrating labeling by SpyCatcher, microscopy can give information about the proportion of bacteria labeled as well localization of the protein (polar vs. whole cell fluorescence signal distribution, for example). Example images for labeling of bacteria expressing SpyTagged intimin are shown in Fig. 2b.

1. Perform intimin expression and labeling by following **steps 1–14** as in Subheading 3.2. Note that experimental triplicates are not needed unless quantification is planned.
2. During the last wash step, prepare agarose pad slides: add 1 mL of molten agarose and spread across the slide using the pipette tip. Allow to set for a few minutes.
3. Once the bacteria are labeled, drop 5 μL of the samples onto a pad and apply a coverslip.
4. View the bacteria using phase contrast mode and image.
5. Then image in epifluorescence mode.
6. Overlay images using either microscope software or other software (e.g., Fiji [22]).

4 Notes

1. Anhydrotetracycline is light-sensitive, so stocks should be protected, e.g., by covering tubes with foil. It is also recommended to use a recently prepared stock.
2. The agarose solution can be prepared by boiling with a microwave oven shortly before use. The small amount of agarose will cool and solidify on the slide quickly.
3. We have used an Epiphyte3 LEX-10 bioreactor for protein production. However, in the absence of equivalent equipment, the cultures can be grown in flasks in a standard shaking incubator. The rule of thumb is that a flask five times the volume of the liquid should be used, i.e., for 1 L of culture, a 5 L flask should be employed. Where necessary, the cultures can be split into smaller flasks. Antifoam should be added to the cultures irrespective of whether they are grown in a shaker or a bioreactor. As antifoam is very viscous, we simply dip a pipette tip in the antifoam and allow a few drops to drip into the culture medium.
4. The length of sonication depends on the probe size and sonicator model, but 5 × 30 s of sonication with a medium probe is generally sufficient, with 1 min in between on ice to cool the sample down. A freeze-thaw cycle enhances the sonication, as does the addition of 0.1 mg/mL of lysozyme. In addition, we recommend adding magnesium chloride and manganese chloride to 5 mM and a pinch of DNAse I (dip a pipette tip in the DNAse I powder and add to the sample) to help degrade DNA and reduce viscosity. Adding an EDTA-free protease inhibitor cocktail is optional, though we have not found this to be necessary.

5. We have used an Äkta Go system for this protocol step. It may be advisable to add some imidazole (10 mM) to the wash steps to improve removal of impurities; this is easily done using an automated system. For elution, the "upflow" option in this system allows for a more concentrated protein fraction. In the absence of an FPLC unit, gravity flow columns can also be used.
6. Due to the highly specific nature of the SpyCatcher-SpyTag interaction, some impurities will not affect later steps; so to increase the yield of each round of purification, less pure fractions can be included in the pool. However, this will make quantification of the SpyCatcher(EQ)-sfGFP less precise.
7. Other ways to determine the protein concentration, e.g., a Bradford assay, can be used. 1 mg/mL of SpyCatcher-sfGFP is equivalent to 24 μM.
8. If the OD_{600} of the culture is 2.0, you would pellet 5 mL of the culture. Similarly, if the OD_{600} is 0.75, the amount of culture to pellet would be 13.33 mL to get the same number of cells.
9. If a continuous wavelength filter is unavailable, standard GFP or green light (e.g., FITC) filters will suffice.

Acknowledgments

D.L. receives support from the Research Council of Norway, grants 294605, 302723, and 331752. R.D. and J.C.L. are supported by Nottingham Trent University internal funding.

References

1. Doyle MT, Grabowicz M, Morona R (2015) A small conserved motif supports polarity augmentation of *Shigella flexneri* IcsA. Microbiology 161:2087–2097
2. Gunasinghe SD, Shiota T, Stubenrauch CJ et al (2018) The WD40 protein BamB mediates coupling of BAM complexes into assembly precincts in the bacterial outer membrane. Cell Rep 23:2782–2794
3. Rassam P, Copeland NA, Birkholz O et al (2015) Supramolecular assemblies underpin turnover of outer membrane proteins in bacteria. Nature 523:333–336
4. Chauhan N, Hatlem D, Orwick-Rydmark M et al (2019) Insights into the autotransport process of a trimeric autotransporter, *Yersinia* Adhesin A (YadA). Mol Microbiol 111:844–862
5. Rice JJ, Schohn A, Bessette PH et al (2006) Bacterial display using circularly permuted outer membrane protein OmpX yields high affinity peptide ligands. Protein Sci 15:825–836
6. Oberhettinger P, Leo JC, Linke D et al (2015) The inverse autotransporter intimin exports its passenger domain via a hairpin intermediate. J Biol Chem 290:1837–1849
7. Casasanta MA, Yoo CC, Smith HB et al (2017) A chemical and biological toolbox for Type Vd secretion: characterization of the phospholipase A1 autotransporter FplA from *Fusobacterium nucleatum*. J Biol Chem 292:20240–20254
8. Oberhettinger P, Schütz M, Leo JC et al (2012) Intimin and invasin export their C-terminus to the bacterial cell surface using an inverse mechanism compared to classical autotransport. PLoS One 7:e47069
9. Curkić I, Schütz M, Oberhettinger P et al (2016) Epitope-tagged autotransporters as

single-cell reporters for gene expression by a *Salmonella* Typhimurium *wbaP* mutant. PLoS One 11:e0154828

10. Sun J, Lin X, He Y et al (2023) A bacterial outer membrane vesicle-based click vaccine elicits potent immune response against *Staphylococcus aureus* in mice. Front Immunol 14: 1088501
11. Van Der Ley P, Heckels JE, Virji M et al (1991) Topology of outer membrane porins in pathogenic *Neisseria* spp. Infect Immun 59:2963–2971
12. Newton SM, Klebba PE, Michel V et al (1996) Topology of the membrane protein LamB by epitope tagging and a comparison with the X-ray model. J Bacteriol 178:3447–3456
13. Jose J, von Schwichow S (2004) "Cystope tagging" for labeling and detection of recombinant protein expression. Anal Biochem 331: 267–274
14. Keeble AH, Howarth M (2020) Power to the protein: enhancing and combining activities using the Spy toolbox. Chem Sci 11:7281–7291
15. Zakeri B, Fierer JO, Celik E et al (2012) Peptide tag forming a rapid covalent bond to a protein, through engineering a bacterial adhesin. Proc Natl Acad Sci U S A 109:4347–4348
16. Hatlem D, Trunk T, Linke D et al (2019) Catching a SPY: using the SpyCatcher-SpyTag and related systems for labeling and localizing bacterial proteins. Int J Mol Sci 20:2129
17. Wrobel A, Saragliadis A, Pérez-Ortega J et al (2020) The inverse autotransporters of *Yersinia ruckeri*, YrInv and YrIlm, contribute to biofilm formation and virulence. Environ Microbiol 22:2939–2955
18. Whelan R, McVicker G, Leo JC (2020) Staying out or going in? The interplay between Type 3 and Type 5 secretion systems in adhesion and invasion of enterobacterial pathogens. Int J Mol Sci 21:4102
19. Leo JC, Oberhettinger P, Schütz M et al (2015) The inverse autotransporter family: intimin, invasin and related proteins. Int J Med Microbiol 305:276–282
20. Leo JC, Oberhettinger P, Yoshimoto S et al (2016) Secretion of the intimin passenger domain is driven by protein folding. J Biol Chem 291:20096–20112
21. Studier FW (2005) Protein production by auto-induction in high-density shaking cultures. Protein Expr Purif 41:207–234
22. Schindelin J, Arganda-Carreras I, Frise E et al (2012) Fiji: an open-source platform for biological-image analysis. Nat Methods 9: 676–682

Chapter 5

In Vitro Reconstruction of Bacterial β-Barrel Membrane Protein Assembly Using *E. coli* Microsomal (Mid-Density) Membrane

Eriko Aoki, Edward Germany, and Takuya Shiota

Abstract

The in vitro reconstruction assay enables us to evaluate in detail the insertion and proper protein folding (together termed assembly) of β-barrel membrane proteins. Here, we introduce an in vitro reconstitution experiments using isolated membrane fractions from *Escherichia coli* (*E. coli*). Membrane fractions isolated from *E. coli* cells and disrupted by sonication, which we have termed *E. coli* microsomal (mid-density) membrane (EMM), are ideal for biochemical experiments, as they can be harvested by high-speed centrifugation and do not require ultra-centrifugation. EMM pretreated with detergent can assemble externally supplemented β-barrel membrane proteins via intact β-barrel assembly machinery (BAM) complex retained in EMM. This method not only allows assembly analysis with inexpensive equipment but it also can be applied to drug screening using assembly as an indicator with high reproducibility. In this chapter, we introduce our method of evaluating assembled β-barrel membrane proteins by demonstrating four representative β-barrel membrane proteins: *E. coli* major porins OmpA and OmpF; enterohemorrhagic *E. coli* (EHEC) autotransporter EspP, and *Haemophilus influenzae* (*H. influenzae*) adhesin Hia.

Key words β-barrel membrane protein assembly, OMPs, In vitro reconstitution, *E. coli*, BAM complex

1 Introduction

Bacterial outer membrane proteins (OMPs) are characterized by a β-barrel transmembrane domain [1]. Newly synthesized OMPs cross the inner membrane (IM) through the Sec translocon [2], followed by the removal of the signal sequence attached to the N-terminal side of the OMPs by signal peptidases. The mature OMPs traverse the periplasm from IM to outer membrane (OM) with assistance from chaperons such as SurA or Skp [3]. At the OM interface, the BAM complex is then responsible for the assembly of the majority of OMPs [4].

Raffaele Ieva (ed.), *Transmembrane β-Barrel Proteins: Methods and Protocols*, Methods in Molecular Biology, vol. 2778, https://doi.org/10.1007/978-1-0716-3734-0_5,

To analyze the OMP assembly via the BAM complex in detail, it is necessary to evaluate the assembly efficiency by in vitro reconstitution experiments. In the 1990s, de Cock et al. established the first in vitro reconstitution experimental system utilizing membrane fraction mainly containing OM components isolated from *E. coli* disrupted by a French press [5]. They demonstrated that 0.08% of Triton X-100 stimulated in vitro synthesized PhoE assembly into the OM fraction [6]. We adapted this finding for our EMM assembly assay introduced in this chapter. With the identification of the BAM complex as the primary machinery for OMPs assembly into bacterial OM, Hagan et al. developed a method that utilizes artificial proteoliposomes containing purified BAM complex for assembly [7], and this method has become a major in vitro reconstitution system for studying BAM complex-assisted OMP assembly.

Recently, we invented an alternate in vitro reconstitution system utilizing EMM, a mid-density membrane fraction isolated from *E. coli* disrupted by sonication [8, 9]. The EMM method has various advantages over reconstitution experiments using artificial proteoliposomes. First, EMM can be stored at −80 °C and thus stable experiments can be performed with a single lot via large-scale preparation. Second, many scientists can use this method because EMM preparation requires only sonication and centrifugation, thus removing the need for high-performance liquid chromatography purified BAM complex or the use of an ultracentrifuge harvested membrane fraction. Third, the isolation of EMM from various mutant *E. coli* strains enables us to directly analyze the impact of BAM mutations on its function. Fourth, EMM membranes act as a more physiological condition, containing the asymmetric bilayers composed of LPS and lipids, which are difficult to reconstitute in liposomes. Finally, this method can analyze the OMPs assembly which requires the support of multiple protein complexes. The EMM contains not only the BAM complex but also other integral membrane protein complexes similar to the native stoichiometry. Here, we are introducing EMM preparation, EMM assembly assay, and evaluation examples using four different substrates: *E. coli* major porin OmpA [10] and OmpF [11], EHEC autotransporter EspP [12], and *H. influenzae* adhesin Hia [13].

2 Materials

Prepare all solutions using purified water (prepared by purifying water to attain a sensitivity >5 MΩ-cm at 25 °C) and special reagent grade. Prepare all reagents at room temperature; storage temperature is described below.

2.1 EMM Preparation

1. L medium (800 mL): Dissolve 8 g of tryptone, 8 g of NaCl, and 4 g of yeast extract in purified water and then make up to 800 mL. Autoclave and store at room temperature (*see* **Note 1**).
2. 1 M Tris–HCl, pH 7.5 (500 mL): Dissolve 60.55 g of Tris in 400 mL purified water and adjust pH to 7.5 with HCl. Fill with purified water to 500 mL and store at room temperature.
3. 4 M NaCl (500 mL): Dissolve 116.9 g of NaCl in purified water and then fill with purified water to 500 mL. Autoclave and store at room temperature.
4. 0.5 M EDTA, pH 8.0 (500 mL): Dissolve 93 g of Na_2EDTA $2H_2O$ in 400 mL purified water and adjust pH to 8.0 with 5 N NaOH. Make up to 500 mL. Store at room temperature.
5. 2 M Sucrose (200 mL): Dissolve 136.92 g of sucrose in purified water and then make up to 200 mL. Store at 4 °C.
6. 5 N KOH (100 mL): Dissolve 28.05 g of KOH in purified water and then make up to 100 mL. Store at room temperature.
7. 1 M 3-Morpholinopropanesulfonic acid (MOPS)-KOH, pH 7.2 (250 mL): Dissolve 52.3 g of MOPS in 175 mL of purified water and adjust pH to 7.2 with 5 N KOH. Make up to 250 mL and store at room temperature.
8. Sonication buffer (500 mL): 50 mM Tris–HCl pH 7.5, 150 mM NaCl, and 5 mM EDTA pH 8.0. Mix 25 mL 1 M Tris–HCl pH 7.5, 18.75 mL 4 M NaCl, 5 mL 0.5 M EDTA pH 8.0, and 451.25 mL purified water. Store at 4 °C.
9. SEM buffer (500 mL): 250 mM sucrose, 10 mM MOPS-KOH pH 7.2, 1 mM EDTA. Mix 62.5 mL 2 M sucrose, 5 mL 1 M MOPS-KOH pH 7.2, 1 mL 0.5 M EDTA pH 8.0, and 431.5 mL purified water. Store at 4 °C.
10. 0.6% (w/v) Sodium dodecyl sulfate (SDS) (50 mL): Dissolve 300 mg SDS in purified water and then make up to 50 mL. Store at room temperature.
11. Probe type sonicator (*see* **Note 2**).
12. Liquid nitrogen.

2.2 Synthesis of Substrate OMPs

2.2.1 *In Vitro* Transcription

1. Commercially available nuclease-free water. Store at room temperature.
2. Linearized DNA for a template of substrate OMP: Amplify the DNA fragment which contains the indicated elements in the following order: (1) SP6 promoter, (2) start codon, (3) mature part of OMPs, (4) polyA tail, by high fidelity PCR enzyme. Purify PCR fragment using nuclease-free water for elution (*see* **Note 3**).

3. Commercially available SP6 RNA polymerase kit. For example, Clonetech's SP6 RNA Polymerase contains SP6 RNA polymerase, 10 × transcription buffer, 0.1% BSA, and 100 mM DTT. Store at −20 °C.
4. Commercially available RNase inhibitor cocktail. Store at −20 °C.
5. 10 mM G-cap [m7G(5′)ppp(5′)G]: Add 99 μL nuclease-free water to 1.0 μmol of m7G(5′)ppp(5′)G. Store at −20 °C (*see* **Note 4**).
6. 5 mM γNTPs mixture: Mix 5 μL of each ATP, CTP, GTP, and UTP to 80 μL nuclease-free water. Store at −20 °C.

2.2.2 *In Vitro Translation*

1. Commercially available nuclease-free water.
2. Amino acid mixture minus methionine.
3. Commercially available RNase inhibitor cocktail. Store at −20 °C.
4. RNA-coding substrate OMP: This was produced in the in vitro transcription reaction (Subheading 3.2.1). Store at −20 °C (*see* **Note 5**).
5. [^{35}S]-labeled methionine and cysteine, specific activity 1175 Ci/mmol, 10 mCi/mL. Store at −20 °C (*see* **Notes 6** and 7).
6. Commercially available rabbit reticulocyte lysate. Store at −80 °C (*see* **Note 8**).

2.3 EMM Assembly Assay

1. 1 M MOPS-KOH pH 7.2: *See* as in Subheading 2.1.
2. 1 M KH_2PO_4-K_2HPO_4 pH 7.2 (250 mL): Dissolve 10.9 g of KH_2PO_4 in purified water and then make up to 80 mL. Dissolve 34.8 g of K_2HPO_4 in purified water and then make up to 200 mL. Mix 70.75 mL of KH_2PO_4 solution and 179.25 mL of K_2HPO_4 solution. Adjust the pH to 7.2 with either KH_2PO_4 solution or K_2HPO_4 solution, and store at room temperature.
3. 2 M sucrose (200 mL): Dissolve 136.9 g of sucrose in purified water and then make up to 200 mL. Store at 4 °C.
4. 2.5 M KCl (50 mL): Dissolve 9.32 g of KCl in purified water and then make up to 50 mL. Store at room temperature.
5. 1 M $MgCl_2$ (50 mL): Dissolve 10.2 g of $MgCl_2$ $6H_2O$ in purified water and then make up to 50 mL. Store at room temperature.
6. 1 M dithiothreitol (DTT): Dissolve 770 mg of DTT in purified water and then make up to 5 mL. Store aliquot at −20 °C.

7. 200 mM methionine (5 mL): Dissolve 150 mg of methionine in the 10 mM HCl and then make up to 5 mL. Store aliquot at −20 °C.
8. 200 mM cysteine (5 mL): Dissolve 176 mg of cysteine in the 10 mM HCl and then make up to 5 mL. Store aliquot at −20 °C.
9. 20% (v/v) Triton X-100 (50 mL): Dissolve 10 mL of Triton X-100 in 30 mL purified water. Shake overnight to mix completely and make up to 50 mL. Store at room temperature.
10. 10% (w/v) Bovine serum albumin (BSA): Dissolve 100 mg of BSA in purified water and then make up to 1 mL. A fresh solution should be prepared before each experiment.
11. Assembly assay buffer (1 mL): 10 mM MOPS-KOH pH 7.2, 2.5 mM KH_2PO_4-K_2HPO_4 pH 7.2, 250 mM sucrose, 15 mM KCl, 5 mM $MgCl_2$, 4 mM methionine, 4 mM cysteine, 5 mM DTT, 1% (w/v) BSA, 0.09% (v/v) TritonX-100. Mix 10 μL 1 M MOPS-KOH pH 7.2, 2.5 μL 1 M KH_2PO_4-K_2HPO_4 pH 7.2, 125 μL 2 M sucrose, 6 μL 2.5 M KCl, 5 μL 1 M $MgCl_2$, 20 μL 200 mM methionine, 20 μL 200 mM cysteine, 5 μL 1 M DTT, 100 μL 10% (w/v) BSA, 4.5 μL 20% (v/v) Triton X-100, and 702 μL purified water. A fresh solution should be prepared before each experiment.
12. EMM. Kept in aliquots at −80 °C.
13. 30 μM peptide 23: Purchase peptide (H-INTDNIVALGLV YQF -OH) and dissolve it in DMSO to 30 mM. Store aliquot at −20 °C. This peptide corresponds to the last 15 residues of the OmpC containing the bacterial β-signal sequence.
14. [^{35}S]-labeled substrate OMPs protein prepared via in vitro translation reaction. Prepare fresh when prior to starting EMM assembly assay.

2.4 Analysis of Assembled OMPs

2.4.1 Analysis of OmpA Assembly

1. SEM buffer.
2. SDS-polyAcrylamide gel electrophoresis (PAGE) sample buffer (100 mL): 125 mM Tris–HCl, pH 6.8, 2 mM EDTA, 2% (w/v) SDS, 8% (v/v) glycerol, 0.03% (w/v) bromophenol blue, 2% (v/v) β-mercaptoethanol. Dissolve 1.51 g of Tris, 74.4 mg Na_2EDTA $2H_2O$, 2 g SDS, 30 mg bromophenol blue in 60 mL purified water, and adjust pH to 6.8 with HCl. Mix 8 mL 100% (v/v) glycerol and make up to 100 mL with purified water. Store at room temperature. Add 20 μL β-mercaptoethanol to 1 mL SDS-PAGE sample buffer immediately prior to use.

2.4.2 Analysis of EspP Assembly

1. SEM buffer.
2. 5 mg/mL Proteinase K (PK) (5 mL): Dissolve 25 mg of PK in the SEM buffer and then make up to 5 mL. Store aliquot at −20 °C.
3. Dimethyl sulfoxide (DMSO).
4. 2-Propanol.
5. 1 M Phenylmethylsulfonyl fluoride (PMSF) (5 mL): Dissolve 870 mg of PMSF in the 100% DMSO and then make up to 5 mL. Store aliquot at −20 °C. Dilute to 100 mM with 2-propanol when it is used (*see* **Note 9**).
6. SDS-PAGE sample buffer.

2.4.3 Analysis of Hia Assembly

1. SEM buffer.
2. 1 M PMSF.
3. SDS-PAGE sample buffer.

2.4.4 Analysis of Hia Assembly by Carbonate Extraction

1. EMM. Kept in aliquots at −80 °C.
2. Assembly assay buffer.
3. SEM buffer.
4. 1 M PMSF.
5. 100 mM Na_2CO_3: Dissolve 53 mg of Na_2CO_3 in 4 mL of pure water and mix 50 μL of 100 mM PMSF. Make up to 5 mL by purified water. A fresh solution should be prepared before each experiment.
6. Micro ultracentrifuge (e.g., himac CS 120GXL, Hitachi koki).
7. Ultracentrifuge tubes (1.5 mL) (e.g., himac 1.5 mL micro tube, Eppendorf Himac Technologies).
8. 3× SDS-polyacrylamide gel electrophoresis (PAGE) sample buffer (100 mL): 375 mM Tris–HCl, pH 6.8, 6 mM EDTA, 6% (w/v) SDS, 24% (v/v) glycerol, 0.09% (w/v) bromophenol blue. Dissolve 4.53 g of Tris, 223.2 mg $Na_2EDTA\ 2H_2O$, 6 g SDS, 90 mg bromophenol blue in 60 mL purified water and adjust pH to 6.8 with HCl. Mix 24 mL 100% (v/v) glycerol and make up to 100 mL with purified water. Store at room temperature.
9. 100% (v/v) β-mercaptoethanol. Store at room temperature.

2.4.5 Analysis of OmpF Assembly

1. SEM buffer.
2. n-Dodecyl-β-D-maltopyranoside (DDM): store at 4 °C (*see* **Note 10**).
3. 1 M PMSF.

4. 2× Blue Native (BN)-PAGE lysis buffer (100 mL): 50 mM imidazole-HCl, pH 7.0, 100 mM NaCl, 100 mM 6-aminohexianoic acid, 2 mM EDTA, 15% (w/v) glycerol. Dissolve 340 mg imidazole, 600 mg NaCl, 1.3 g 6-aminohexianoic acid, 74.4 mg $Na_2EDTA\ 2H_2O$ in 50 mL purified water and adjust pH to 7.0 with HCl. Mix 15 mL 100% (v/v) glycerol and make up to 100 mL with purified water. Store at 4 °C.
5. BN-PAGE sample buffer (50 mL): 4.0%(w/v) Coomassie Brilliant Blue (CBB) G-250, 100 mM 6-aminohexanoic acid. Dissolve 2 g CBB G-250 and 656 mg 6-aminohexianoic acid in purified water and then make up to 50 mL. Store at 4 °C.
6. 1.5% (w/v) DDM BN-PAGE lysis buffer (1 mL): Dissolve 15 mg of DDM in 500 μL 2× BN-PAGE lysis buffer and mix 10 μL of 100 mM PMSF. Make up to 1 mL with purified water.

3 Method

3.1 EMM Preparation

E. coli cells are grown in Erlenmeyer flasks with baffled bottom in liquid culture under constant orbital shaking (approximately 120 rpm). While the optimal incubation temperature for *E. coli* is 37 °C, the temperature and medium should be optimized depending on the character of *E. coli* strain (and the mutant derivatives) and growth conditions. Therefore, it is advisable to examine the conditions under which *E. coli* strain of interest can grow sufficiently in a small-scale culture in advance.

1. Culture the *E. coli* strain of interest in 200 mL of L medium for mid to late log phase (OD_{600} = 1.0 to 1.2). Take care that culture does not reach the stationary phase (OD_{600} > 2.0). In this chapter, we utilized BL21(DE3)* (*see* **Notes 11** and **12**).
2. Harvest cells into the 50 mL conical tube at 3,000 × *g* for 7 min at 4 °C (*see* **Note 13**).
3. For the following steps, all samples should be kept on ice.
4. Resuspended the cell pellet in 15 mL sonication buffer.
5. Disrupt cells by probe-type sonicator (we use Brason sonifier 450, 3 mm diameter probe) at power 5 with constant sonication for 7 s allowed cells to chill on ice >1 min between sonication; repeat 20 times (*see* **Note 14**).
6. Centrifuge the sonicated sample at 2,800 × g for 7 min at 4 °C to spin down cell debris. Transfer the supernatant to a fresh 50 mL conical tube (*see* **Note 15**).
7. Centrifuge at 15,000 × *g* for 10 min at 4 °C to harvest EMM. Discard the supernatant.

8. Gently resuspend the pellet in 2 mL of SEM buffer by pipette.
9. Measure protein concentration in EMM. Take 10 μL of resuspended EMM solution and mix with 990 μL of 0.6% (w/v) SDS solution. Boil the mixture for 10 min and centrifuge at 15,000 × *g* for 5 min at room temperature. Measure protein absorbance at 280 nm. To calculate the amount of protein per EMM, assume an EMM protein concentration of 10 mg/mL at $A_{280} = 0.21$ (*see* **Note 16**).
10. Make aliquots of 200 μL, immediately snap-freeze in liquid nitrogen, and then store at −80 °C (*see* **Notes 17** and **18**).

3.2 Synthesis of Substrate OMPs

3.2.1 In Vitro Transcription

1. To make 40 μL of transcription mixture, mix 1 μL of the linearized DNA, 4 μL of 10 × transcription buffer, 4 μL of 0.1% BSA, 4 μL of 100 mM DTT, 1 μL of RNase inhibitor cocktail, 2 μL of 5 mM γNTP mixture, 1 μL of 10 mM G-cap, 1 μL of SP6-RNA polymerase, and 16 μL of nuclease-free water.
2. Incubate for 90 min at 37 °C.
3. Store at −20 °C.

3.2.2 In Vitro Translation

1. To make 100 μL of translation products, mix 5 μL of transcription mixture produced as above, 1 μL of amino acid mixture minus methionine, 2 μL of RNase inhibitor cocktail, 35 μL of rabbit reticulocyte lysate, and 5 μL of [^{35}S]-labeled methionine and cysteine and make up to 100 μL with nuclease-free water (*see* **Notes 19** and **20**).
2. Incubate for 90 min at 30 °C.
3. Transfer on ice (*see* **Note 21**).

3.3 EMM Assembly Assay

The standard EMM assembly assay protocol monitors the assembly of substrate OMPs into EMM by co-incubating EMM and [^{35}S]-labeled substrate OMPs. The ratio of EMMs to substrate OMPs needs to be optimized as the efficiency of assembly depends on the substrate, and the efficiency of detection depends on the analysis method. Here, to introduce four different methods to analyze assembly efficiency, we used four different substrate OMPs, OmpA, EspP, Hia, and OmpF.

1. Defrost and transfer the appropriate amount of EMM to the two new tubes. For each substrate, the amounts of EMM required are as follows: OmpA, 225 μg; EspP, 600 μg; Hia, 280 μg; OmpF, 450 μg. One tube is for peptide 23, the other tube is prepared for DMSO control. Harvest EMM by centrifugation at 10,000 × g for 5 min at 4 °C.
2. Resuspend the pellet by pipetting with appropriate volume of assembly assay buffer: OmpA, 300 μL; EspP, 800 μL; Hia, 400 μL; OmpF, 600 μL.

3. Add peptide 23 to final concentration 100 μM (e.g., add 1.0 μL of 30 mM peptide 23 stock into 300 μL of reaction mixture for OmpA). In a negative control reaction, add equal volume of DMSO.
4. Incubate on ice for 5 min (*see* **Note 22**).
5. Transfer to a 30 °C block incubator and pre-incubate for 2 min (*see* **Note 23**).
6. Add appropriate volume of translated product of substrate OMPs: OmpA, 30 μL; EspP, 80 μL; Hia, 40 μL; OmpF, 60 μL, respectively. Mix gently by pipetting.
7. Continue to incubate at 30 °C. To monitor assembly reaction, withdraw aliquots of equal volume and incubate on ice to halt the assembly reaction. Each time point for each substrate and the amount of mixture to be collected at that time are as follows: OmpA: 1, 3, 10 min, 300 μL; EspP: 5, 10, 20, 40 min, 200 μL; Hia: 3, 5, 10, 30 min, 100 μL; OmpF: 10, 30, 90 min, 200 μL (*see* **Note 24**).
8. Analyze the EMM containing assembled OMPs as described in the following section.

3.4 Analysis of Assembled OMPs

3.4.1 Analysis of OmpA Assembly Reactions

The assembly of OmpA can be assessed by "heat modifiability", as folded OmpA is SDS-resistant at room temperature (non-boiled sample) thereby migrating faster on SDS-PAGE than the denatured form (boiled sample) [14] (*see* also Chap. 9).

1. Harvest EMM by centrifugation at 10,000 × *g* for 5 min at 4 °C. Discard the supernatant.
2. Gently resuspend pellets in 100 μL of SEM buffer containing 1 mM PMSF.
3. Harvest EMM by centrifugation as above. Discard the supernatant.
4. Add 40 μL of SDS-PAGE sample buffer to EMM and resuspend using a bath sonicator (*see* **Note 25**).
5. Transfer 20 μL into new tubes (Split half).
6. Incubate the sample tubes for 10 min at room temperature or 98 °C.
7. Apply all samples onto SDS-PAGE and analyze them by radioimaging (Fig. 1).

3.4.2 Analysis of EspP Assembly Reactions

EspP is a 1300 aa protein and contains two domains, a C-terminal β-barrel domain and a N-terminal passenger domain. Assembled EspP via the BAM complex removes the passenger domain by self-cleavage and only the β-barrel domain stays in the outer membrane. Assembly of EspP can be assessed by the amount of fragmented EspP after protease treatment as the membrane-embedded β-barrel domain becomes protease-resistant [15].

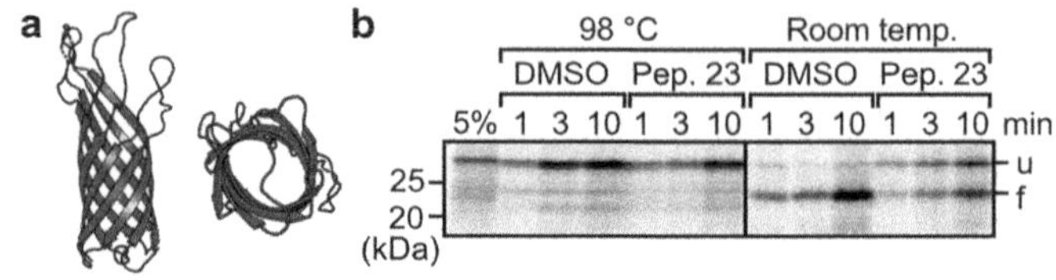

Fig. 1 Analysis of OmpA assembly. (**a**) Structure of the β-barrel domain of OmpA (PDB: 1BXW), side view (left) and bottom view (right). (**b**) In vitro–synthesized OmpA containing [^{35}S]-labeled methionine and cysteine were incubated with EMM in the presence or absence of peptide 23 (Pep. 23) for indicated times at 30 °C. Assembly reaction was halted by placing each reaction aliquot on ice and the indicated times and then EMM were harvested. After washing EMM, EMM proteins were solubilized with SDS-PAGE sample buffer and incubated at 98 °C or room temperature. The samples were analyzed by SDS-PAGE and detected by autoradiography. 5% shows an equivalent of 5% of the radiolabeled OmpA used in each reaction. (u) and (f) indicate unfolded and folded forms of OmpA, respectively. The addition of a chemical amount of the peptide 23 impaired the assembly of OmpA, indicating the assembly reaction occurred via the BAM complex

1. To split half the assembly mixture of each time point, transfer 100 μL of assembly mixture into new tubes.
2. Add 2 μL of 5 mg/mL PK or SEM buffer to the halted reactions and mix by tapping the tubes by hand. Incubate on ice for 20 min and then add 2 μL of 100 mM PMSF (final concentration is 1 mM) to stop the digestion.
3. Harvest EMM by centrifugation at 15,000 × *g* for 5 min at 4 °C. Discard the supernatant and add 100 μL of SEM buffer containing 1 mM PMSF, and gently shake the tubes by hand.
4. Harvest EMM by centrifugation as above. Discard the supernatant.
5. Add 20 μL of SDS-PAGE sample buffer to EMM and resuspend using a bath sonicator.
6. Incubate the sample tubes for 10 min at 98 °C.
7. Apply all samples onto SDS-PAGE and analyze them by radio-imaging (Fig. 2).

3.4.3 Analysis of Assembled Hia Trimers

Hia is trimeric autotransporter adhesin that forms a 12-stranded single pore β-barrel by four strands originating from one polypeptide. The assembled Hia trimer becomes stable and remains in trimeric form even when boiled in the presence of SDS. Therefore, assembly of Hia can be assessed by the amount of trimer of the boiled sample run by SDS-PAGE.

1. Harvest EMM by centrifugation at 10,000 × *g* for 5 min at 4 °C. Discard the supernatant.

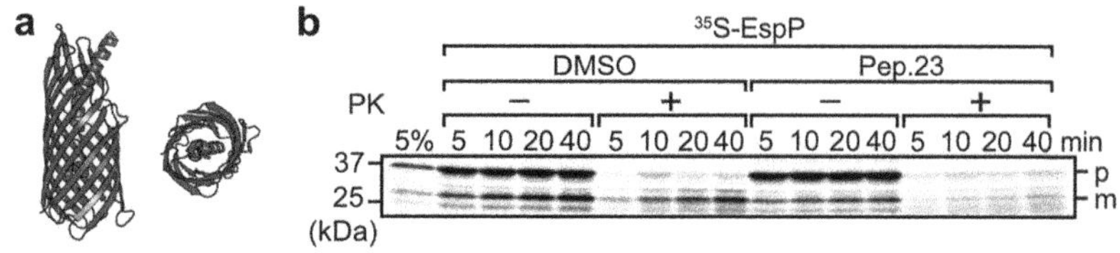

Fig. 2 Analysis of EspP assembly. (**a**) Structure of the β-barrel domain of EspP (PDB: 3SLJ), side view (left) and bottom view (right). (**b**) In vitro–synthesized EspP (974–1300 aa) containing [^{35}S]-labeled methionine and cysteine were incubated with EMM in the presence or absence of peptide 23 (Pep. 23) for indicated times at 30 °C. Assembly reaction was halted by moving the reaction aliquots on ice, and then treated with or without PK. PK treatment was stopped by the addition of PMSF. EMM were harvested and washed. EMM proteins were analyzed by SDS-PAGE and detected by autoradiography. 5% shows an equivalent of 5% of the radiolabeled EspP used per lane. (p) and (m) indicate the precursor and mature forms of EspP, respectively. The assessment of peptide 23 is same as in Fig. 1b

2. Gently resuspend pellets by pipetting with 100 μL of SEM buffer containing 1 mM PMSF.
3. Harvest EMM by centrifugation at 15,000 × *g*. Discard the supernatant.
4. Add 100 μL of SEM buffer containing 1 mM PMSF to the pellet and then mix by tapping the tube.
5. Spin down the EMM by centrifugation at 10,000 × *g* for 10 min at 4 °C. Discard the supernatant.
6. Resuspend the EMM in 20 μL of SDS-PAGE sample buffer. Incubate the sample tubes for 5 min at 95 °C.
7. Apply all samples onto SDS-PAGE and analyze them by radio-imaging (Fig. 3).

3.4.4 Analysis of Hia Assembly by Carbonate Extraction

The assembly of OMPs can also be analyzed by monitoring the amount of protein inserted into the membrane phase. To this end, we use carbonate extraction as follows. Here we demonstrate carbonate extraction of Hia.

1. Resuspend the 140 μg EMM pellet in 200 μL of the assembly assay buffer by pipette.
2. Preincubate the EMM at 30 °C for 3 min.
3. Add 20 μL of the in vitro synthesized [^{35}S]-labeled Hia.
4. Incubate for 90 min at 30 °C. Transfer on ice.
5. After assembly assay, centrifuge at 10,000 × *g* at 4 °C for 5 min to harvest Hia-assembled EMM.
6. Centrifuge at 10,000 × *g* at 4 °C for 5 min to harvest EMM. Discard the supernatant.

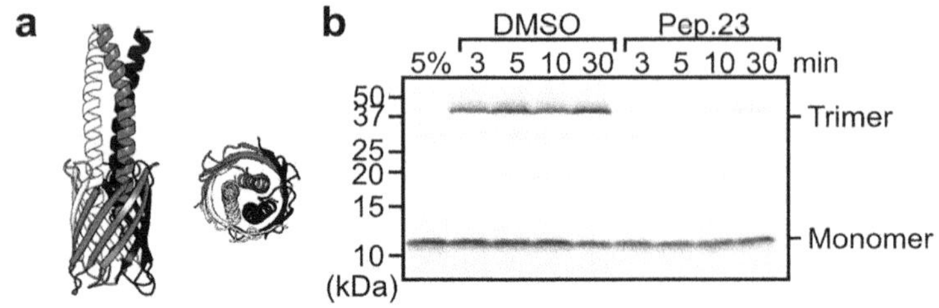

Fig. 3 Analysis of Hia assembly. (**a**) Structure of the part of the autotransporter domain and the β-barrel domain of Hia (PDB: 2GR7), side view (left) and bottom view (right). (**b**) In vitro–synthesized Hia (998–1098 aa) containing [^{35}S]-labeled methionine and cysteine were incubated with EMM in the presence or absence of peptide 23 (Pep. 23) for indicated times at 30 °C. Assembly reaction was halted by shifting on ice and then EMM were harvested. After washing EMM, EMM proteins were solubilized with SDS-PAGE sample buffer and incubated at 98 °C. The samples were analyzed by SDS-PAGE and detected by autoradiography. 5% shows an equivalent of 5% of the radiolabeled Hia used per lane. Monomer and trimer of Hia were indicated. The assessment of peptide 23 is same as in Fig. 1b

7. Resuspend the pellet in 100 μL of SEM buffer containing 1 mM PMSF by pipette.
8. Centrifuge at 10,000 × *g* at 4 °C for 5 min. Discard the supernatant.
9. Resuspend the pellet in 100 μL of SEM buffer containing 1 mM PMSF by tapping the bottom of the tube.
10. Centrifuge at 10,000 × *g* at 4 °C for 10 min. Discard the supernatant.
11. Resuspend the pellet in 25 μL of 100 mM Na_2CO_3 by pipette.
12. Incubate for 15 min on ice.
13. Collect the membrane by ultracentrifugation at 100,000 × *g* (Eppendorf Himac Technologies rotor S55A2) at 4 °C for 60 min.
14. Resuspend the pellet in 100 μL of SEM buffer containing 1 mM PMSF by pipette.
15. Collect the membrane by ultracentrifugation at 100,000 × *g* at 4 °C for 20 min.
16. Transfer the supernatant into a new micro tube. Add 13 μL 3× SDS-PAGE sample buffer without β-mercaptoethanol and 2.4 μL β-mercaptoethanol. Resuspend the membrane pellet in 40 μL of SDS-PAGE sample buffer.
17. Split the samples in half. Heat one aliquot for 5 min at 95 °C and incubate the other at room temperature.
18. Analyze sample by SDS-PAGE and radioimaging or immunoblotting using anti-OmpA or anti-SurA antibodies (Fig. 4; *see* **Note 26**).

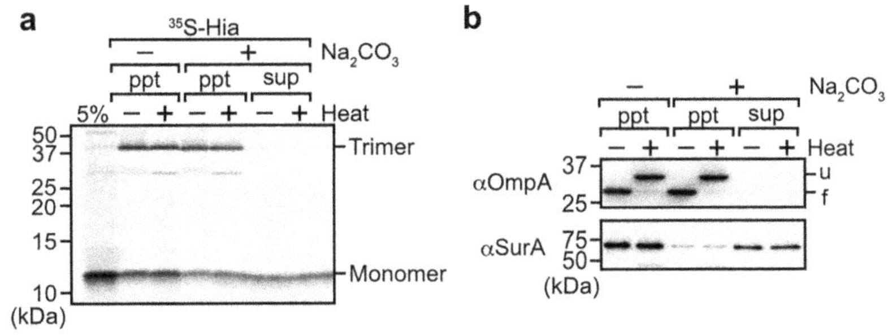

Fig. 4 Analysis of membrane insertion of Hia. (**a**) After performing the assembly reaction of Hia as in Fig. 3b, EMM were treated with or without carbonate (Na_2CO_3 +/−). Membrane-embedded proteins and extracted proteins were separated by ultra-centrifugation and subjected to SDS-PAGE after incubating at 98 °C or room temperature (Heat +/−). Hia was analyzed by radio imaging. (**b**) Endogenous OmpA and SurA were analyzed by immunoblotting using antibodies against OmpA or SurA, respectively. Hia trimer was observed only in the pellet (ppt) fraction after carbonate extraction, similar to OMP control, OmpA. However, the monomer was observed in both ppt and supernatant (sup) samples. Model soluble protein SurA, the periplasmic chaperone, was observed in the supernatant fraction. These results suggest that the Hia trimer was correctly assembled into the membrane

3.4.5 Analysis of OmpF Assembly

A unit of the assembled OmpF is a homo-trimer composed of three mono-polypeptide single pore barrels. While this unit can be observed by heat modifiability on SDS-PAGE, this unit appears in multiple bands due to sample temperature or kinds of lipids surrounding the OmpF. To measure the trimer OmpF simply, we utilize BN-PAGE, because OmpF trimer is observed single band on BN-PAGE [8].

1. Harvest EMM by centrifugation at 10,000 × *g* for 5 min at 4 °C. Discard the supernatant.
2. Gently resuspend pellets by pipetting with 200 μL of SEM buffer containing 1 mM PMSF.
3. Harvest EMM by centrifugation as above. Discard the supernatant.
4. Gently resuspend pellets by pipetting with 30 μL of 1.5% DDM BN-PAGE lysis buffer. Incubate for 20 min on ice.
5. Centrifuge at 13,000 × *g* for 10 min at 4 °C, and transfer supernatant to the new tube.
6. Add 2 μL of BN-PAGE sample buffer and mix them by tapping the tube. Incubate for 5 min on ice.
7. Spin down by centrifugation at 10,000 × *g* for 1 min at 4 °C, and apply all supernatants onto BN-PAGE. Analyze the gel by radioimaging (Fig. 5; *see* **Note 27**).

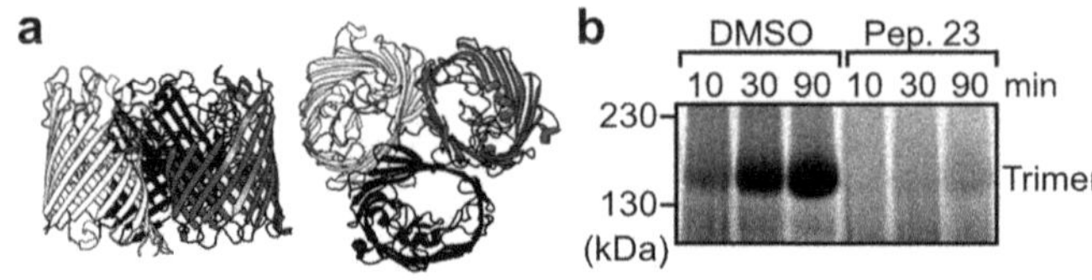

Fig. 5 Analysis of OmpF assembly. (**a**) Structure of OmpF (PDB: 3O0E), side view (left) and bottom view (right). (**b**) In vitro–synthesized OmpF containing [^{35}S]-labeled methionine and cysteine were incubated with EMM in the presence or absence of peptide 23 (Pep. 23) for indicated times at 30 °C. Assembly reaction was halted by shifting on ice and then EMM were harvested. After washing EMM, EMM proteins were solubilized with 1.5% DDM BN-PAGE lysis buffer. The samples were analyzed by BN-PAGE and detected by autoradiography. Trimer of OmpF was indicated. The assessment of peptide 23 is same as in Fig. 1b

4 Notes

1. We usually use L medium; however, any medium can be used as long as *E. coli* can be recovered. When isolating EMM from mutants or from *E. coli*-harboring plasmids, the addition of antibiotics to the medium as appropriate is not a problem for EMM isolation.
2. We use Branson sonifier 450, 3 mm diameter probe. Other sonicator can also be used.
3. We usually clone OMPs-encoding open reading frames excluding the segment encoding the signal sequence into pTnT vector (Promega) downstream of the SP6 RNA polymerase binding site and upstream of the polyA tail. To improve assembly efficiency via the BAM complex, we remove the signal sequence. Because EMM contains the inner membrane fraction as well as the outer membrane fraction, the Sec translocon also exists in EMM. Sec translocon traps the substrate OMPs by binding to the signal sequence. Signal sequence-less OMPs can exclude this possibility and increase assembly efficiency. When we use the OMPs containing additional large domains such as autotransporter (EspP) or adhesin (Hia), we remove these additional non-β-barrel domains of them. We removed 1–973 amino acids of EspP or 1–997 amino acids of Hia, respectively.
4. We add G-cap during in vitro translation to synthesize mRNA capped at 5′-end. The cap structure enhances translation initiation in vitro by improving mRNA stability, protecting from exonuclease degradation, and promoting the formation of translation initiation complexes.
5. Avoid multiple freezing and defrosting of the transcription kit mixture, because this process significantly decreases transcription impairing protein yield.

6. Instructions recommend storing at −80 °C but storing at −20 °C does not affect translation efficiency. The radioisotope will decay but can be used without problems for up to two half-lives (174 days).
7. Diligently follow standards for permission of the use of radioisotopes in individual countries when using or disposing of radiolabeled reagents.
8. Avoid multiple freezing and defrosting of reticulocyte lysate. Aliquot the samples appropriately according to the experimental plan and snap-freeze them with liquid nitrogen to store them.
9. PMSF diluted with 2-propanol should be freshly prepared before use.
10. DDM tends to absorb moisture easily, so shift to room temperature prior to opening the vial.
11. 200 mL is the minimum cell culture scale for EMM preparation for use in the assembly assay. The culture scale can be increased if larger quantities of EMM are needed.
12. The assembly efficiency of a particular substrate OMP can be assessed using EMMs isolated from both wild-type and mutant strains.
13. Regardless of culture volume, each 50 mL conical tube should only contain 200 mL of *E. coli* culture.
14. The state of disruption can be evaluated by absorbance at 600 nm. The OD_{600} of *E. coli* suspended in the sonication buffer before disruption was about 14.38 (calculated by the result of the measurement of ten-fold dilution sample), and the OD_{600} of *E. coli* after disruption was about 1.71.
15. Because the cell debris pellet is soft, do not transfer the supernatant of the bottom of the tube to the new tube to avoid sucking the cell debris.
16. Use quartz cuvette or UV transmittable cuvettes.
17. EMMs are harvested by centrifugation prior to EMM assembly assay, so you do not need to adjust the concentration prior to freezing. Rapid freezing is preferred because the time between crushing and freezing of *E. coli* affects the OMP assembly activity of the EMM.
18. Electron microscopic analysis of the sample immediately after crushing and the EMM fraction sample confirms the removal of cell debris and the enrichment of the membrane in the EMM fraction (Fig. 6).
19. Reducing the amino acid mixture to one-tenth of the amount as recommended in the instruction does not affect translation efficiency. It is advisable to consider the conditions for saving amino acid mixture.

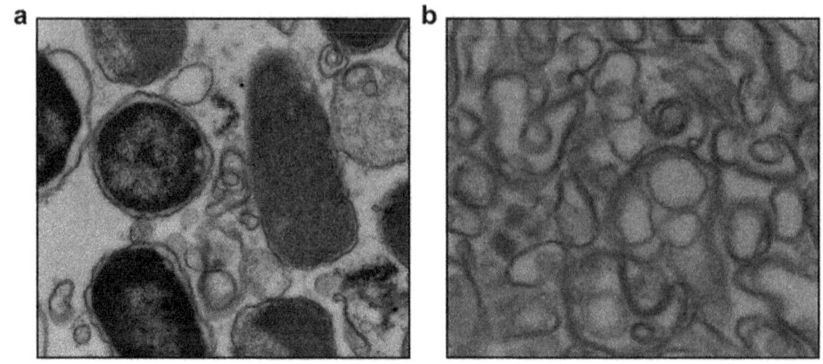

Fig. 6 Transmission electron microscopy images of disrupted cell (**a**) or EMM fraction (**b**)

20. [^{35}S]-labeled methionine and cysteine can be added up to 2× (10 μL for this protocol) depending on decay.
21. It is advisable to prepare the assembly assay buffer and EMM during the incubation for in vitro translation to shorten the time of storage on ice.
22. Note that incubation longer than 1 h will decrease assembly efficiency.
23. When comparing multiple conditions, shift the timing such as every 20 s.
24. When analyzing multiple time courses, dividing the total volume of translational product and assembly assay buffer into on-ice tubes will result in an insufficient sample for the last time course because of residual solution on the tip, etc. Therefore, transfer the amount of assembly assay buffer to a tube on ice.
25. Be careful that the sonication process does not cause the sample to become too hot, as this may lead to protein degradation and poor EMM quality.
26. OmpA is one of the OMPs known as integral membrane protein, and SurA is the periplasmic chaperone known as soluble protein. They were utilized as controls for membrane protein and soluble protein, respectively.
27. To avoid temperature rise, electrophoresis should be performed in a refrigerator or cold room.

Acknowledgments

We thank Yoshiteru Goto (Division of Electron Microscopy in Frontier Science Research Center, University of Miyazaki). We thank the members of the Shiota lab for discussions and critical comments on the manuscript. This work was supported by JST FOREST Program to TS (JPMJFR2064) JSPS KAKENHI to TS (21KK0126), and EA (22K15463).

References

1. Horne JE, Brockwell DJ, Radford SE (2020) Role of the lipid bilayer in outer membrane protein folding in Gram-negative bacteria. J Biol Chem 295:10340–10367. https://doi.org/10.1074/jbc.REV120.011473
2. Rapoport TA (2007) Protein translocation across the eukaryotic endoplasmic reticulum and bacterial plasma membranes. Nature 450: 663–669. https://doi.org/10.1038/nature06384
3. Wang X, Peterson JH, Bernstein HD (2021) Bacterial outer membrane proteins are targeted to the bam complex by two parallel mechanisms. MBio 12. https://doi.org/10.1128/mBio.00597-21
4. Benn G, Mikheyeva IV, Inns PG et al (2021) Phase separation in the outer membrane of Escherichia coli. Proc Natl Acad Sci U S A 118:e2112237118. https://doi.org/10.1073/pnas.2112237118
5. de Cock H, Hendriks R, de Vrije T, Tommassen J (1990) Assembly of an in vitro synthesized Escherichia coli outer membrane porin into its stable trimeric configuration. J Biol Chem 265:4646–4651
6. de Cock H, van Blokland S, Tommassen J (1996) In vitro insertion and assembly of outer membrane protein PhoE of Escherichia coli K-12 into the outer membrane. Role of Triton X-100. J Biol Chem 271:12885–12890. https://doi.org/10.1074/jbc.271.22.12885
7. Hagan CL, Kim S, Kahne D (2010) Reconstitution of outer membrane protein assembly from purified components. Science 328:890–892. https://doi.org/10.1126/science.1188919
8. Gunasinghe SD, Shiota T, Stubenrauch CJ et al (2018) The WD40 protein BamB mediates coupling of BAM complexes into assembly precincts in the bacterial outer membrane. Cell Rep 23:2782–2794. https://doi.org/10.1016/j.celrep.2018.04.093
9. Thewasano N, Germany EM, Maruno Y et al (2023) Categorization of Escherichia coli outer membrane proteins by dependence on accessory proteins of the β-barrel assembly machinery complex. J Biol Chem 299: 104821. https://doi.org/10.1016/j.jbc.2023.104821
10. Pautsch A, Schulz GE (1998) Structure of the outer membrane protein A transmembrane domain. Nat Struct Biol 5:1013–1017. https://doi.org/10.1038/2983
11. Phale PS, Philippsen A, Kiefhaber T et al (1998) Stability of trimeric OmpF porin: the contributions of the latching loop L2. Biochemistry 37:15663–15670. https://doi.org/10.1021/bi981215c
12. Barnard TJ, Dautin N, Lukacik P et al (2007) Autotransporter structure reveals intra-barrel cleavage followed by conformational changes. Nat Struct Mol Biol 14:1214–1220. https://doi.org/10.1038/nsmb1322
13. Meng G, Spahich N, Kenjale R et al (2011) Crystal structure of the Haemophilus influenzae Hap adhesin reveals an intercellular oligomerization mechanism for bacterial aggregation. EMBO J 30:3864–3874. https://doi.org/10.1038/emboj.2011.279
14. Ohnishi S, Kameyama K, Takagi T (1998) Characterization of a heat modifiable protein, Escherichia coli outer membrane protein OmpA in binary surfactant system of sodium dodecyl sulfate and octylglucoside. Biochim Biophys Acta 1375:101–109. https://doi.org/10.1016/s0005-2736(98)00145-x
15. Roman-Hernandez G, Peterson JH, Bernstein HD (2014) Reconstitution of bacterial autotransporter assembly using purified components. elife 3. https://doi.org/10.7554/eLife.04234

Chapter 6

Examining Protein Translocation by β-Barrel Membrane Proteins Using Reconstituted Proteoliposomes

Minh Sang Huynh, Jiaming Caitlyn Xu, and Trevor F. Moraes

Abstract

β-barrel membrane proteins populate the outer membrane of Gram-negative bacteria, mitochondria, and chloroplasts, playing significant roles in multiple key cellular pathways. Characterizing the functions of these membrane proteins *in vivo* is often challenging due to the complex protein network in the periplasm of Gram-negative bacteria (or intermembrane space in mitochondria and chloroplasts) and the presence of other outer membrane proteins. *In vitro* reconstitution into lipid-bilayer-like environments such as nanodiscs or proteoliposomes provides an excellent method for examining the specific function and mechanism of these membrane proteins in an isolated system. Here, we describe the methodologies employed to investigate Slam, a 14-stranded β-barrel membrane protein also known as the type XI secretion system that is responsible for translocating proteins across the outer membrane of many bacterial species.

Key words Proteoliposome reconstitution, β-barrel membrane protein, Gram-negative bacteria, outer membrane proteins, Surface lipoprotein translocation, *In vitro* translocation, Spheroplast-released translocation, Denatured protein translocation, Type XI Secretion System

1 Introduction

Integral membrane proteins play vital roles in a variety of cellular pathways including signal transduction, adhesion, and ion and macromolecules transport [1]. Membrane proteins are structurally divided into two distinct classes based on secondary structure: α-helical and β-barrel. While the α-helical membrane proteins are commonly found in the bacterial inner membrane and membranes of eukaryotic cells, β-barrel membrane proteins are exclusively found in the outer membrane of mitochondria, chloroplasts, and Gram-negative bacteria [2, 3].

Approximately 3% of the Gram-negative bacterial genome encodes for outer membrane proteins (OMPs) that fulfil a variety of important roles including serving as adhesion factors for virulence, channels for nutrients uptake, receptors, or enzymes [4, 5]. Many OMPs have been identified and extensively studied

Raffaele Ieva (ed.), *Transmembrane β-Barrel Proteins: Methods and Protocols*, Methods in Molecular Biology, vol. 2778,
https://doi.org/10.1007/978-1-0716-3734-0_6,

such as BamA, a β-barrel membrane protein component of the β-barrel assembly machinery (BAM) that inserts most membrane proteins into the outer membrane of Gram-negative bacteria [6]. Many OMPs form transient or stable complexes with other OMPs or OM-associated lipoproteins in the outer membrane prior to being properly inserted in the OM for their optimal activity. For example, the assembly of the type V secretion system, such as EspP produced by some pathogenic strains of *Escherichia coli*, requires BamA, a component of the BAM complex for proper insertion of its transmembrane domain before translocating its passenger domain to the surface [7]. Another prominent example is LptD, a 26-stranded β barrel protein that forms a complex with the lipoprotein LptE in order to insert lipopolysaccharide into the outer leaflet of the outer membrane [8]. Characterizing the cellular function of a novel β-barrel membrane protein is often a challenging task, as *in vivo* assays cannot completely rule out the influence of the surrounding environment on protein activity. Thus, an *in vitro* reconstitution assay with minimal components provides a method to examine the explicit function of novel β-barrel membrane proteins.

In this study, we describe an *in vitro* liposome reconstitution assay to characterize the function of a surface lipoprotein assembly modulator (Slam), a β-barrel membrane protein that translocates transferrin binding lipoprotein B (TbpB) to the bacterial cell surface [9]. The methodologies described in this paper are derived and modified from our recent publication on the reconstitution of Slam into liposomes [10]. Here, we describe how to generate liposomes from *Escherichia coli* lipid extracts, insert purified β-barrel membrane proteins into liposomes, conduct sucrose flotation gradient assays to test liposome integrity, and perform a full *in vitro* assay to characterize the translocation of substrate TbpB into the lumen of the Slam-containing proteoliposomes. These methodologies can be applied to reconstitute other similar β-barrel membrane proteins into liposomes and examine their functions in an isolated environment.

2 Materials

2.1 Liposome Generation

1. Lipid powder: *E. coli* polar lipid extract from Avanti—100 mg.
2. Chloroform: 99.8% purity.
3. Nitrogen gas tank and dispenser.
4. Heated water bath.
5. Glass vials, glass pipettes.
6. Resuspension buffer: 50 mM Tris–HCl pH 8, 200 mM NaCl (*see* **Note 1**).

7. Liquid nitrogen.
8. Extrusion kit (Avanti): 2 × 1 mL glass syringes, 0.2 μm Whatman membrane.
9. Whatman filter papers (10 mm diameter).

2.2 Membrane Protein Insertion

1. Purified membrane protein in detergent micelles (preferably n-dodecyl β-D-maltoside, DDM). *See* **Note 2**.
2. SM2 BioBeads (Biorad).
3. SDS-PAGE gel and apparatus.
4. Coomassie blue stain.

2.3 Sucrose Flotation Assay

1. 50 mL of 60% sucrose buffer: 50 mM Tris–HCl pH 8, 200 mM NaCl, 60% (w/v) sucrose.
2. 50 mL of 30% sucrose buffer: 50 mM Tris–HCl pH 8, 200 mM NaCl, 30% (w/v) sucrose.
3. 50 mL of 10% sucrose buffer: 50 mM Tris–HCl pH 8, 200 mM NaCl, 10% (w/v) sucrose.
4. 50 mL of 100% TCA: dissolve 71.4 g of trichloroacetic acid in 50 mL of distilled water, store at room temperature.
5. Special equipment: swinging bucket rotor (Beckman–SW41).
6. Beckman ultracentrifuge.
7. 13.2 mL, thin wall polypropylene tubes (for ultracentrifugation).
8. 100% Trichloroacetic acid.
9. 20% ice-cold acetone.
10. Benchtop refrigerated centrifuge.

2.4 Spheroplast Release Assay

1. ZY media: 1% (w/v) bio-tryptone, 0.5% (w/v) yeast extract. Autoclave media for sterilization.
2. 20 × NPS: 1 M Na_2HPO_4, 1 M KH_2PO_4, 0.5 M $(NH_4)_2SO_4$. 1 L of 20 × NPS stock is prepared as follows: in sequence, add and resuspend 268 g of Na_2HPO_4, 136 g of KH_2PO_4, and 66 g of $(NH_4)_2SO_4$ in 1 L of distilled H2O. Autoclave media for sterilization. The final pH of the solution is approximately 6.75.
3. 50 × 5052: 25% (v/v) glycerol, 2.5% (w/v) glucose, and 10% (w/v) α-lactose. 1 L of 50 × 5052 is prepared as follows: add and resuspend 250 mL of 100% glycerol stock, 25 g of glucose, and 100 g of α-lactose in a final volume of 1 L of distilled H_2O, then autoclave for sterilization. Lactose may take 1–2 h to fully resuspend and this process is facilitated by heating at above 60 °C.

4. 1 M $MgSO_4$ stock solution: dissolve 24.65 g of $MgSO_4$ in 100 mL of distilled H_2O and then autoclave for sterilization.
5. 50 mL autoinduction media: mix 46.5 mL of ZY media, 2.5 mL of 20 × NPS, 1 mL of 50 × 5052, 50 μL of $MgSO_4$, and appropriate amount of antibiotic. This recipe for 50 mL of autoinduction media was adapted from the protocol outlined by Studier, FW [11].
6. Spheroplast resuspension buffer: 50 mM Tris–HCl pH 8 (stock of 1 M), 0.5 M sucrose. For 4 mL of buffer, mix 200 μL of stock Tris–HCl pH 8 with 1 mL of sucrose stock and 2.8 mL of distilled water. Leave on ice.
7. Spheroplast conversion buffer: 50 mM Tris–HCl pH 8, 16 mM EDTA, 0.1 mg/mL lysozyme. For 4 mL of buffer, mix 200 μL of Tris–HCl pH 8 stock with 128 μL of EDTA stock and 20 μL of lysozyme stock in 3652 μL of distilled water. Leave on ice.
8. 10 × M9 minimal salts: For 1 L of stock solution, resuspend 64 g of Na_2HPO_4, 15 g of KH_2PO_4, 2.5 g of NaCl, and 5 g of NH4Cl resuspended in 1 L of distilled H2O. Autoclave to sterilize media.
9. 20% glucose stock solution: For 100 mL of stock solution, resuspend 20 g of glucose in 100 mL of distilled H_2O and autoclave to sterilize.
10. 2 M sucrose stock solution: For 100 mL of stock solution, resuspend about 68.46 g of sucrose in 100 mL of distilled H_2O and autoclave to sterilize.
11. M9 minimal media: 1 × M9 minimal salts, 2% glucose, and 0.25 M sucrose.
12. 1 M Isopropyl β-D-1-thiogalactopyranoside (IPTG): 238 mg in 1 mL dH_2O.
13. Purified *E. coli* LolA protein (required for use in Sect. 3.4.1 only).
14. Benchtop refrigerated centrifuge.

2.5 Proteinase K Digestion Assay

1. Tris–HCl pH 8 buffer: 50 mM Tris–HCl pH 8, 200 mM NaCl
2. 2 mg/mL proteinase K stock solution: dissolve 2 mg of proteinase K in 1 mL of resuspension buffer. Aliquot to 50 μL each and store at −20 °C (short term).
3. 200 mM PMSF stock solution: dissolve 17.4 mg of PMSF in 500 μL of 100% isopropanol, aliquot to 50 μL each, and store at −20 °C.
4. 10% Triton X-100 solution: Add 1 mL of 100% Triton X-100 to 9 mL of distilled water.
5. Rabbit IgG α-FLAG antibody.

6. Mouse anti-poly-His primary antibody.
7. Horseradish peroxidase-conjugated anti-rabbit and anti-mouse secondary antibodies.
8. SDS-PAGE gel apparatus.
9. PVDF membrane for Western blot.
10. Western blot apparatus.
11. Skim milk powder
12. PBS-T buffer: 1X PBS buffer + 0.05% Tween 20.
13. Western developing solutions (Bio-Rad).
14. ChemiDoc Imager (Bio-Rad).

3 Methods

3.1 Liposome Preparation

1. Prepare glass vials (2–10 mL) and pre-equilibrate them with nitrogen gas.
2. Add 5 mL of chloroform to 100 mg of *E. coli* polar lipid extract powder or lipid powder of choice and gently resuspend the lipid. The lipid concentration should be approximately 20 mg/mL (*see* **Note 3**).
3. Aliquot 0.5 mL of resuspended lipids into 2 mL glass vials. As only ~10 mg (1 × 2mL vial) of lipid is required in the following steps, the remaining aliquots can be stored at −20 °C for 1–2 months (Fig. 1, **steps 1** and **2**).
4. Evaporate chloroform: open the nitrogen gas tank and adjust the gas flow gently. Tilt the glass vial at an angle of 45° and allow the gas to dry the upper wall of the vial and rotate the vial as illustrated in Fig. 1. An indicator that the glass vial was treated with nitrogen gas is that the vial should feel colder. Multiple layers of lipids will form on the side of the vials (Fig. 1, **step 3**). If the remaining chloroform at the bottom of the vial forms yellow droplets/layers, submerge the vial in a hot (50–60 °C) water bath to liquify the droplets/layers. Repeat **step 3** until most of the lipid forms layers on the side of the vial.
5. Forming multilamellar vesicles (MLVs): add 1 mL of resuspension buffer into the vial. This will give the solution a final concentration of 10 mg/mL lipids. Leave the lipids in the buffer solution at room temperature for 30–60 min. Use a 200 μL micropipette tip to gently resuspend the MLVs from the side of the glass vial. Resuspend until the solution is homogenously cloudy white. Avoid making bubbles. At the end, transfer the solution to a 1.5 mL tube.

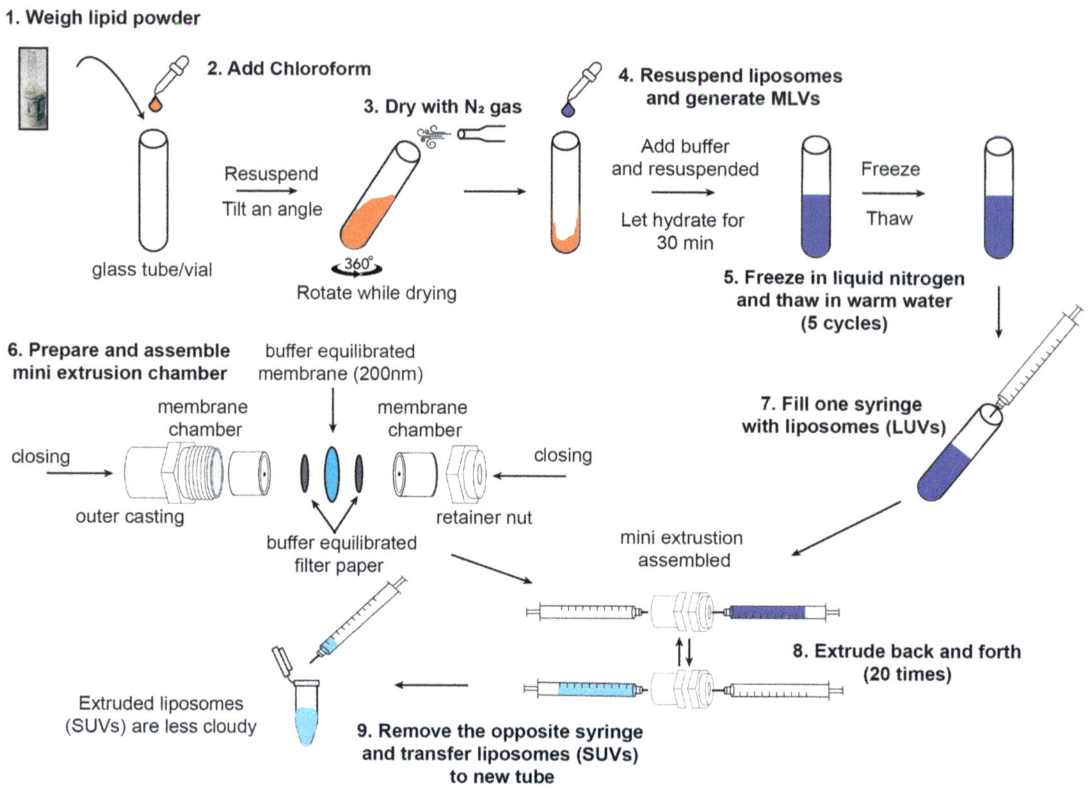

Fig. 1 Generation of liposomes from *E. coli* lipid extract. Details are described in Subheading 3.1

6. Forming large unilamellar vesicles (LUVs): prepare separate liquid nitrogen and warm water (30–37 °C) baths. Freeze the MLV solution in liquid nitrogen and then thaw in the warm water bath. Slowly thawing a frozen MLV allows the outer layer to bud off and form into a unilamellar vesicle (single bilayer). Repeat the freeze-thaw cycle at least 5 times.
7. Assembling the extrusion kit: while waiting for the freeze-thaw step, equilibrate one 0.2 μm Whatman membrane and two filter papers in resuspension buffer (the same buffer that was used for the liposomes). Assemble the chamber as illustrated in Fig. 1, **step 6**. Wash the two syringes with resuspension buffer two to three times. Tighten the syringe needles to avoid leakage. Plug "syringe 1" into one end of the chamber. Fill "syringe 2" with 1 mL of resuspension buffer. Plug "syringe 2" into the other end of the chamber. Slowly push all the buffer to "syringe 1." Take out "syringe 1" and remove the air bubbles. Refill the buffer in "syringe 1" to 1 mL. Plug "syringe 1" back into the chamber. Slowly push all the buffer to "syringe 2." Take out "syringe 2" and remove the air bubbles. Refill the buffer in "syringe 2" and plug it back into the chamber. Slowly push all the buffer back and forth between the two syringes. There

should be no air bubbles in either of the syringes. Push back and forth five to six times and check if there is any leakage, which can be detected by observing the volume of buffer in the syringe after every cycle. If there is no leakage, leave the two syringes (both filled with 0.5 mL of buffer) in the chamber until the freeze-thaw cycles are done and liposomes are ready to be extruded (*see* **Note 4**).

8. Empty the buffer in both syringes once the freeze-thaw cycles are finished. Plug one syringe into one side of the chamber (either side of the chamber can be used and this will be referred to as the "output" chamber). Fill the other syringe with the LUV solution (from the freeze-thaw cycles). Remove any bubbles if necessary. Plug the LUV syringe into the other side of the chamber ("input" chamber).
9. Gently push the solution from one side to the other. Repeat for at least 20 times back and forth. The LUVs will initially be a cloudy-white solution that will become homogenously white-clear after multiple passages. At the end of the extrusion, push the solution to the "output" chamber and remove the syringe to avoid any debris deposited on the filter paper in the "input" chamber.
10. Transfer 1 mL of the extruded liposomes to a 1.5 mL tube and leave at room temperature for membrane protein insertion. Otherwise, store liposomes at 4 °C up to 3–4 days (*see* **Note 5**).

3.2 Membrane Protein Insertion

Membrane proteins in detergent (DDM preferred) that are freshly purified or thawed (on ice) from −80 °C storage can be inserted into liposomes. Here, we will describe the insertion of Slam, a β-barrel membrane protein from *Moraxella catarrhalis.* In the following section, we will examine Slam's function as a translocon using a proteoliposome reconstitution assay (Fig. 2).

1. Thaw 100 μL of 1 mg/mL of Slam-DDM protein on ice.
2. Prewash SM2 BioBeads (Biorad): before insertion, wash 200 mg SM2 biobeads twice with 1 mL of resuspension buffer (same buffer that was used to make the liposomes).
3. Split the liposomes into 2 × 1.5 mL tubes so that each tube has 500 μL of liposomes. Add approximately 10 mg of pre-washed BioBeads into each tube. Label the tube and add protein/ buffer as follows:
   (a) Empty liposomes: 100 μL of DDM buffer + 500 μL of liposomes.
   (b) Slam proteoliposomes: 100 μL of Slam-DDM + 500 μL of liposomes (*see* **Note 6**).

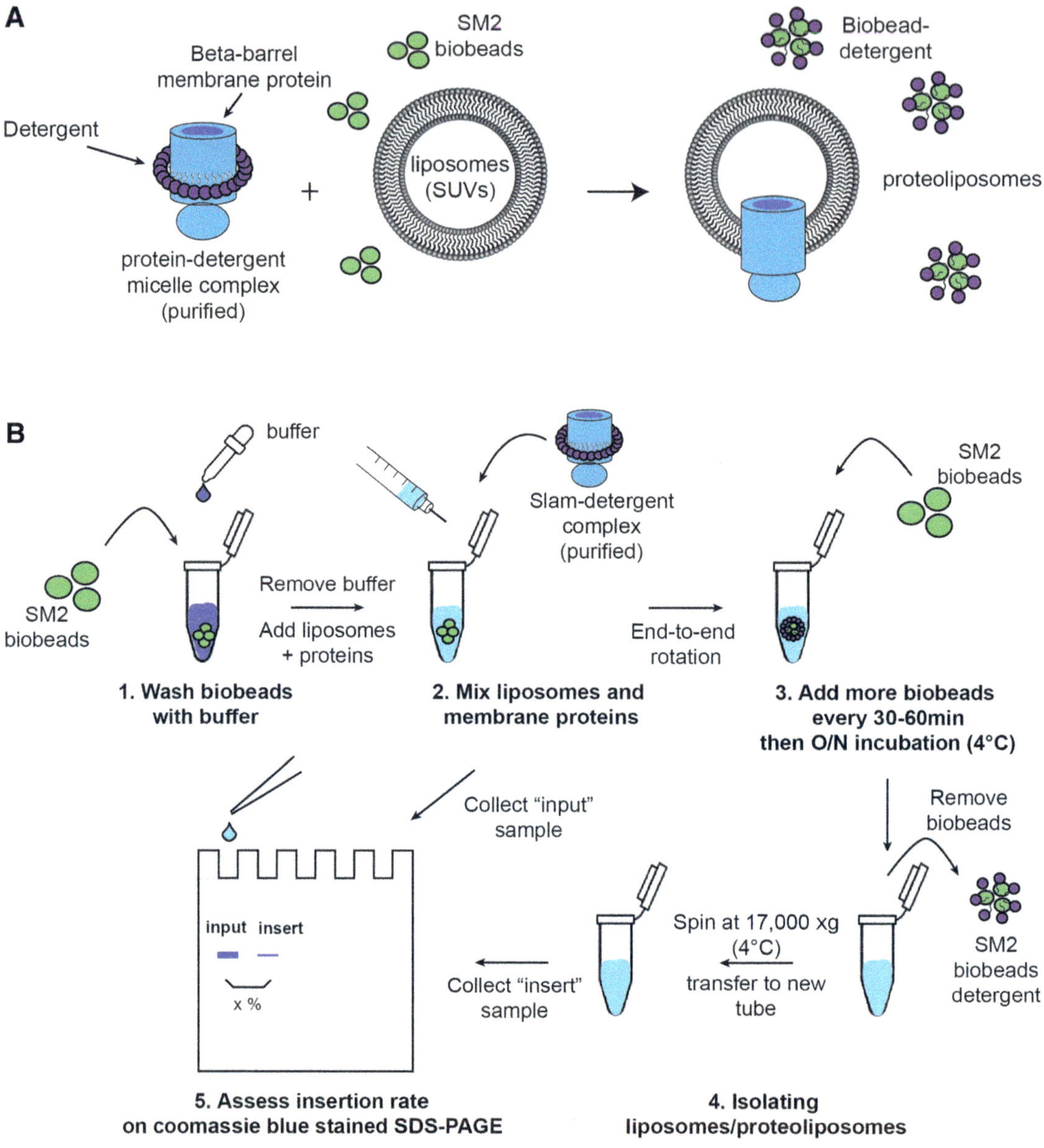

Fig. 2 (**a**) Schematic protocol of membrane protein insertion into liposomes. SM2 BioBeads are a strong absorbent for non-ionic detergents like Triton X-100 and DDM. In this preparation, adding biobeads removes the DDM detergent and forces the β-barrel membrane protein to insert into liposomes or precipitate out of solution. (**b**) An illustration of membrane protein insertion into liposomes. Details are described in Subheading 3.2

4. Removing DDM detergent: leave the tubes at room temperature with end-to-end rotation. Add 10 mg pre-washed BioBeads into each tube every 30 min or 1 h for a total of 2 h. Once 50 mg BioBeads are added into each tube, leave tubes overnight at 4 °C with end-to-end rotation (*see* **Note 7**).

5. Isolating liposomes/proteoliposomes: separate liposomes/proteoliposomes from the biobeads by aspirating the solution using gel loading tip. Transfer to new tubes. Take samples for SDS-PAGE gel and label them as "input." Spin down liposomes/proteoliposomes at 17,000 × *g* at 4 °C. Transfer the supernatant containing proteoliposomes-incorporated membrane proteins to new 1.5 mL tubes. Take samples for SDS-PAGE gel and label them as "inserted." There will be precipitation as not all proteins will be incorporated into the liposomes.
6. Run "input" and "inserted" samples on SDS-PAGE gel side by side and stain with Coomassie blue to estimate the percentage of proteins retained in the supernatant after centrifugation (i.e., the amount of protein that was inserted into the floated liposomes present in the soluble fraction).
7. Leave empty liposomes and proteoliposomes at room temperature to warm up for at least 15 min before performing the following translocation assay. Otherwise, leave the liposomes on ice or at 4 °C.

3.3 Sucrose Flotation Assay (Optional)

We recommend performing a sucrose flotation assay to assess proteoliposome integrity after membrane protein insertion at least once before further examining the function of the proteoliposomes (Fig. 3).

1. Add 1 mL of 60% sucrose buffer to a 13 mL polypropylene tube (specialized tubes for a swinging bucket rotors may have different volumes). Leave on ice for at least 10 min. Solutions may start to solidify on ice.
2. Tilt the tube at an angle of 30–45° and gently add 1 mL at a time for up to 10 mL of 30% sucrose buffer to the side of the tube (avoid disturbing and mixing with the bottom layer). Leave on ice for another 10 min.
3. Tilt the tube again and gently add 1 mL of 10% sucrose buffer solution. Leave the tube on ice or 4 °C for at least 10 min.
4. Tilt the tube again and gently add 500 μL of liposomes/proteoliposomes on top.
5. Assemble the tubes in the SW41 swinging bucket rotor.
6. Spin for at least 18 h at 90,000 × *g* at 4 °C.
7. Collecting fractions after ultracentrifugation: collect and transfer 500 μL of the top layer (liposomes layer) to a 1.5 mL tube. Collect the next 1 mL layer (10% sucrose layer), mix and transfer 500 μL to a 1.5 mL tube. Collect the next 2 mL layer (30% sucrose layer), mix and transfer 500 μL to a 1.5 mL tube. Continue collecting 0.5 mL for every 2 mL fractions for the

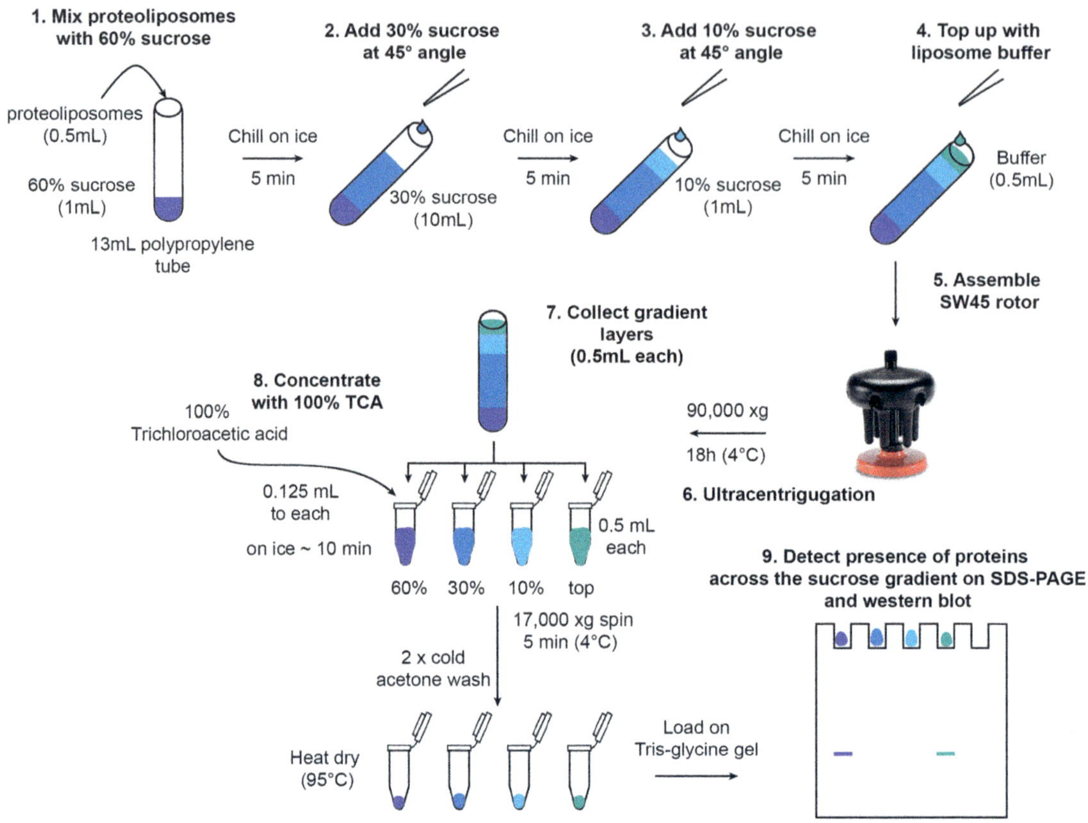

Fig. 3 A step-by-step visual of the sucrose flotation assay for proteoliposomes. Intact proteoliposomes float up to the very top of the sucrose gradient, while a majority of precipitate and free, non-incorporated proteins localize to the bottom (60% sucrose) and middle (30–10% sucrose) of the gradient respectively. Details are described in Subheading 3.3

entire 30% sucrose layer. And finally collect the bottom 1 mL layer (60% sucrose layer), mix and transfer 500 μL to a 1.5 mL tube.

8. Concentrate the protein using trichloroacetic acid (TCA): add to each tube 125 μL of 100% TCA solution (1:5 ratio) and incubate on ice for 10 min.
9. Spin down at 17,000 rcf at 4 °C for 5 min.
10. Discard the supernatant and leave the pellet intact (pellet should be whitish and fluffy).
11. Wash the pellet with 200 μL of ice-cold acetone.
12. Spin down at 17,000 RCF for 5 min at 4 °C and discard the acetone.
13. Repeat the wash step >2 additional times.

14. Dry the pellet by placing tubes in a 95 °C heat block for 10 min or until all the acetone evaporates.
15. Add to each tube 50 μL of 1× SDS loading buffer.
16. Load onto an SDS-PAGE gel and run for 45 min at 200 V.
17. Stain the gel with Coomassie blue dye buffer or perform a Western blot analysis to transfer the proteins to PVDF membrane.
18. For Western blotting, we block the membrane for 1 h with 5% skim milk powder in PBST, followed by overnight incubation at 4 °C with mouse IgG α-His antibody. Incubation with a secondary α-mouse IgG antibody conjugated to horseradish peroxidase is used to detect our Slam (His-tagged) protein.

3.4 (Lipo)protein Translocation by Reconstituted Proteoliposomes

Below, we describe two methods to prepare substrates for the *in vitro* proteoliposomes translocation assay to examine the function of β-barrel outer membrane proteins. The substrate that we used for the translocation assay is the surface lipoprotein known as transferrin binding protein B from *M. catarrhalis* (Mcat TbpB) with a C-terminal FLAG tag.

3.4.1 Substrates Released Directly from Spheroplasts (Semi–In Vitro Translocation Assay, Fig. 4)

1. Growing and expressing TbpB: grow C43 (DE3) *E. coli* cells transformed with plasmid that contains Mcat TbpB in 3 mL of autoinduction media overnight at 37 °C (*see* **Note 8**).
2. Measure cell density and adjust OD_{600} to 1.0. Each translocation experiment requires at least 400 μL of cells at OD_{600} ~ 1.0. The following setup is described for two treatments: empty liposomes and Slam proteoliposomes.
3. Harvest 800 μL of cells at 800 × *g* for 5 min at room temperature.
4. Gently resuspend cell pellets in 200 μL of ice-cold spheroplast resuspension buffer and incubate on ice for 5 min.
5. Add 200 μL of ice-cold spheroplast conversion buffer to the solution and gently mix, followed by incubation on ice for 20 min.
6. Spin down the solution at 10,000 × *g* for 10 min at 4 °C. Remove and discard supernatant.
7. Resuspend the pellet (now spheroplast) in 200 μL of M9 minimal media. The spheroplast pellet is sticky and hard to resuspend. Avoid vigorous pipetting that may cause lysis and generate bubbles.
8. Add in 10 μM of purified *E. coli* LolA and 0.1 mM IPTG to the spheroplast solution.
9. Split the spheroplast solution into 2 × 1.5 mL tubes, 100 μL each.

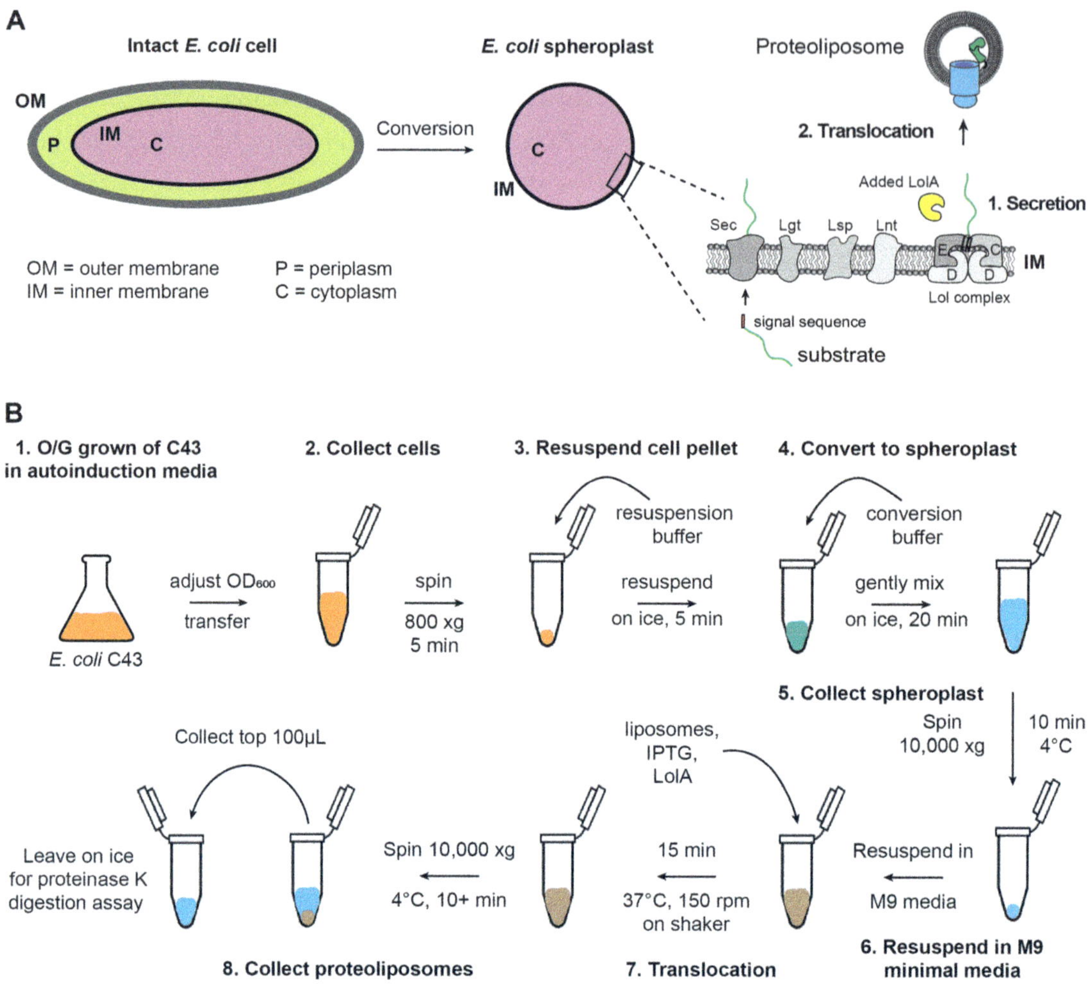

Fig. 4 (**a**) Schematic protocol of the spheroplast release translocation assay. The substrate for this assay (a lipoprotein – TbpB) was expressed overnight. Once treated with lysozyme and EDTA, the outer membrane becomes permeable. Adding purified LolA, a triacyl-lipid binding protein triggers the release of mature TbpB that populates the inner membrane directly into the media, bypassing the outer membrane. Isolated media rich with secreted TbpB is used as the substrate for the spheroplast-secreted translocation assay. (**b**) A step-by-step illustration of the proteoliposome translocation assay using a spheroplast-secreted substrate. Details are described in Subheading 3.4.1

10. Add to each tube either 100 μL of empty liposomes or Slam proteoliposomes.
11. Incubate the tubes in a 37 °C shaker set to 150 rpm for 30 min (*see* **Note 9**).
12. Spin down the spheroplasts at 10,000 × *g* for 10 min at 4 °C. Spin longer if the spheroplasts do not pellet.
13. Collect 100 μL of the top supernatant for proteinase K digestion assay (proceed to Subheading 3.5).

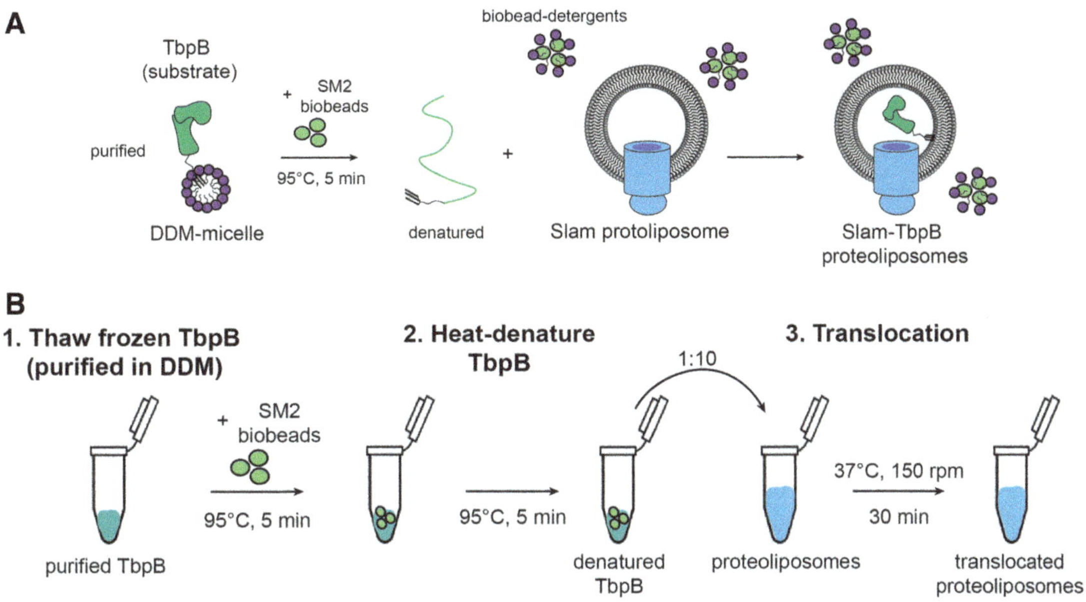

Fig. 5 (**a**) Schematic protocol of the proteoliposome translocation assay using purified substrate. TbpB lipoprotein is purified in DDM and treated with SM2 BioBeads to remove the detergent. TbpB is unfolded by heating at 95 °C before incubating with Slam proteoliposomes for translocation. (**b**) A step-by-step visual of the proteoliposome translocation assay using purified substrate. Details are described in Subheading 3.4.2

3.4.2 Purified Substrates (Full In Vitro *Translocation Assay, Fig. 5)*

1. Thaw 10 μL of TbpB (stock of 20 μM) on ice.
2. Add 10 mg of prewashed SM2 BioBeads into TbpB and incubate for 15–30 min at room temperature to remove DDM from the solution.
3. Incubate the sample at 95 °C for 5 min to denature the protein.
4. Add 5 μL of heat-unfolded TbpB into 95 μL of either empty liposomes or Slam proteoliposomes (1:20 ratio selected to lower the TbpB concentration to 1 μM) (*see* **Note 10**).
5. Incubate the reactions at 37 °C for 15–20 min (or optimal time for translocation) with shaking set to 150 rpm.
6. Leave on ice and proceed to proteinase K digestion.

3.5 Assessing (Lipo) protein Translocation by Proteinase K Digestion Assay (Fig. 6)

1. Prepare three reactions for each liposome treatment (empty and Slam proteoliposomes).
   (a) No PK: 5 μL of Tris–HCl pH 8 buffer.
   (b) PK: 4 μL of Tris–HCl pH 8 buffer + 1 μL of proteinase K (2 mg/mL).
   (c) PK+TritonX: 3 μL of Tris–HCl pH 8 buffer + 1 μL of proteinase K (2 mg/mL) + 1 μL of Triton X-100 (10%) (*see* **Note 11**).

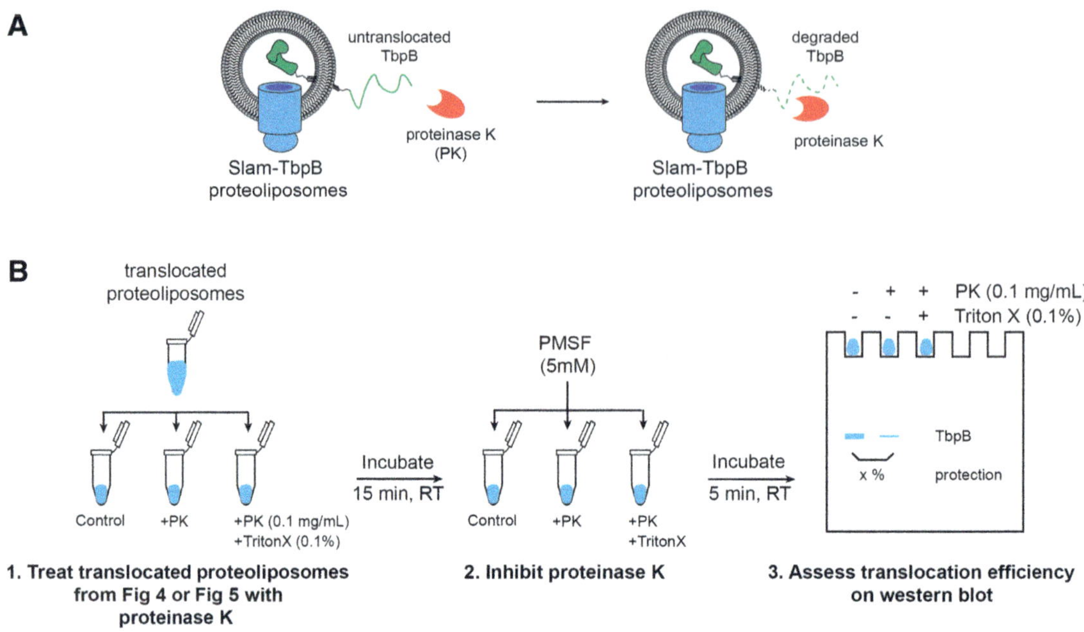

Fig. 6 (**a**) Schematic protocol of the proteinase K digestion assay. Proteinase K is a broad-spectrum serine protease that degrades all proteins remaining outside of the Slam proteoliposomes post-translocation. TbpB in the lumen of the liposomes is protected from proteinase K digestion and used to estimate the efficiency of Slam translocation. (**b**) A step-by-step illustration of the proteinase K digestion assay. Details are described in Subheading 3.5

2. To each reaction, add 15 μL of liposomes following the translocation events in Subheadings 3.4.1 or 3.4.2 above.
3. Incubate reactions at room temperature for 15 min.
4. Add 1 μL of PMSF to each tube and incubate for another 5 min to inhibit proteinase K activity.
5. Add about 5.5 μL of 5 × SDS loading buffer to each tube and boil samples at 95 °C for 5 min. ×
6. Analyze samples by SDS-PAGE in the following order: No PK; PK; PK + Triton X-100 for empty liposomes and Slam proteoliposomes.
7. Run for 45 min at 200 V.
8. Perform a Western blot analysis by first transfering samples onto PDVF.
9. Block the blot with 10 mL of 5% skim milk powder in PBST buffer for at least 1 h at room temperature.
10. Incubate the blot with rabbit IgG α-FLAG antibody (1:5000 in 1% milk PBST buffer) for at least 1 h at room temperature or overnight at 4 °C.

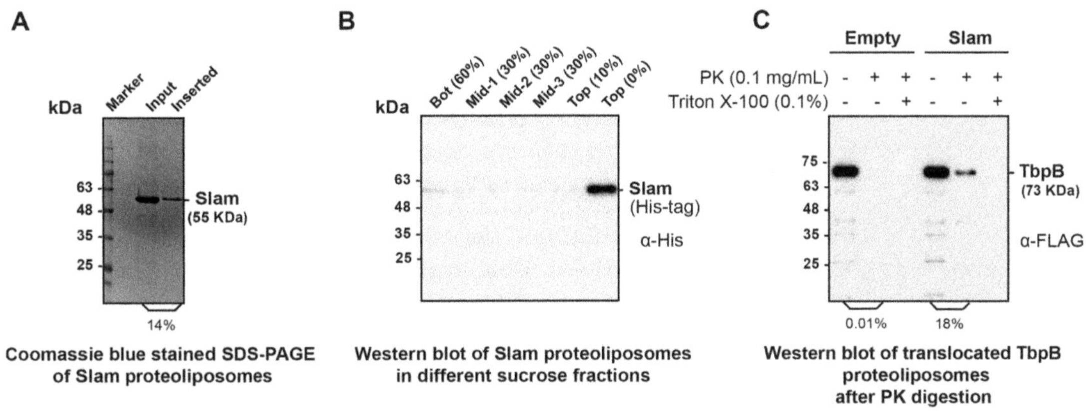

Fig. 7 Example of Slam-dependent translocation for TbpB. (**a**) Insertion of the membrane protein, Slam, into liposomes. The insertion percentage was calculated using the densitometry ratio of insert: input. (**b**) Western blot using α-His antibody to detect His-tagged Slam protein in the sucrose fraction. Most Slam was found in the top fraction (intact liposomes floated to the top of the sucrose gradient). (**c**) Representative Western blot of the Slam proteoliposomes translocation assay for heat-denatured TbpB using α-FLAG antibody. A fraction of TbpB was protected from proteinase K digestion. Triton X-100 was used as a negative control as it dissolves liposomes leaving TbpB exposed for degradation. Percentage of protection was calculated using the densitometry ratio of +PK sample over -PK sample

11. Wash the blots three times with 10 mL of PBST buffer for 5 min each time.
12. Incubate the blot with α-rabbit IgG antibody (1:5000 in 1% milk PBST buffer) for at least 1 h at room temperature.
13. Wash the blot three times with 10 mL of PBST buffer for 5 min each time before applying 300 μL of developing solution.
14. Develop the blot using chemiluminescence function of the Chemi-Imager (Biorad).
15. Quantify the amount of translocation by dividing the intensity of the TbpB band in the PK treatment by the intensity of TbpB band in the no PK treatment.

Examples of the entire experiment including protein insertion, sucrose flotation, and proteinase K digestion for Slam proteoliposomes translocation assay using heat denatured TbpB are shown in Fig. 7.

4 Notes

1. Resuspension buffer should be the same buffer used to purify the membrane protein, but without detergent. This is to maintain the same buffer content inside and outside of the liposome lumen to avoid osmotic pressure.

2. Our membrane protein is purified and stable in DDM. SM2-BioBeads are known to be the best and most efficient mehanism to remove detergents with low critical micelle concentrations such as DDM [13].
3. Lipids from Avanti are usually freshly purified and often have a 1 year expiration.
4. Batch sonication is the desired method to generate a large volume of SUVs. Both techniques have their own advantages and disadvantages. Extrusion is often chosen to control the size of the SUVs by changing the size of the membrane pores. For more information on the extrusion kit or how to assemble all the components of the extrusion kit for the first time, please visit Avanti's website: https://avantilipids.com/divisions/equipment-products/mini-extruder-assembly-instructions.
5. Liposomes in buffers should not be stored at 4 °C for too long. Liposomes taken out of 4 °C should be spun down at 10,000 × *g* at 4 °C for 5 min to remove any precipitation before protein incorporation.
6. For our experiment, we used a 1:100 protein:lipid ratio (w/w). However, this ratio may need to be optimized for different membrane proteins that may behave differently during liposome insertion due to unique properties. We also used a 1:5 protein:liposomes ratio (v/v) to lower the DDM concentration (0.03% initially) to 0.005%, which is less than its critical micelle concentration (CMC ~ 0.00875%) in order to coax Slam out of the detergent micelles. We also slowly increase the amount of biobeads to slowly absorb DDM detergent and to minimize Slam precipitation. Different proteins may have different requirement for optimal lipid insertion; thus it is recommended to test for various insertion conditions and protein:lipid ratios for your own protein.
7. Cold temperatures are not optimal for liposomes and thus membrane protein insertion. However, Slam is extremely unstable if left at room temperature for long periods of time, and since we cannot perform the follow-up experiments on the same day, our proteoliposomes had to be incubated overnight at 4 °C.
8. Growing C43 cells containing a TbpB plasmid in Lysogeny broth (LB) media and inducing TbpB expression with IPTG for a few hours would also work. However, we had more consistent TbpB expression when we used autoinduction media. The main goal is to have TbpB (the substrate) expressed and incorporated into the inner membrane of *E. coli* cells before converting them to spheroplasts, so that upon addition of purified *E. coli* LolA, lipoproteins are released from the LolCDE complex in the inner membrane [12].

9. For Slam and TbpB, we have experimentally confirmed that efficient translocation occurs within 15–30 min. Different proteins may have different translocation activities and, therefore, this requires adjusting the length of the translocation reaction.
10. From our main study (Huynh et al. 2022), Slam can only translocate its substrate TbpB in an unfolded manner, thus we had to denature TbpB prior to incubating with Slam proteoliposomes. Originally, we used 8 M urea to unfold the proteins. However, urea-unfolded TbpB often had very low efficiency. Recently, we have found an alternative method in which TbpB (in DDM buffer + biobeads) was denatured at 95 °C for 5 min and immediately diluted to the proteoliposomes for translocation. The efficiency was much higher and experiments are reproducible. Figure 7c is the result of heat-denatured TbpB translocation by empty and Slam proteoliposomes. This modification was developed by Caitlyn Xu.
11. Triton X-100 is a strong non-ionic detergent that was used to solubilize the liposomes in this experiment. Dissolving the liposomes would expose any translocated TbpB to the proteinase K. This is the second negative control that is included to demonstrate that protease protection is the result of liposome incorporation.

References

1. Cournia Z, Allen TW, Andricioaei I, Antonny B et al (2015) Membrane protein structure, function, and dynamics: a perspective from experiments and theory. J Membr Biol 248(4):611–640
2. Cavalier-Smith T (2000) Membrane heredity and early chloroplast evolution. Trends Plant Sci 5:174–182
3. Sachs JN, Engelman DM (2006) Introduction to the membrane protein reviews: the interplay of structure, dynamics, and environment in membrane protein function. Annu Rev Biochem 75:707–712
4. Rollauer SE, Sooreshjani MA, Noinaj N, Buchanan SK (2015) Outer membrane protein biogenesis in Gram-negative bacteria. Philos Trans R Soc Lond Ser B Biol Sci 370(1679): 20150023
5. Wimley WC (2003) The versatile beta-barrel membrane protein. Curr Opin Struct Biol 13(4):404–411
6. Wang X, Peterson JH, Bernstein HD (2021) Bacterial outer membrane proteins are targeted to the Bam complex by two parallel mechanisms. MBio 12(3):e00597–e00521
7. Ryoo D, Rydmark MO, Pang YT, Lundquist KP, Linke D, Gumbart JC (2020) BamA is required for autotransporter secretion. Biochim Biophys Acta Gen Subj 1864(7):129581
8. Wu T, McCandlish AC, Gronenberg LS, Chng SS, Silhavy TJ, Kahne D (2006) Identification of a protein complex that assembles lipopolysaccharide in the outer membrane of Escherichia coli. Proc Natl Acad Sci U S A 103(31): 11754–11759
9. Hooda Y, Lai CC, Judd A et al (2016) Slam is an outer membrane protein that is required for the surface display of lipidated virulence factors in Neisseria. Nat Microbiol 1:16009
10. Huynh MS, Hooda Y, Li YR, Jagielnicki M, Lai CC, Moraes TF (2022) Reconstitution of surface lipoprotein translocation through the Slam translocon. elife 11:e72822
11. Studier FW (2005) Protein production by auto-induction in high density shaking cultures. Protein Expr Purif 41(1):207–234
12. Yokota N, Kuroda T, Matsuyama S, Tokuda H (1999) Characterization of the LolA-LolB system as the general lipoprotein localization mechanism of Escherichia coli. J Biol Chem 274(43):30995–30999
13. Schubert R (2003) Liposome preparation by detergent removal. Methods Enzymol 367: 46–70

Chapter 7

In Vivo Disulfide-Bond Crosslinking to Study β-Barrel Membrane Protein Interactions, Dynamicity, and Folding Intermediates

Matthew Thomas Doyle

Abstract

Membrane-embedded β-barrels are the major building blocks of the Gram-negative outer membrane and are involved in antibiotic resistance, virulence, and the maintenance of bacterial cell physiology. The increased frequency of multidrug resistant Gram-negative infections warrants the sharing of accessible methods for the study of β-barrels. One such method is "in vivo disulfide-bond crosslinking" which is a highly informative and cost-effective approach to study the structure, topology, dynamicity, and function of β-barrels in situ. The approach can also be used to identify and finely map both stable or transient interactions between β-barrels and other interacting proteins. In this chapter, I describe the conceptual basis of in vivo disulfide-bond crosslinking and the potential pitfalls in experimental design. I also provide a general protocol for high-efficiency in vivo disulfide-bond crosslinking and modified protocols as examples for how the method can be adapted to different scenarios.

Key words Outer membrane protein, β-barrel, Disulfide bond, Interaction-mapping, Dynamicity, Protein folding, AlphaFold

1 Introduction

Outer membrane β-barrels usually contain few cysteine residues (or none at all). Researchers have taken advantage of this characteristic to introduce pairs of cysteines at specific locations within β-barrels which are then monitored for their potential to form disulfide bonds in vivo. The disulfide bonds are easily identified and quantified using modern Western immunoblotting procedures. The information generated is highly specific and structurally informative because cysteine residue thiol groups must come within ~2–3 Å of each other to form a disulfide bond. Because the disulfide-bond oxidation reaction can be efficiently catalyzed in vivo,

Raffaele Ieva (ed.), *Transmembrane β-Barrel Proteins: Methods and Protocols*, Methods in Molecular Biology, vol. 2778, https://doi.org/10.1007/978-1-0716-3734-0_7,

disulfide crosslinking is also a highly useful method to complement other structural approaches that rely on purified components such as X-ray crystallography and cryo-electron microscopy.

Several examples of the use of in vivo disulfide-bond crosslinking are displayed in important studies that characterized the function of the homologous bacterial β-barrel proteins BamA and TamA that are involved in the insertion and folding of other β-barrel proteins [1–9]. Some of these studies used intra-molecular disulfide bonding to artificially stabilize the β-barrel domains of BamA and TamA (among other inter-domain locations) and showed that their first and last strands must separate as a requirement for their function [1, 7, 8]. Several studies have also used inter-molecular disulfide crosslinking to provide insight into protein-protein interactions between bacterial β-barrels, partner periplasmic proteins, and substrate proteins [1, 2, 4–6]. For example, in vivo inter-molecular disulfide crosslinking was first used to show that BamA forms amazing hybrid-barrel structures that are required intermediates in the β-barrel folding mechanism [6]. After an interface is mapped by in vivo disulfide-bond crosslinking, the method can be adapted to other experiments that are commonly used to characterize bacterial β-barrel proteins. For example, the dynamicity of the intermediate hybrid-barrel structures formed between BamA and its substrates was initially characterized by disulfide-bonding the relevant interfaces and then observing the folding states using a heat-modifiable electrophoretic mobility shift assay (see adapted protocol) [6]. In vivo intermolecular disulfide crosslinking has also been used to monitor the secretion path of several virulence factors across the bacterial outer membrane including the so-called type 5a autotransporters and the type 5b two-partner secretion systems [4, 10–13]. I also note that disulfide crosslinking is not only applicable to the study of bacterial β-barrels but has also been used in the study of mitochondrial β-barrels with a couple of exquisite examples reported recently [14, 15].

Although the protocol itself is extremely simple, the successful execution of in vivo disulfide-bond crosslinking hinges on several experimental design considerations and intrinsic factors of the proteins of interest. As mentioned, a critical factor is the number of endogenous cysteines within the protein of interest. When considering the use of this approach, the first action is to inspect the sequence of your protein of interest. Proteins containing many endogenous cysteines might not be good candidates for characterization by disulfide crosslinking. If there are few endogenous cysteines in the protein(s) of interest, they can be substituted for serine. Alternatively, if prior experiments (e.g., crystal structures) suggest that the endogenous cysteines are located distally from the site of investigation, or are known to form natural disulfide bonds, they might not interfere with the analysis (this must be empirically determined). Disruption of natural disulfide bonds should be carefully considered due to their potential involvement in protein folding, stability, and function. For example, surface loop 6 of *E. coli* BamA contains two endogenous cysteines that form a disulfide in vivo which do not interfere with disulfide-crosslinking experiments designed to investigate the distally located first and last strands of the BamA β-barrel [6, 16].

Another consideration is the design of the expression system for the production of the protein(s) of interest. It is especially important that outer membrane β-barrels are expressed at levels that avoid a backlog in their biogenesis to the outer membrane. In such a scenario, significant levels of nonspecific crosslinking might occur due to misfolding and aggregation in the periplasm. The expression of potentially toxic gene products (such as those that encode for mutant β-barrels that fold slowly and enable characterization of protein folding intermediates) must also be tightly regulated. Tunable low copy-number expression plasmids (like the rhamnose inducible plasmid pSCRhaB2 [17]) have proven very useful [2, 9], while other systems, such as the pET series of plasmids, may not be a great choice. For the investigation of hetero-oligomeric complexes, each subunit should be expressed at levels that are consistent with the physiological stoichiometry of the complex. This can be achieved by the construction of plasmids containing natural or synthetic operons. Such an approach was required for the successful expression and characterization of the BAM complex which contains BamA and four accessory lipoproteins [18]. Depending on the protein system and hypotheses under investigation, a multi-plasmid expression system may be constructed wherein each plasmid encodes a protein with a single engineered cysteine. For these approaches, careful consideration of the relative expression levels is important to reduce the occurrence of nonspecific oxidation reactions which can make subsequent data interpretation difficult.

Perhaps the most critical consideration is the location of engineered cysteine substitution sites. Whether disulfide-bond oxidation occurs is dependent on the proximity and orientation of the thiol groups of the introduced cysteine pair. Consider the scenario presented in Fig. 1 which is a structure of a hybrid barrel based on recent studies [2, 6]; highly efficient disulfide-bond formation was observed when single cystines were introduced at luminally oriented positions in the β-barrel protein EspP at positions 1293, 1295, 1297, and 1299 and paired with cysteines introduced into immediately adjacent luminally oriented positions in BamA at positions 431, 429, 427, 425, respectively. Although disulfide-bond formation was high for the cystine pair EspP(S1299C)-BamA (S425C), the cystine pair EspP(S1297C)-BamA(S425C) produced low levels of disulfide bonds which shows the distance specificity of the experiment. Showing the orientation specificity of the

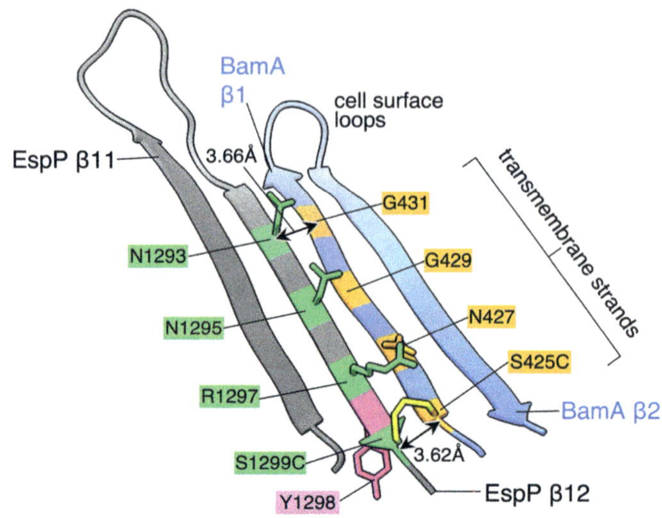

Fig. 1 Cysteine proximity and orientation specificity in disulfide bond formation. The structure shows the interface between the β-barrel protein BamA and its substrate β-barrel EspP which creates a hybrid-barrel structure (see PDBID: 7TT5 [2]). The following BamA(orange):EspP(green) cysteine pairs form disulfide bonds in vivo at 80–90% efficiency: N1293C:G431C, N1295C:G429C, R1297C: N427C, and S1299C:S425C. The disulfide bond between S1299C:S425C is shown in yellow. The side of the β-sheet facing out of the page is the luminal side of the hybrid barrel

experiment, when the lipid-facing EspP(Y1298C) was paired with either BamA(S425C) or BamA(N427C), very little disulfide binding was observed. As an aside, the distance between β-strands in a β-sheet is serendipitously very similar to the backbone-backbone distance within a disulfide bond indicating that such bonds are likely to have very little impact on the native structure of the protein under investigation.

How can in vivo disulfide crosslinking experiments be reliably designed if disulfide-bond formation requires such high precision of cystine placement? Previously solved protein structures in the Protein Data Base (PDB) are particularly helpful for the identification of target positions. The advent of highly reliable structure prediction software (e.g., AlphaFold2 [19] or ESMFold [20]) has also made the design of potentially disulfide-bond forming cysteine pairs substantially easier for proteins without solved structures. Several versions of AlphaFold2 are now available that can predict structures of complex heterooligomers [21]. Moreover, the current generation of protein structure prediction tools is particularly successful in the prediction of bacterial membrane embedded β-barrels. From a different perspective, in vivo disulfide-bond crosslinking is an excellent method to directly test such predictions. However, if a reliable protein structure prediction is unattainable, or if the subject of the analysis is a protein folding intermediate or a transient conformational state, a library containing many cysteine-pairs positioned widely across many different locations of protein sequence must be screened until a disulfide bond is observed. Unless one has the available resources to make and screen a saturated cysteine-pair substitution library, the researcher must instead rely on prior observations and their "best guess" while designing the initial library of cysteine substitutions. This might appear to be a gamble, but it is important to note that just one observed disulfide bond between a particular cysteine pair subsequently enables additional cysteine pairs to be designed to allow the identified interface to be finely mapped.

2 Materials

2.1 Reagents

1. Milli-Q water, autoclaved.
2. Lysogeny broth (LB): 10 g/L tryptone, 5 g/L yeast extract, 5 g/L NaCl. Autoclaved.
3. LB agar: LB broth as above plus 15 g/L agar.
4. Glycerol stock media: 30% (v/v) glycerol, 1% (w/v) peptone. Filter sterilized.
5. SOC media: 20 g/L tryptone, 5 g/L yeast extract, 0.5 g/L NaCl, 0.18 g/L KCl, 10 mM $MgCl_2$, 10 mM $MgSO_4 \bullet 7H_2O$. Autoclaved.
6. Phosphate buffered saline (PBS) and PBS supplemented with 0.05% (v/v) Tween 20 (PBST) for room temperature membrane washes. Keep an additional PBS at 4 °C and precool further on ice before assays.
7. Thiol-specific oxidizer: 4-DPS (4,4′-dipyridyl disulfide): working stock is 100 mM diluted in 200 proof ethanol. Kept at 4 °C.
8. Phenylmethanesulfonyl fluoride (PMSF): working stock is 100 mM diluted in 200 proof ethanol. Kept at 4 °C.
9. Trichloroacetic acid (TCA). Kept at 4 °C.
10. Acetone. Kept at 4 °C.
11. 2 × SDS protein gel loading solution: 12.11 g/L Tris base, 40 g/L SDS, 2 g/L bromophenol blue, 20% (v/v) glycerol.
12. Precast tris-glycine-SDS polyacrylamide (8–16% gradient) mini gels.
13. iBlot2 nitrocellulose mini transfer stacks.
14. Immunoblocking buffer: 50% (v/v) Intercept blocking buffer (PBS) (Licor), .50% PBS, 0.05% Tween 20.

15. Protein size standard compatible with fluorescent blots, e.g., Chameleon Duo Prestained Protein Ladder (Licor).
16. Primary antibodies generated against protein of interest or tags.
17. Secondary antibodies with fluorescent conjugate, e.g., 800CW IRDyes (Licor).
18. Proteinase K (PK) stock solution: 10 mg/mL of PK in PK resuspension buffer .
19. PK resuspension buffer: 5 mM $CaCl_2$, 50 mM Tris–HCl, pH 8. Filter sterilized.
20. BugBuster Master Mix (Merck).

2.2 Equipment

1. 90 mm petri dishes.
2. Nichrome loop (1 μL).
3. 50 mL conical centrifuge tubes.
4. 1.6 mL semimicro spectrophotometer cuvettes.
5. 0.2 cm gap electroporation cuvettes.
6. 1.5 mL microcentrifuge tubes.
7. 2 mL cryo-tubes.
8. Microbiology incubator.
9. Microbiology shaking incubator.
10. Visible spectrophotometer.
11. Refrigerated swing-bucket centrifuge.
12. Electroporator device.
13. Vortex.
14. Rotator wheel.
15. Microcentrifuge.
16. Mini centrifuge.
17. Thermomixer.
18. Protein gel electrophoresis equipment and power supply.
19. iBlot2 transfer device.
20. 2D rocking platform.
21. Trays for Western immunoblot incubations.
22. Imager capable of scanning fluorescent blots (Amersham Typhoon 5 imager (GE Healthcare) or Odyssey DLx imager (Licor)).

3 Methods

3.1 Bacterial Disulfide Crosslinking

This is a standard method to monitor the formation of disulfide bonds between engineered pairs of cysteines within proteins of interest that are expressed in bacteria. Use this method to screen a cysteine pair library, to build a disulfide-bond map, or to probe secondary structure. Two convenient modifications of the method to generate disulfide-bond formation kinetics information, probe the membrane topology of interactions, and observe folded states and dynamicity of disulfide-bonded species, are described in subsequent subsections. The protocol assumes that the pitfalls and other considerations outlined in the "Introduction" have been taken into account and that the required plasmids, and their cystine substitution mutant variants, have been constructed.

3.1.1 Preparation of Test Bacteria

1. Use the streak-plating method to isolate single colonies of the host strain (from glycerol stock) on LB agar or equivalent (*see* **Note 1**). Incubate overnight at 37 °C or as necessary.
2. Inoculate a single colony into a 50 mL conical centrifuge tube containing 10 mL of LB. Culture with shaking overnight at 37 °C/250 rpm or as necessary.
3. Inoculate a fresh 50 mL conical centrifuge tube containing 10 mL of LB with the overnight culture at a 1/100 dilution. Culture with shaking until mid- to late-log phase (OD_{600} of ~0.5).
4. Cool cultures on ice for 5 min. Conduct **steps 5–9** on ice.
5. Using a precooled (4 °C) centrifuge containing a swing-bucket rotor, centrifuge the culture at 3000 × *g* for 10 min. Carefully pour off the waste supernatant.
6. Resuspend the bacteria in 10 mL of ice-cold sterile Milli-Q water.
7. Repeat **steps 6–5** once. Then repeat **step 5** again.
8. Resuspend bacterial pellet into 200 μL of ice-cold sterile Milli-Q water. Aliquot 100 μL of washed bacteria into a 0.2 cm gap electroporation cuvette that was precooled on ice.
9. Add 1 μL of plasmid DNA to the bacteria and mix by tapping the cuvette gently. Incubate cuvette on ice for 10 min.
10. Use an electroporation device to electro-shock the bacteria according to the requirements for the test species.
11. Add 900 μL of SOC media to the cuvette, mix by pipetting, and transfer to a sterile 1.5 mL microcentrifuge tube.
12. Incubate bacteria at 37 °C on a rotating wheel for 30–60 min.

13. Use the spread-plating method to inoculate 50–100 μL of the bacteria to an LB agar plate containing necessary selection antibiotics. Pellet the remaining bacteria at 3000 × *g* for 2 min using a microcentrifuge, resuspend the pellet into 100 μL of fresh SOC by pipetting, and spread the entire volume onto a second plate. Incubate overnight at 37 °C or as necessary.
14. Using a sterile loop, take a single transformant colony and streak a lawn on a fresh LB plate containing the selection antibiotic. Incubate overnight at 37 °C or as necessary to produce a lawn of the transformed strain.
15. Using a sterilized nichrome loop (1 μL calibration), scrape a loop-full of the transformed strain. Transfer the bacteria to a 2 mL cryotube containing 750 μL of sterile glycerol stock storage media (*see* **Note 2**). Vortex to remove clumps. Store at −80 °C.
16. If doubly transformed bacteria (*see* **Note 3**) are necessary, repeat **steps 1–15** using the newly created singly transformed stock. Maintain the selection antibiotic for the first plasmid for **steps 1–3**. Include both selection antibiotics for each of the two plasmids in **steps 13** and **14**.

3.1.2 Oxidation Assay

1. Use the streak-plating method to isolate single colonies of the test bacteria that contain plasmids for the controlled production of the cysteine-containing protein(s) under investigation (from glycerol stock) on LB agar or equivalent containing appropriate selection antibiotics (maintain the selection antibiotics through all culturing steps). Incubate overnight at 37 °C or as necessary.
2. Inoculate a single colony into a 50 mL conical centrifuge tube containing 10 mL of LB. Culture with shaking overnight at 37 °C/250 rpm or as necessary.
3. Inoculate another 50 mL conical centrifuge tube containing 10 mL of sterile LB with an appropriate volume of the overnight culture to give a starting sub-culture OD_{600} of 0.02–0.05. Culture with shaking under desired conditions until an OD_{600} of 0.5. Induce the production of the cysteine containing proteins using conditions empirically determined for each subject (*see* **Note 4**).
4. Pipette 1 mL aliquots of the induced culture to one 1.6 mL semimicro spectrophotometer cuvette and to two 1.5 mL microcentrifuge tubes precooled on ice (complete **steps 5–13** on ice). Immediately measure and record the OD_{600} of the cuvette sample and discard.

5. Using a microcentrifuge (precooled to 4 °C), pellet the two 1 mL aliquots at 10,000 × *g*, for 2 min. Carefully remove the media by pipetting.
6. Add 1 mL of ice-cold PBS to each bacterial pellet. Resuspend the bacteria by using a vortex.
7. After vortexing, collect the entire volume at the bottom of the tube by a 1–2 s spin in a mini centrifuge (*see* **Note 5**).
8. Add oxidizer 4-DPS to one tube at a final concentration of 0.2 mM (2 μL of a 100 mM stock precooled on ice). Add 2 μL of 200 proof ethanol (precooled on ice) to the other tube (this will be the mock-treated sample). Mix the bacteria by a 1–2 s vortex and repeat **step 7** (*see* **Note 6**). Incubate tubes on ice for 30 mins.
9. Repeat **step 5**, add 0.5 mL of ice-cold PBS, vortex to resuspend the bacteria, and repeat **step 7**.
10. Add PMSF to each tube to a final concentration of 4 mM, vortex for 1 s, and repeat **step 7**.
11. Add TCA to each tube to a final concentration of approximately 10% (v/v), vortex for 1 s, and repeat **step 7**. Incubate tubes on ice for 10 min to allow cell lysis and protein precipitation.
12. Pellet precipitated proteins by using a microcentrifuge (precooled to 4 °C) at 20,817 × *g* for 10 min (*see* **Note 7**).
13. Wash precipitated protein by adding 0.6 mL of ice-cold 100% acetone to the tubes. Roughly homogenize the pelleted protein by scrapping the tube wall with a p1000 pipette tip and pipetting. Repeat **step 12**.
14. Remove all acetone by pipetting. Transfer tubes to a tube rack and air dry the tubes at 37 °C with the lids left open. Close lids and store the precipitated protein at −20 °C.

3.1.3 SDS-PAGE

1. Add 2 × SDS protein gel-loading solution to the tubes containing precipitated protein samples at a volume normalized to the final OD_{600} reading of the culture from which the sample was taken. A good formula is: volume (μL) = $OD_{600} \times 200$). Do not attempt to homogenize at this step (*see* **Note 8**).
2. Using a Thermomixer, heat samples at 95 °C while vortexing at maximum rpm (~1400 rpm). Allow samples to cool to room temperature. Collect condensation by a 1–2 s spin in a mini centrifuge, vortex for 1–2 s, and then collect the sample by another 1–2 s spin.
3. Set up a precast tris-glycine-SDS polyacrylamide mini gel (8–16% gradient) in an appropriate tank and buffer at room temperature (*see* **Note 9**).

4. For a 12-well gel, load 5 μL of protein samples to wells. Into one well at both sides of the gel, add an appropriate protein ladder that emits in the infrared (*see* **Note 10**). Add 5 μL of 2 × SDS protein gel-loading solution to any spare well.
5. Electrophorese samples at 150 V for an amount of time empirically determined to adequately resolve all proteins species under investigation (*see* **Note 11**).

3.1.4 Western Immunoblotting

1. Remove the gel and place into a clean tray containing milli-Q water.
2. Use an iBlot2 transfer device on the "P0" setting to transfer separated proteins to a nitrocellulose membrane (*see* **Note 12**).
3. After the transfer, use tweezers to rapidly submerge the membrane into 5 mL of immunoblocking buffer to avoid drying. Block with gentle 2D rocking at room temperature overnight.
4. Pour off buffer and remove excess buffer using a vacuum aspirator.
5. Add primary antibodies to the membrane (diluted into 5 mL of immunoblocking buffer) and incubate with rocking for between 1 and 18 h.
6. Pour off the antibody solution and wash the membrane three times with 10 mL of PBST (5 min rock for each wash). Repeat **step 4**.
7. Add secondary antibodies that contain an infrared emitting conjugate (or other fluorophore of choice) to the membrane (diluted according to the manufacturer's specifications into 5 mL of immunoblocking buffer) and incubate with rocking for between 1 and 2 h.
8. Repeat **step 6**. Then, wash the membrane twice with 10 mL of PBS (5 min rock for each wash). Repeat **step 4**.
9. Fold a kimwipe in half and place flat onto the bottom of a dry tray. Use tweezers to place the membrane on top of the kimwipe with protein side facing up. Air dry the membrane at 37 °C for 20 min (*see* **Note 13**). Protect membrane from light until imaged.

3.1.5 Imaging, Quantitation, and Calculation of Crosslinking Levels

1. Use an Amersham Typhoon 5 imager (GE Healthcare) outfitted with 785 nm and 685 nm lasers and IRlong 825BP30 and IRShort 720BP20 filters or an Odyssey DLx imager (Licor) to image the membrane using an empirically determined intensity setting that results in peak intensities within the dynamic range (*see* **Note 14**). Store images as raw uncompressed tiff.

2. Use an image analysis program (such as the open source program FIJI image J [22]) to identify bands that correspond to disulfide-bonded species (*see* **Note 15**). Quantify the pixel intensities within each lane using the area-under-curve approach (in FIJI go to Analyze > Gels).
3. Calculate the proportion of the disulfide-bonded band of interest relative to other bands in the same lane that correspond to non-crosslinked species. For instance, fraction crosslinked = intensity of disulfide-bonded band divided by the sum of band intensities in the same lane.

3.2 Kinetics of Disulfide-Bond Formation

This modification is used to monitor the accumulation of disulfide-bonded species for a cysteine pair of interest. The information generated can indicate the stability or dynamicity of an interface/interaction under investigation. For an example of the use of this method, see Fig. 6B of this reference [6].

3.2.1 Time-Resolved Oxidation Assay

1. Prepare tubes labeled "0.5, 2, 5, 15, 30, 60, and 90" on ice containing 50 μL of TCA and 20 μL of 100 mM PMSF.
2. Conduct **steps 1–3** as in Subheading 3.1.2.
3. Pipette a 1 mL aliquot of the induced culture to a 1.6 mL semimicro cuvette and immediately measure and record the OD_{600}.
4. Using a precooled centrifuge (4 °C), centrifuge the remaining culture in a swing-bucket rotor at 5000 × *g* for 5 min. Carefully pour off the waste supernatant.
5. Add 10 mL of ice-cold PBS to the pellet and resuspend by vortexing. Repeat **step 4**. Resuspend the pellet in 5 mL of ice-cold PBS and place the tube on ice.
6. Add 4-DPS to the bacteria at a final concentration of 0.2 mM, mix by a quick swirl of the tube by hand, and begin a timer.
7. Aliquot 0.5 μL samples of the bacteria to the preprepared tubes at 0.5, 2, 5, 15, 30, 60, and 90 min time points. Vortex for 1–2 s and then spin in a mini centrifuge1–2 s immediately after aliquoting each timepoint sample. After the last time point sample, incubate tubes on ice for a further 10 min.
8. Conduct **steps 12–14** as in Subheading 3.1.2.
9. Conduct steps of Subheadings 3.1.3 to 3.1.5 as required.

3.3 Disulfide-Bond Assay Coupled with Protein Cell-Surface Accessibility to Protease Treatment

This modification is used to digest cell surface accessible portions of proteins prior to, or after, disulfide-bond formation. For an example for how the treatment of bacteria with protease prior to disulfide-bond formation can be used, see Fig. 5D of this reference [2]. For an example of how the treatment of bacteria with protease after the formation disulfide bonds can be used, see Fig. 6B of this reference [4].

3.3.1 Digestion-Oxidation Assays

Conduct all steps as in Subheading 3.1.2 except take four aliquots of the culture to 1.5 mL tubes at **step 4**, and between **steps** 7 and **8** add the following steps:

1. Add proteinase K (PK) to two of the four tubes at a final concentration of 200 μg/mL (20 μL from a 10 mg/mL stock solution). To the other two, add 20 μL of PK resuspension buffer (mock-treated controls). Mix all tubes by vortexing for 1–2 s and collect the entire volume at the bottom of the tube by a 1–2 s spin in a mini centrifuge.
2. Incubate bacteria on ice for the desired time interval.
3. Continue with protocol in Subheading 3.1.2 from **step 5** noting that now four tubes will be processed with the following treatment combinations: PK(−)/4-DPS(−), PK(+)/4-DPS (−), PK(−)/4-DPS(+), PK(+)/4-DPS(+).
4. Conduct steps of Subheadings 3.1.3 to 3.1.5 as required.

3.3.2 Oxidation-Digestion Assays

Conduct all steps as in Subheading 3.1.2 except take four aliquots of the culture to 1.5 mL tubes at **step 4**, and between **steps 8** and **9** add the following steps:

1. Conduct this step as in Subheading 3.3.1, **step 1**.
2. Incubate bacteria on ice for the desired time interval.
3. Continue with protocol in Subheading 3.1.2 from **step 5** noting that now four tubes will be processed with the following treatment combinations: PK(−)/4-DPS(−), PK(+)/4-DPS (−), PK(−)/4-DPS(+), PK(+)/4-DPS(+).
4. Conduct steps of Subheadings 3.1.3 to 3.1.5 as required.

3.4 Disulfide-Bond Assay Coupled with Heat-Induced Mobility Shift

This modification is used to observe folded states of membrane-embedded β-sheets or β-barrels. For an example of how it is used to observe multiple conformations of an outer membrane-embedded hybrid-barrel structure, see Fig. 6C of this reference [6].

3.4.1 Oxidation Assay with Modified Lysis

Conduct all steps as in Subheading 3.1.2 up to and including **step 9**, except take four aliquots of the culture to 1.5 mL tubes at **step 4**. Then add the following steps:

1. Using a microcentrifuge (precooled to 4 °C), pellet the 1 mL aliquots at 10,000 × *g* for 2 min. Carefully remove the media by pipetting.
2. Add BugBuster Master Mix at a volume normalized to the final OD_{600} reading of the culture from which the sample was taken. A good formula is: volume (μL) = $OD_{600} \times 100$). Immediately pipette to homogenize. Incubate on ice for 3 min to allow bacteria to lyse.

3. Take two 30 μL aliquots of each lysate to new 1.5 ml microcentrifuge tubes. To each, add 10 μL of 2 × SDS protein gel loading solution and mix by tapping the tube (final SDS concentration is 1%) (*see* **Note 16**). Keep one tube on ice while the other is heated to 99 °C for 10 min.
4. Immediately proceed to steps of Subheading 3.4.2.

3.4.2 Cold-SDS-PAGE

1. Set up a gel as in **step 3** of Subheading 3.1.3 except that the gel tank is set up within a 4 °C room and embedded into packed ice within an ice bucket. Additionally, the electrophoresis buffer is precooled on ice to <4 °C.
2. Conduct **steps 4–5** as in Subheading 3.1.3 (*see* **Note 17**).
3. Conduct steps of Subheadings 3.1.4 to 3.1.5 as required.

4 Notes

1. This method would likely work for many different species. These protocols have been used successfully in *E. coli* and more recently in *Pseudomonas aeruginosa* [23].
2. A glycerol stock of each cysteine pair strain and any strains that contain single cysteine variant proteins should always be stored. Various positions often become extremely useful at later stages of projects, and storage of glycerol stocks saves substantial time, especially when testing one site against many.
3. For double plasmid systems, we have had fantastic experience with combining pTrc99a with pSCrhaB2 [2, 4, 6, 9].
4. For instance, we screened several induction time points and IPTG concentrations for the expression of single-cysteine variants of BamA from pTrc99a and several induction time points and rhamnose concentrations for the production of a stalled version of EspP from pSCRhaB2. 1 h induction with 0.4 mM IPTG followed by 45 min induction with 0.2% rhamnose produced ideal levels without significant degradation products. In this case, we also reduced the culture temperature to reduce protein degradation and to slow the protein folding process [6].
5. It is important to collect the bacterial suspension in the bottom of the tube to ensure that it is being efficiently cooled. Droplets on the tube lid will not be in contact with the ice and therefore might experience higher temperatures during the experiment.
6. It is important to conduct the pipetting, mixing, and spin procedure of this step rapidly in order to have a consistent 30 min oxidation period between experiment repetitions and between test cysteine pairs.

7. The protein pellet will often appear as a diffuse smear along the tube sidewall.
8. The protein clumps usually require heat to fully homogenize into solution. Pipetting the sample at this stage will result in a blocked pipette tip.
9. The use of precast gradient gels is highly recommended due to their consistency which is helpful when quantification is necessary. Gradient gels are also required due to the very large shifts in electrophoretic migration of disulfide-crosslinked species relative to the reduced protein bands.
10. The Chameleon Duo Prestained Protein Ladder is very reliable.
11. The electrophoresis time should be consistent across replicates. This results in higher precision when quantifying.
12. The use of traditional wet blotting procedures is absolutely not recommended as the variation in transfer efficiency across the membrane is highly variable.
13. This step significantly increases the consistency in background signal across the membrane due to even drying. The kimwipe raises the membrane allowing drying from all sides.
14. Use of other systems is fine as well. We have observed that infrared detection systems are significantly more sensitive, with a better signal to noise ratio which is particularly important for the identification of low levels of disulfide bonding.
15. Assess your Western blot; intermolecular disulfide-bonded species are often detected as high-molecular-weight bands that correspond to protein-protein adducts and are detected by both antibodies when probed simultaneously for each protein in the adduct. Intramolecular disulfide bonds result in less pronounced, but still readily detectible, shifts in electrophoretic migration. In this case, comparison to single cysteine and WT controls is important.
16. Not all β-barrels are equally resistant to unfolding by SDS. It might be necessary to titrate the final concentration of SDS across samples to identify an ideal SDS concentration to observe folded states.
17. Note that the dye front may take over 2 h to run to the bottom of the gel at these temperatures.

References

1. Stubenrauch CJ, Bamert RS, Wang J et al (2022) A noncanonical chaperone interacts with drug efflux pumps during their assembly into bacterial outer membranes. PLoS Biol 20(1):e3001523
2. Doyle MT, Jimah JR, Dowdy T et al (2022) Cryo-EM structures reveal multiple stages of bacterial outer membrane protein folding. Cell 185(7):1143–1156 e13

3. Wu R, Bakelar JW, Lundquist K et al (2021) Plasticity within the barrel domain of BamA mediates a hybrid-barrel mechanism by BAM. Nat Commun 12(1):7131
4. Doyle MT, Bernstein HD (2021) BamA forms a translocation channel for polypeptide export across the bacterial outer membrane. Mol Cell 81(9):2000–2012 e3
5. Tomasek D, Rawson S, Lee J et al (2020) Structure of a nascent membrane protein as it folds on the BAM complex. Nature 583(7816):473–478
6. Doyle MT, Bernstein HD (2019) Bacterial outer membrane proteins assemble via asymmetric interactions with the BamA beta-barrel. Nat Commun 10(1):3358
7. Gu Y, Li H, Dong H et al (2016) Structural basis of outer membrane protein insertion by the BAM complex. Nature 531(7592):64–69
8. Noinaj N, Kuszak AJ, Balusek C et al (2014) Lateral opening and exit pore formation are required for BamA function. Structure 22(7): 1055–1062
9. Peterson JH, Doyle MT, Bernstein HD (2022) Small molecule antibiotics inhibit distinct stages of bacterial outer membrane protein assembly. MBio 13(5):e0228622
10. Baud C, Guerin J, Petit E et al (2014) Translocation path of a substrate protein through its Omp85 transporter. Nat Commun 5:5271
11. Sauri A, Ten Hagen-Jongman CM, van Ulsen P et al (2012) Estimating the size of the active translocation pore of an autotransporter. J Mol Biol 416(3):335–345
12. Leyton DL, Sevastsyanovich YR, Browning DF et al (2011) Size and conformation limits to secretion of disulfide-bonded loops in autotransporter proteins. J Biol Chem 286(49): 42283–42291
13. Junker M, Besingi RN, Clark PL (2009) Vectorial transport and folding of an autotransporter virulence protein during outer membrane secretion. Mol Microbiol 71(5):1323–1332
14. Takeda H, Busto JV, Lindau C et al (2023) A multipoint guidance mechanism for beta-barrel folding on the SAM complex. Nat Struct Mol Biol 30:176–187
15. Hohr AIC, Lindau C, Wirth C et al (2018) Membrane protein insertion through a mitochondrial beta-barrel gate. Science 359(6373): eaah6834
16. Rigel NW, Ricci DP, Silhavy TJ (2013) Conformation-specific labeling of BamA and suppressor analysis suggest a cyclic mechanism for beta-barrel assembly in Escherichia coli. Proc Natl Acad Sci U S A 110(13):5151–5156
17. Cardona ST, Valvano MA (2005) An expression vector containing a rhamnose-inducible promoter provides tightly regulated gene expression in *Burkholderia cenocepacia*. Plasmid 54(3):219–228
18. Roman-Hernandez G, Peterson JH, Bernstein HD (2014) Reconstitution of bacterial autotransporter assembly using purified components. elife 3:e04234
19. Jumper J, Evans R, Pritzel A et al (2021) Highly accurate protein structure prediction with AlphaFold. Nature 596(7873):583–589
20. Lin Z, Akin H, Rao R et al (2022) Evolutionary-scale prediction of atomic level protein structure with a language model. bioRxiv:2022.07.20.500902
21. Evans R, O'Neill M, Pritzel A et al (2022) Protein complex prediction with AlphaFold-Multimer. bioRxiv:2021.10.04.463034
22. Schindelin J, Arganda-Carreras I, Frise E et al (2012) Fiji: an open-source platform for biological-image analysis. Nat Methods 9(7): 676–682
23. Hanson SE, Doyle MT, Bernstein HD (2023) The patatin-like protein PlpD forms novel structurally dynamic homodimers in the *Pseudomonas aeruginosa* outer membrane. bioRxiv:2023.04.17.537245

Chapter 8

Site-Specific Photocrosslinking to Investigate Toxin Delivery Mediated by the Bacterial β-Barrel Assembly Machine

Emily M. Bouzan and Christine L. Hagan

Abstract

Contact-dependent inhibition (CDI) is a mechanism of interbacterial competition in Gram-negative organisms that relies on a specific interaction between a CdiA protein on the surface of one cell and a β-barrel protein on the surface of a neighboring cell. This interaction triggers the transport of a protein toxin into the neighboring cell where it exerts its lethal activity. Several classes of CdiA proteins that bind to different β-barrel receptors have been identified, but the molecular mechanism by which they deliver their toxins across the outer membranes of their target cells is poorly understood. Here we describe the use of site-specific photocrosslinking to characterize the interaction between a CdiA protein and its receptor. We describe the method for an *E. coli* CdiA that utilizes BamA as its receptor. BamA's central role in assembling β-barrel proteins in the outer membrane makes its role in CDI particularly intriguing; it suggests that these two different protein transport processes might share mechanistic features. Our in vitro photocrosslinking method is useful in elucidating early steps in the CDI mechanism, but it could be adapted to study later steps or to study other CdiA-receptor pairs.

Key words Contact-dependent inhibition, CdiA, BamA, Unnatural amino acid, *p*BPA, Photocrosslinking

1 Introduction

The β-barrel assembly machine (Bam) plays an essential role in the folding, insertion, and secretion of proteins in the outer membranes of Gram-negative bacteria, but certain bacteriophages, bacteriocins, and contact-dependent inhibition (CDI) systems also utilize Bam as a receptor to infect or intoxicate these cells [1–5]. These phages and interbacterial competition systems bind to BamA, the central β-barrel component of Bam, on the extracellular surface of the bacteria they target. They exhibit high selectivity for certain species (or strains) of bacteria by binding to non-conserved regions of BamA's extracellular loops [4–6]. A detailed

Raffaele Ieva (ed.), *Transmembrane β-Barrel Proteins: Methods and Protocols*, Methods in Molecular Biology, vol. 2778,
https://doi.org/10.1007/978-1-0716-3734-0_8,

understanding of how these phages, bacteriocins, and CDI systems interact with BamA may reveal commonalities in their mechanisms. We have focused on a particular CDI system from *E. coli* strain EC93, which targets BamA in other *E. coli* cells [3, 7]. This CDI system delivers a protein toxin across the target outer membrane without an external energy source and thus may serve as a simple model for protein translocation from the outside of the cell into the periplasm [8].

CDI systems are widespread in Gram-negative species and consist of three proteins: (1) CdiA, a large filamentous protein with a cleavable C-terminal toxin domain; (2) CdiB, a β-barrel of the Omp85 family that assembles CdiA on the producing (inhibitor) cell's surface; and (3) CdiI, an immunity protein that protects the producing cell from self-intoxication by inhibiting the CdiA toxin [9–11]. The receptor-binding domain (RBD) of CdiA, which interacts with BamA, is located in the middle of the protein sequence [12]. CdiA is initially assembled on the inhibitor cell's surface in a hairpin structure such that the RBD is poised at the distal end of the hairpin ready to interact with BamA on neighboring cell [13]. The C-terminal toxin domain is initially retained within the periplasm of the inhibitor cell, but, upon binding to a BamA, a large conformational change occurs, which extracts the toxin domain from the inhibitor cell and delivers it across the outer membrane of the target cell [13]. The molecular mechanism of the toxin delivery across the target cell's outer membrane and BamA's role in that process is, however, largely unknown.

Site-specific crosslinking methods have been highly informative in elucidating the mechanisms of other protein transport processes in the outer membrane. For example, photocrosslinking and cysteine crosslinking methods have been used to study the assembly of β-barrel proteins [14, 15], the assembly of autotransporters and secretion of their passenger domains [16–18], and the import of a bacteriocin [19]. These methods capture weak or transient interactions that might not be otherwise observable and identify individual amino acids at the protein-protein interface.

Site-specific photocrosslinking methods rely on the incorporation of an unnatural, photoreactive amino acid at a designated position in the protein under investigation. Schultz and colleagues developed and optimized an amber codon suppression system that uses an orthogonal tRNA and aminoacyl-tRNA synthetase (aaRS) pair from *Methanocaldococcus jannaschii* to accomplish this task [20, 21]. The aaRS was engineered to aminoacylate the tRNA with *p*-benzoylphenylalanine (*p*BPA), and the tRNA was mutated to recognize amber nonsense codons. When employing this system, *E. coli* cells are co-transformed with two plasmids: one encoding the engineered aaRS and tRNA and one encoding the protein of interest with a single codon in its gene sequence mutated to an amber nonsense (TAG) codon. During protein expression, the mutated tRNA incorporates *p*BPA into the growing polypeptide at the site of the amber codon. Ultraviolet irradiation of the synthesized protein then causes the benzophenone side chain of *p*BPA to form a biradical, which can react with nearby C-H bonds within approximately 3 Å [22, 23]. If the *p*BPA is located at a site on the protein that interacts with another protein, interprotein crosslinks can be formed. Typically, *p*BPA substitutions at a large number of amino acid positions in the protein of interest are evaluated in order to map the protein-protein interface.

We have applied the *p*BPA-incorporation technique to identify specific sites of interaction between BamA and the $CdiA^{EC93}$ receptor-binding domain (CdiA RBD) in a purified system [24]. Here we describe the separate expression and purification of the CdiA RBD and *p*BPA-substituted BamA proteins, the co-purification and photocrosslinking of BamA-CdiA RBD complexes, and the subsequent SDS-PAGE and immunoblotting analysis. Our initial studies indicated that the CdiA RBD binds to BamA near the lateral gate in its β-barrel, suggesting that the gate may play a role in the subsequent toxin translocation steps. We expect that the photocrosslinking method described here will be useful in probing those subsequent steps in the CDI mechanism.

2 Materials

2.1 Protein Expression Strain and Plasmids

1. *E. coli* BL21(DE3) competent cells.
2. pEVOL-pBpF plasmid encoding the orthogonal tRNA and aaRS for *p*BPA incorporation (Addgene plasmid #31190) [20, 21] (*see* **Note 1**).
3. Plasmid encoding the CdiA RBD (amino acids R1358-F1646) with an N-terminal His-tag under an IPTG-inducible promoter (pCH441) [24] (*see* **Note 2**).
4. Plasmid encoding the mature sequence of BamA (amino acids E22-W810) with an amber codon (TAG) mutation at the desired site of *p*BPA substitution (e.g., pCH531, pCH607) [24] (*see* **Note 2**).

2.2 Bacterial Growth, Protein Expression, and Cell Lysis

1. Sterilized Luria Broth (LB) Media.
2. 50 mg/mL carbenicillin in ultrapure water. Store at −20 °C. Dilute 1000-fold into growth media as needed.
3. 50 mg/mL kanamycin in ultrapure water. Store at −20 °C. Dilute 1000-fold into growth media as needed.
4. 30 mg/mL chloramphenicol in ethanol. Store at −20 °C. Dilute 1000-fold into growth media as needed.
5. 0.9 M *p*BPA: Add 1 mL 1 M NaOH to 0.2693 g *p*-benzoylphenylalanine (Bachem) and vortex until fully dissolved. Store at −20 °C wrapped in aluminum foil.

6. 1 M isopropyl β-D-1-thiogalactopyranoside (IPTG). Store at −20 °C.
7. 20% (w/v) glucose. Sterile filter and store at 22 °C.
8. 20% (w/v) arabinose. Store at −20 °C.
9. 10 mg/mL deoxyribonuclease I (Millipore Sigma) dissolved in 50% (v/v) glycerol. Store at −20 °C.
10. 10 mg/mL lysozyme dissolved in 50% (v/v) glycerol. Store at −20 °C.
11. 100 mM phenylmethylsulfonyl fluoride (PMSF) dissolved in isopropanol. Store at −20 °C. Warm to room temperature and vortex to dissolve PMSF prior to use.

2.3 Buffers

Store all buffer solutions at 4 °C unless noted otherwise.

1. 25× Tris-Buffered Saline (TBS), pH 8.0: 0.5 M Tris base, 3.75 M NaCl, adjusted to pH 8.0 with concentrated HCl. Store at 22 °C.
2. 4 M Imidazole, pH 8.0: pH adjusted to 8.0 with concentrated HCl. Store at 22 °C wrapped in aluminum foil.
3. TBS, pH 8.0: Dilute 25× TBS, pH 8.0 stock solution with ultrapure water to working concentration of 20 mM Tris, pH 8.0, 150 mM NaCl.
4. TBS, pH 8.0, 5% (v/v) glycerol for size-exclusion chromatography. Sterile filtered (through 0.2 μm membrane).
5. Glycerol wash buffer: TBS, pH 8.0, 5% (v/v) glycerol, 20 mM Imidazole.
6. Glycerol elution buffer: TBS, pH 8.0, 5% (v/v) glycerol, 200 mM Imidazole.
7. 8 M urea: Prepared fresh on the day of use in ultrapure water.
8. BamA folding buffer: TBS, pH 8.0, 0.5% (w/v) lauryldimethylamine-N-oxide
9. DDM wash buffer: TBS, pH 8.0, 0.05% (w/v) n-dodecyl-β-D-maltopyranoside, 20 mM Imidazole.
10. DDM elution buffer: TBS, pH 8.0, 0.05% (w/v) n-dodecyl-β-D-maltopyranoside, 200 mM Imidazole.

2.4 Protein Purification and Quantification

1. French Press G-M with standard pressure cell (Glen Mills).
2. Ni-NTA Superflow Resin (Qiagen).
3. Econo-Pac chromatography columns (Bio-Rad).
4. Bio-Spin chromatography columns (Bio-Rad).
5. Superdex 200 Increase 10/300 GL Prepacked Tricorn Column (Cytiva).

6. Amicon Ultra 10 kDa molecular weight cutoff filters (Millipore).
7. Bugbuster Protein Extraction Reagent (Millipore).
8. Detergent compatible (DC) Protein Assay (Bio-Rad).

2.5 Photocross-linking

1. 96-well clear, polystyrene plates with flat-bottomed wells.
2. UVP 100 W high-intensity UV inspection lamp (Analytik Jena Model B-100AP).

2.6 SDS-PAGE and Immunoblotting

1. 2× SDS sample buffer: 125 mM Tris-HCl, pH 6.8, 4% (w/v) sodium dodecyl sulfate, 20% (v/v) glycerol, 0.04% (w/v) bromophenol blue, 5% (v/v) β-mercaptoethanol.
2. SDS-PAGE running buffer: 25 mM Tris, pH 8.3, 192 mM glycine, 0.1% (w/v) SDS.
3. 10% TGX mini-gels (Bio-Rad).
4. 1× Transfer buffer: Dilute 5× Trans-Blot Turbo Transfer Buffer stock (Bio-Rad) with ultrapure water and ethanol according to the manufacturer's instructions.
5. Trans-Blot Turbo Transfer System (Bio-Rad).
6. PVDF membrane (Bio-Rad).
7. Blocking buffer: Dilute 10× stock solution (Millipore) with ultrapure water.
8. Penta-His HRP conjugate (Qiagen).
9. TBS-T Wash Buffer: 20 mM Tris-HCl, pH 8.0, 150 mM NaCl, 0.1% (v/v) Tween 20.
10. Clarity Western ECL substrate (Bio-Rad).
11. ChemiDoc MP Imaging System (Bio-Rad).

3 Methods

3.1 Expression and Purification of His-CdiA Receptor-Binding Domain (CdiA RBD)

The CdiA RBD is typically purified in a large quantity and stored at −80 °C for use in numerous photocrosslinking experiments with different *p*BPA-substituted BamA proteins.

1. Transform BL21(DE3) cells with a plasmid encoding the His-tagged CdiA RBD (pCH441) by heat shock at 42 °C. Select transformants by plating on LB agar containing 50 μg/mL kanamycin and incubate at 37 °C overnight.
2. Select a colony and inoculate a 50 mL starter culture in LB media containing 50 μg/mL kanamycin. Incubate overnight at 37 °C with shaking at 220 rpm.
3. The next day, dilute the starter culture 100-fold into new LB media containing 50 μg/mL kanamycin (*see* **Note 3**).

4. Incubate the diluted cultures at 37 °C with shaking at 220 rpm.
5. When the cultures reach an optical density (OD_{600}) of approximately 0.4, add IPTG to a final concentration of 0.1 mM to induce protein expression.
6. Incubate the cultures at 37 °C with shaking at 220 rpm for a further 4 h.
7. Collect the cells by centrifugation at 5250 × *g* for 15 min at 4 °C. Decant the supernatants (*see* **Note 4**).
8. Resuspend the cell pellets in TBS, pH 8.0, 5% (v/v) glycerol. About 15–20 mL of buffer is typically required to resuspend the cells collected from 1 L of culture.
9. Dilute the 10 mg/mL deoxyribonuclease and lysozyme stock solutions into the resuspended cells to yield a final concentration of 0.1 mg/mL of each enzyme. Likewise, dilute the 100 mM PMSF solution into the resuspended cells to a final concentration of 1 mM. Keep the cells on ice.
10. Lyse the cells by passing them through the French pressure cell at 20,000 psi. Collect the lysate on ice and pass it through the press two or three more times to ensure complete lysis (*see* **Note 5**).
11. Centrifuge the cell lysate at 5250 × *g* for 15 min at 4 °C.
12. While the lysate is in the centrifuge, prepare the materials for Ni-NTA chromatography by transferring Ni-NTA Superflow resin into an Econo-Pac column. Use approximately 0.5 mL of packed resin (or 1 mL of 50% Ni-NTA slurry) per liter of bacterial culture (in **Step 3**). Wash the resin with 5 column volumes (CV) of the glycerol wash buffer (*see* **Note 6**).
13. Transfer the cell lysate supernatant to the prepared Ni-NTA column. Collect the flow-through.
14. Reapply the flow-through to the Ni-NTA resin (*see* **Note 7**).
15. Wash the Ni-NTA resin with 10 CV of the glycerol wash buffer. Allow the column to drain completely.
16. Repeat **Step 15**.
17. Elute the His-tagged CdiA RBD by adding 3 CV of the glycerol elution buffer to the Ni-NTA column.
18. Concentrate the eluate in an Amicon Ultra 10 kDa filter by centrifugation at 4000 × *g* at 4 °C (*see* **Note 8**). Typically at this stage, the protein purified from 3 L of bacterial culture can be concentrated to a volume of approximately 1 mL (*see* **Note 9**).
19. Transfer the concentrated protein to a microcentrifuge tube and centrifuge at 21,130 × *g* for 10 min at 4 °C to remove any precipitated protein.

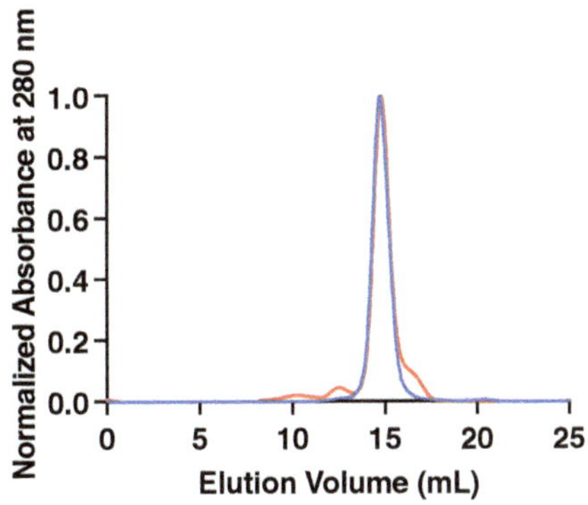

Fig. 1 Size-exclusion chromatography (SEC) purification of the His-tagged CdiA receptor-binding domain (RBD). The His-CdiA RBD is initially purified by Ni-NTA affinity chromatography and then subjected to SEC on a Superdex 200 Increase 10/300 GL column in TBS, pH 8.0, 5% (v/v) glycerol. The crude eluate from the Ni-NTA chromatography (red trace) contains small quantities of contaminating proteins, which elute on SEC just prior to and just after the His-CdiA RBD peak at 14.7 mL. When the central fractions from the His-CdiA RBD peak are collected and subjected to a second round of SEC (blue trace), these contaminating peaks are no longer observed

20. Subject the supernatant to size-exclusion chromatography on a Superdex 200 Increase 10/300 GL column that has been equilibrated in TBS, pH 8.0, 5% (v/v) glycerol buffer. Collect 0.5 mL (or smaller) fractions.
21. The His-tagged CdiA RBD protein elutes at ~14.7 mL. Collect the fractions from this peak and centrifuge them in an Amicon Ultra 10 kDa filter at 4000 × *g* at 4 °C (*see* Fig. 1 and **Note 10**). The protein can be concentrated to ~20 mg/mL without any precipitation.
22. Determine the concentration of the purified protein using the DC Protein Assay according to the manufacturer's instructions.
23. Flash freeze the purified protein in liquid nitrogen and store at −80 °C until use.

3.2 Expression and Purification of BamA with pBPA Substitutions

1. By performing a heat shock in a 42 °C water bath, transform BL21(DE3) competent cells with two plasmids: (1) pEVOL-pBpF and (2) a plasmid encoding the mature BamA sequence with an amber stop codon mutation at the desired *p*BPA location (e.g., pCH531 or pCH607) (*see* **Notes 11** and **12**).
2. Select transformants by plating on LB agar with 50 μg/mL carbenicillin and 30 μg/mL chloramphenicol. Incubate at 37 °C overnight.

3. Select a colony and inoculate a 5 mL starter culture in LB media containing 50 μg/mL carbenicillin, 30 μg/mL chloramphenicol, and 0.02% (w/v) glucose. Incubate overnight at 37 °C with shaking at 220 rpm.
4. The next day, prewarm a flask containing 50 mL of LB media to 37 °C. Add 50 μg/mL carbenicillin and 30 μg/mL chloramphenicol to the media.
5. Prepare the 0.9 M *p*BPA solution in 1 M NaOH. Ensure the *p*BPA is fully dissolved. Swirl the prewarmed media vigorously while adding 50 μL of the 0.9 M *p*BPA to the 50 mL of media (*see* **Note 13**).
6. Add 0.5 mL of the starter culture to the flask and incubate at 37 °C with shaking at 220 rpm.
7. When the cultures reach an optical density (OD_{600}) of approximately 0.4, add arabinose to a final concentration of 0.05% (w/v) and IPTG to a final concentration of 0.2 mM.
8. Incubate the cultures at 37 °C with shaking at 220 rpm for a further 4 h.
9. Collect the cells by centrifugation at 5250 × *g* for 15 min at 4 °C. Decant the supernatants.
10. Resuspend the cell pellet in 1 mL Bugbuster protein extraction reagent and transfer to a microcentrifuge tube. Add 10 μL of each of the deoxyribonuclease, lysozyme, and PMSF stock solutions to final concentrations of 0.1 mg/mL, 0.1 mg/mL, and 1 mM, respectively.
11. Incubate the resuspended cells on a rocker at 22 °C for approximately 30 min (*see* **Note 14**).
12. Collect the BamA inclusion bodies by centrifugation at 10,000 × *g* for 5 min at 4 °C. Carefully remove the supernatant by vacuum aspiration.
13. Wash the inclusion body pellet by resuspending it in 1 mL TBS, pH 8.0. Centrifuge again at 10,000 × *g* for 5 min for 4 °C. Carefully remove the supernatant by vacuum aspiration.
14. Resuspend the inclusion body pellet in 250 μL of 8 M urea. Incubate on a rocker at 22 °C for approximately 30 min to allow the inclusion bodies to dissolve and the solution to clarify.
15. Centrifuge at 21,130 × *g* for 10 min at 4 °C to remove any undissolved material or debris. Carefully transfer the supernatant to a new microcentrifuge tube.
16. Store the urea-denatured protein at −80 °C until use in the subsequent steps.

3.3 Purification of BamA-CdiA RBD Complexes

In a typical experiment, several (6 or 7) different BamA mutants containing *p*BPA substitutions at different locations will be purified with the CdiA RBD according to the procedure below. The small-scale Ni-NTA column purifications are easily performed in parallel.

1. Dilute 100 μL of each urea-denatured BamA protein (containing a *p*BPA substitution) into 900 μL of BamA folding buffer. The final concentration of each BamA protein should be approximately 25–30 μM. Incubate these solutions overnight (~16 h) on a rocker at 22 °C.
2. Centrifuge the BamA solutions at 21,130 × *g* for 10 min at 4 °C to pellet any precipitated material.
3. Carefully transfer the supernatants to new microcentrifuge tubes and determine the protein concentration in the solutions using the Bio-Rad DC Protein Assay according to the manufacturer's instructions.
4. Prepare 500 μL solutions containing 5 μM CdiA RBD and 20 μM BamA in refolding buffer. Incubate these mixtures on ice for 10 min.
5. Prepare the Ni-NTA columns by loading 400 μL of 50% Ni-NTA slurry into Bio-Spin chromatography columns. Allow the columns to drain by gravity filtration (*see* **Note 6**). There should be 200 μL of packed resin in the columns.
6. Equilibrate the Ni-NTA resin by adding 5 CV (1 mL) of DDM wash buffer to each column.
7. Load each BamA-CdiA RBD mixture onto an equilibrated Ni-NTA column.
8. Collect the solution that flows through each column and reapply it to the top of the Ni-NTA resin.
9. Add 10 CV (2 mL) of DDM wash buffer to each column and allow to drain completely.
10. Repeat **Step 9**.
11. Transfer the columns to microcentrifuge tubes and add 1 CV (200 μL) of DDM elution buffer to each. Allow to drain completely.
12. Repeat **Step 11**. There should now be 400 μL of eluate in each microcentrifuge tube.
13. Cap the microcentrifuge tubes and invert them a few times to mix the solutions. Place the tubes on ice and cover with aluminum foil.

3.4 Photocrosslinking of BamA-CdiA RBD Complexes

1. Turn on the high-intensity UV lamp and allow it to warm up for 5–10 min.

2. Transfer 125 μL of each BamA-CdiA RBD eluate (from **Step 13** in the previous section) to a well of 96-well plate. Keep the remainder of the eluate solutions on ice and cover with aluminum foil.
3. Place the 96-well plate on ice. Remove the top cover from the plate and place the UV lamp directly on top of the plate. Center the lamp over the samples (*see* Fig. 2 and **Note 15**).
4. Irradiate the samples for 60 min (*see* **Note 16**).
5. Transfer 100 μL of each of the irradiated samples to microcentrifuge tubes. Separately, transfer 100 μL of each of the unirradiated samples to microcentrifuge tubes. Add 100 μL of 2× SDS sample buffer to each and boil for 10 min.

3.5 SDS-PAGE and Immunoblotting

1. Load 20 μL of each unirradiated and irradiated sample in adjacent lanes on a 10% TGX mini-gel and run at 200 V for 40 min in SDS running buffer.
2. While the gel is running, wet a PVDF membrane with methanol and submerge in 1× transfer buffer along with filter paper stacks.
3. Remove the gel from the cassette and incubate it in transfer buffer for a few (~2) min (*see* **Note 17**).

Fig. 2 Experimental setup for UV irradiation of BamA-CdiA RBD complexes. Purified protein complexes are transferred to a 96-well plate and placed on ice. The high-intensity UV inspection lamp is then placed directly on top of the 96-well plate for 1 h

4. Assemble the transfer cassette with the sandwich of filter papers, PVDF membrane, and gel according to the Trans-Blot Turbo instructions.
5. Run the gel transfer on the Trans-Blot Turbo at constant current of 1.3 A for 7 min.
6. Disassemble the transfer sandwich and place the PVDF membrane in 10 mL of blocking buffer. Incubate on a rocker at 22 °C for 1 h.
7. Pour the blocking buffer solution off the membrane and replace it with 10 mL of fresh blocking buffer containing 2 μL of penta-His HRP conjugate (1:5000 dilution). Incubate overnight on a rocker at 4 °C.
8. Pour off the penta-His HRP conjugate solution and briefly wash the membrane with approximately 10 mL of TBS-T wash buffer three times, pouring off the wash buffer each time.
9. Incubate the membrane with approximately 10 mL of TBS-T wash buffer for 10 min on a rocker at 22 °C.
10. Pour off the wash buffer and repeat **Step 9** two more times.
11. Pour off the final wash and remove the residual solution from the membrane by blotting the edges of the membrane on paper towels.
12. Prepare the Clarity Western ECL substrate solution according to the manufacturer's directions and incubate with the membrane for 5 min.
13. Remove the substrate solution from the membrane by blotting the edges of the membrane on paper towels.
14. Image the blot on a Bio-Rad ChemiDoc MP or similar system using the chemiluminescence filter (*see* Fig. 3).

4 Notes

1. The pEVOL-pBpF plasmid contains two copies of the *M. jannaschii* aaRS gene; one is constitutively expressed and the other is driven by an arabinose-inducible promoter [21].
2. The CdiA- and BamA-encoding plasmids are available on request from our laboratory. They were generated by Gibson assembly [25] of the indicated *cdiA* and *bamA* gene fragments into pET expression vectors (Novagen). Amber codon mutations were then introduced at the desired locations in *bamA* by site-directed mutagenesis. The pCH531 and pCH607 plasmids encode BamA with *p*BPA-substitutions at amino acids D685 and D560, respectively.
3. Each liter of culture yields approximately 2.5 mg of purified protein at the end of the procedure. The typical culture volume

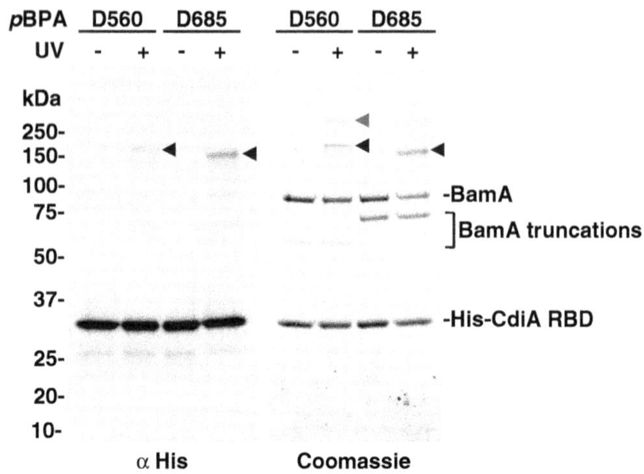

Fig. 3 Photocrosslinking results with *p*BPA substitutions at amino acids D560 and D685 in BamA's extracellular loops. Irradiated (+UV) and unirradiated (−UV) samples of BamA-CdiA RBD complexes containing these substitutions were subjected to SDS-PAGE and immunoblotting with anti-His antibodies (left panel). The bands that appear at high molecular weights in the irradiated samples (indicated by the black arrowheads) are crosslinked species that contain the His-CdiA RBD. The crosslinked species can also be observed by Coomassie staining (right panel). Truncated forms of BamA can be observed on the stained gel; these result from termination of protein synthesis at the amber codon rather than *p*BPA incorporation. The irradiated D560 sample also contains a high molecular weight species that is not recognized by anti-His antibodies (indicated by the red arrowhead). This band likely results from crosslinking of two BamA proteins to one another

is 3 L in our laboratory, but it can be scaled up or down at this point as desired.

4. At this point, the cell pellets can be stored at −80 °C if desired. The frozen pellets should then be thawed on ice prior to resuspension and cell lysis.
5. The cell suspension will not clarify as the cells are lysed because it contains a large quantity of inclusion bodies due to the high level of expression of the CdiA RBD. Nevertheless, a significant quantity of the CdiA RBD will be present in the soluble fraction.
6. All Ni-NTA chromatography steps are performed by gravity filtration. Allow the column to drain completely before proceeding to the next step.
7. Alternatively, this binding step could be performed by capping the column and incubating the resin with the lysate on a rocker at 4 °C for 2 h. This incubation may, however, result in more contaminating proteins in the eluate.
8. Although the calculated molecular weight of the His-tagged CdiA RBD is 32.4 kDa, 30 kDa MWCO filters should not be

used to concentrate the protein. The His-tagged CdiA RBD passes through these filters.

9. The glycerol in the buffer makes concentrating the protein a slow process. Periodically stopping the centrifuge and mixing the protein solution by pipetting up and down helps to speed it along.
10. Small peaks containing contaminating proteins and CdiA RBD truncations typically appear in the chromatogram before and just after the CdiA RBD peak at 14.7 mL. Collect and concentrate only the central fractions from the 14.7 mL peak. When these fractions are subjected to a second round of size-exclusion chromatography, a single peak appears in the chromatogram (*see* Fig. 1).
11. If the transformation with both plasmids fails, the following two-step procedure can be employed. First, transform BL21 (DE3) cells with pEVOL-pBPF and select transformants on agar plates containing 30 μg/mL chloramphenicol. Prepare competent cells from a colony of the transformed cells according to the Inoue method [26, 27]. Include 30 μg/mL chloramphenicol in the growth media to maintain the pEVOL-pBpF plasmid. Then transform these competent cells with the plasmids encoding the desired *p*BPA-substituted BamA proteins. The resulting double transformants should be selected on agar plates containing 50 μg/mL carbenicillin and 30 μg/mL chloramphenicol. The BL21(DE3)/pEVOL-pBpF competent cells can be flash frozen and stored at −80 °C for convenient use in subsequent experiments.
12. BamA is expressed without its signal sequence; it consequently accumulates in the cytoplasm in inclusion bodies. These inclusion bodies are produced in large quantities and can be easily isolated and denatured. Diluting the urea-denatured protein into a detergent solution allows it to fold into its native β-barrel structure. This expression and purification strategy allows for purification of a large number of BamA variants containing different *p*BPA substitutions in an efficient manner.
13. The *p*BPA has a tendency to precipitate upon dilution into LB media. Prewarming and swirling the media eliminates this issue.
14. The suspensions will not clarify as the cell lyse, but they will become less viscous and bright white due to the presence of the BamA inclusion bodies.
15. To protect researchers working in the laboratory, UV-irradiation is performed in a fume hood with the sash covered in an opaque material (e.g., aluminum foil).

16. Similar photocrosslinking results can be obtained with shorter (15–20 min) irradiation periods if the 96-well plate is not placed on ice; however, protein degradation is sometimes observed under these conditions.
17. Alternatively, the gel can be stained with Coomassie blue at this point to assess the purity of the samples and the efficiency of the crosslinking (*see* Fig. 3).

References

1. Tomasek D, Kahne D (2021) The assembly of beta-barrel outer membrane proteins. Curr Opin Microbiol 60:16–23
2. Doyle MT, Bernstein HD (2022) Function of the Omp85 superfamily of outer membrane protein assembly factors and polypeptide transporters. Annu Rev Microbiol 76:259–279
3. Aoki SK, Malinverni JC, Jacoby K et al (2008) Contact-dependent growth inhibition requires the essential outer membrane protein BamA (YaeT) as the receptor and the inner membrane transport protein AcrB. Mol Microbiol 70: 323–340
4. Ghequire MGK, Swings T, Michiels J et al (2018) Hitting with a BAM: selective killing by lectin-like Bacteriocins. MBio 9:1–15. https://doi.org/10.1128/mBio.02138-17
5. Smith DL, James CE, Sergeant MJ et al (2007) Short-tailed stx phages exploit the conserved YaeT protein to disseminate Shiga toxin genes among enterobacteria. J Bacteriol 189:7223–7233
6. Ruhe ZC, Wallace AB, Low DA et al (2013) Receptor polymorphism restricts contact-dependent growth inhibition to members of the same species. MBio 4:1–9. https://doi.org/10.1128/mBio.00480-13
7. Aoki SK, Pamma R, Hernday AD et al (2005) Contact-dependent inhibition of growth in Escherichia coli. Science 309:1245–1248
8. Ruhe ZC, Nguyen JY, Beck CM et al (2014) The proton-motive force is required for translocation of CDI toxins across the inner membrane of target bacteria. Mol Microbiol 94: 466–481
9. Aoki SK, Diner EJ, de Roodenbeke CT et al (2010) A widespread family of polymorphic contact-dependent toxin delivery systems in bacteria. Nature 468:439–442
10. Willett JL, Ruhe ZC, Goulding CW et al (2015) Contact-dependent growth inhibition (CDI) and CdiB/CdiA two-partner secretion proteins. J Mol Biol 427:3754–3765
11. Guerin J, Botos I, Zhang Z et al (2020) Structural insight into toxin secretion by contact-dependent growth inhibition transporters. elife 9:e58100
12. Ruhe ZC, Nguyen JY, Xiong J et al (2017) CdiA effectors use modular receptor-binding domains to recognize target bacteria. MBio 8: 1–16. https://doi.org/10.1128/mBio.00290-17
13. Ruhe ZC, Subramanian P, Song K et al (2018) Programmed secretion arrest and receptor-triggered toxin export during antibacterial contact-dependent growth inhibition. Cell 175:921–933
14. Tomasek D, Rawson S, Lee J et al (2020) Structure of a nascent membrane protein as it folds on the BAM complex. Nature 583:473–478
15. Lee J, Tomasek D, Santos TM et al (2019) Formation of a beta-barrel membrane protein is catalyzed by the interior surface of the assembly machine protein BamA. elife 8. https://doi.org/10.7554/eLife.49787
16. Ieva R, Bernstein HD (2009) Interaction of an autotransporter passenger domain with BamA during its translocation across the bacterial outer membrane. Proc Natl Acad Sci U S A 106:19120–19125
17. Ieva R, Tian P, Peterson JH et al (2011) Sequential and spatially restricted interactions of assembly factors with an autotransporter beta domain. Proc Natl Acad Sci U S A 108: E383–E391
18. Doyle MT, Bernstein HD (2021) BamA forms a translocation channel for polypeptide export across the bacterial outer membrane. Mol Cell 81(2000–2012):e3
19. White P, Joshi A, Rassam P et al (2017) Exploitation of an iron transporter for bacterial protein antibiotic import. Proc Natl Acad Sci U S A 114:12051–12056
20. Chin JW, Martin AB, King DS et al (2002) Addition of a photocrosslinking amino acid to

the genetic code of Escherichia coli. Proc Natl Acad Sci U S A 99:11020–11024

21. Young TS, Ahmad I, Yin JA et al (2010) An enhanced system for unnatural amino acid mutagenesis in E. coli. J Mol Biol 395:361–374
22. Kauer JC, Erickson-Viitanen S, Wolfe HR Jr et al (1986) p-Benzoyl-L-phenylalanine, a new photoreactive amino acid. Photolabeling of calmodulin with a synthetic calmodulin-binding peptide. J Biol Chem 261:10695–10700
23. Farrell IS, Toroney R, Hazen JL et al (2005) Photo-cross-linking interacting proteins with a genetically encoded benzophenone. Nat Methods 2:377–384
24. Curley CL, Fedrigoni TP, Flaherty EM et al (2021) Bacterial contact-dependent inhibition protein binds near the open lateral gate in BamA prior to toxin translocation. Biochemistry 60:2956–2965
25. Gibson DG, Young L, Chuang RY et al (2009) Enzymatic assembly of DNA molecules up to several hundred kilobases. Nat Methods 6: 343–345
26. Inoue H, Nojima H, Okayama H (1990) High efficiency transformation of Escherichia coli with plasmids. Gene 96:23–28
27. Green MR, Sambrook J (2020) The Inoue method for preparation and transformation of competent Escherichia coli: "Ultracompetent" cells. Cold Spring Harb Protoc 2020:101196

Chapter 9

Analysis of Transmembrane β-Barrel Proteins by Native and Semi-native Polyacrylamide Gel Electrophoresis

Violette Morales, Luis Orenday-Tapia, and Raffaele Ieva

Abstract

Membrane-embedded β-barrel proteins are important regulators of the outer membrane permeability barrier of Gram-negative bacteria. β-barrels are highly structured domains formed by a series of antiparallel β-strands. Each β-strand is locked in position by hydrogen bonds between its polypeptide backbone and those of the two neighbouring strands in the barrel structure. Some transmembrane β-barrel proteins form larger homo- or hetero-multimeric complexes that accomplish specific functions. In this chapter, we describe native and semi-native polyacrylamide gel electrophoresis (PAGE) methods to characterize the organization of transmembrane β-barrel proteins. We illustrate blue native (BN)-PAGE as an analytical method to assess the formation of protein complexes. Furthermore, we describe a heat-modifiability assay via semi-native PAGE as a rapid method to investigate the folding of transmembrane β-barrels.

Key words Bacterial envelope, BamA, BAM complex, Transmembrane β-barrel folding, Heat modifiability, BN-PAGE

1 Introduction

The outer membrane (OM) surrounding Gram-negative bacteria represents a vital platform of interactions with the surrounding environment. It also works as a semipermeable barrier that facilitates the transport of small metabolites and critical cofactors while limiting the entry of detergents and large molecules, including some antibiotics. The biogenesis and function of the bacterial OM critically depend on proteins that either associate with or span this lipid bilayer. Membrane-associated proteins generally consist of water-soluble domains attached to the membrane thanks to a lipid modification of their N-termini [1]. Integral OM proteins (OMPs) span the lipid bilayer by folding into amphipathic β-barrel structures with the external hydrophobic surface of the barrel that is in contact with the lipids [2]. The lumen of these barrels is generally polar. Its access to water-soluble molecules, however, can be limited

Raffaele Ieva (ed.), *Transmembrane β-Barrel Proteins: Methods and Protocols*, Methods in Molecular Biology, vol. 2778,
https://doi.org/10.1007/978-1-0716-3734-0_9,

by periplasmic or extracellular protein domains. OMP β-barrel structures consist of an anti-parallel arrangement of β-strands, where the polypeptide backbone of each β-strand forms a series of hydrogen bonds with that of the immediate downstream and upstream strands. Typically, the N- and C-terminal β-strands pair together to close the barrel [3].

Transmembrane β-barrels are highly stable structures that can resist denaturation by strong ionic detergents such as sodium dodecyl sulfate (SDS). In the presence of SDS, additional energy, as heating, is required to fully unfold such structures. As a consequence, upon SDS polyacrylamide gel electrophoresis (SDS-PAGE), β-barrel proteins migrate differently depending on whether they have been heat-denatured or not. Typically, non-heated transmembrane β-barrel proteins are characterized by faster mobility compared to fully denatured proteins. This semi-native PAGE assay was described in seminal studies of the protein content of the bacterial envelope [4, 5] and is commonly referred to as "heat modifiability assay."

Often OMPs do not function as individual protein entities in the membrane but associate with other proteins forming larger multi-subunit complexes. This is the case, for instance, of the β-barrel assembly machinery (BAM), a protein complex responsible for folding and inserting the vast majority of β-barrel structures into the bacterial OM. The central and catalytic BamA subunit consists of a C-terminal membrane-embedded β-barrel domain and an N-terminal domain that localizes in the periplasm [6]. The latter domain promotes interactions with 4 lipoproteins, BamBCDE, leading to the formation of a 200 kDa heteropentamer [7]. The BAM complex binds unfolded OMPs in the periplasm, inserting them into the OM as folded barrels [8]. The BAM complex further associates with accessory factors, including inner membrane protein complexes, periplasmic chaperones, and proteases, as well as other homeostasis factors [9–14]. Recently, we discovered one such association, characterizing the interaction of the BAM complex with DolP, a poorly understood OM lipoprotein that is crucial to preserve OM integrity [15]. The expression of both BAM and DolP is upregulated in response to envelope stress conditions [16], for instance, when an excess of unfolded OMPs accumulates in the periplasm.

We have used blue native PAGE (BN-PAGE) to reveal the association of DolP with the BAM complex. Although the precise function of DolP remains to be further detailed, by using a heat-modifiability assay we have obtained evidence that DolP promotes proper folding of the central catalytic BAM subunit BamA, providing one plausible explanation as to how DolP supports OM integrity.

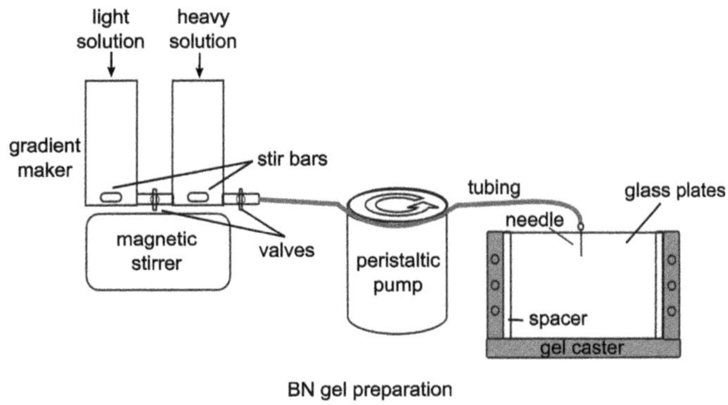

Fig. 1 Schematics of the procedure to cast gradient BN acrylamide gels

In this chapter, we will illustrate how BN-PAGE and heat-modifiability assays have contributed to the characterization of the BAM-DolP interaction. First, we describe the preparation of BN gels (Fig. 1) and the protocol to perform BN-PAGE. This method can be used to resolve the BAM complex extracted from the bacterial envelope upon membrane solubilization with a mild nonionic detergent. In addition, we used BN-PAGE to provide evidence that mixing purified BAM complex and purified DolP, both in detergent micelles, leads to the formation of a larger complex (Fig. 2). Second, we illustrate a heat modifiability assay as a simple and rapid test for transmembrane β-barrel protein folding, which suggested that DolP promotes proper folding of BamA (Fig. 3).

2 Materials

2.1 BN-PAGE Analysis of Transmembrane β-Barrel Protein Complexes

2.1.1 BN Polyacrylamide Gradient Gel Preparation

1. 160 × 180 mm glass plates, with 1 mm spacer and 1 × 25 × 130 mm 15-lane comb.
2. Gel Caster for vertical electrophoresis systems (e.g., SE600 Hoefer).
3. Gradient maker system.
4. Two magnetic stir bars of 1 cm.
5. Magnetic stirrer.
6. Peristaltic pump.
7. Vertical protein electrophoresis system (such as Hoefer SE600).
8. Acrylamide/Bis-acrylamide solution: 49.5% T: 3% C (*see* **Note 1**).
9. 10% (w/v) ammonium persulfate (APS) solution. Store at 4 °C.
10. Tetramethylethylenediamine (TEMED). Store at 4 °C.

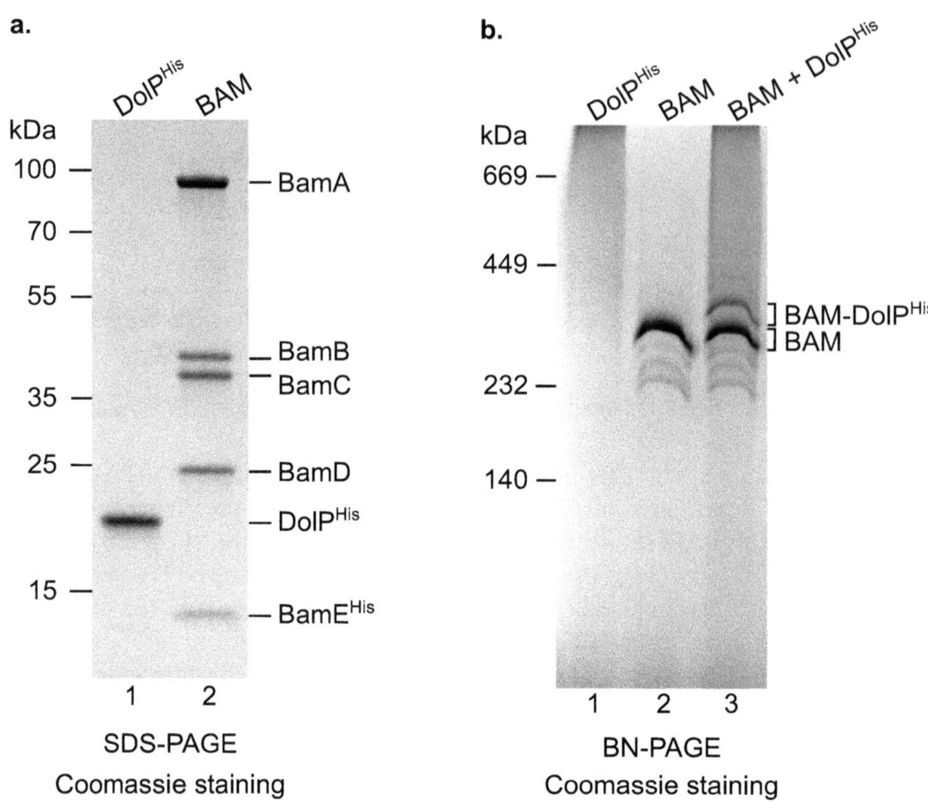

Fig. 2 Migration of purified BAM complex and DolP lipoprotein alone or mixed together on SDS-PAGE (**a**) and on BN-PAGE (**b**), both stained with Coomassie Blue. Migration of standard protein Markers is indicated on the side of the gels

11. 3× BN gel buffer: 200 mM ε-Amino n-caproic acid, 150 mM Bis-Tris-HCl, pH 7.
12. 1× Anode buffer: 50 mM Bis-Tris-HCl, pH 7.0. Prepare a 10× or 40× stock solution, and store at room temperature.
13. 1× Clear cathode buffer: 50 mM Tricine, 15 mM Bis-Tris, pH 7.0. Prepare a 10× or 20× stock solution, and store at room temperature.
14. 1× Blue cathode buffer: 50 mM Tricine, 15 mM Bis-Tris, pH 7.0, 0.02% Coomassie Blue G250 (*see* **Note 2**).

2.1.2 Preparation of BN-PAGE Samples

1. Protein dilution buffer: 20 mM Tris-HCl, pH 7.4, 50 mM NaCl, 0.3% (w/v) digitonin, 10% (w/v) glycerol, freshly supplemented with EDTA-free protease inhibitor cocktail (*see* **Note 3**).
2. Purified proteins of interest in 20 mM Tris-HCl, pH 7.4, 50 mM NaCl, 0.03% DDM, 10% glycerol. Here, we used a preparation of 1 mg/mL DolP^{His} and 2 mg/mL BAM

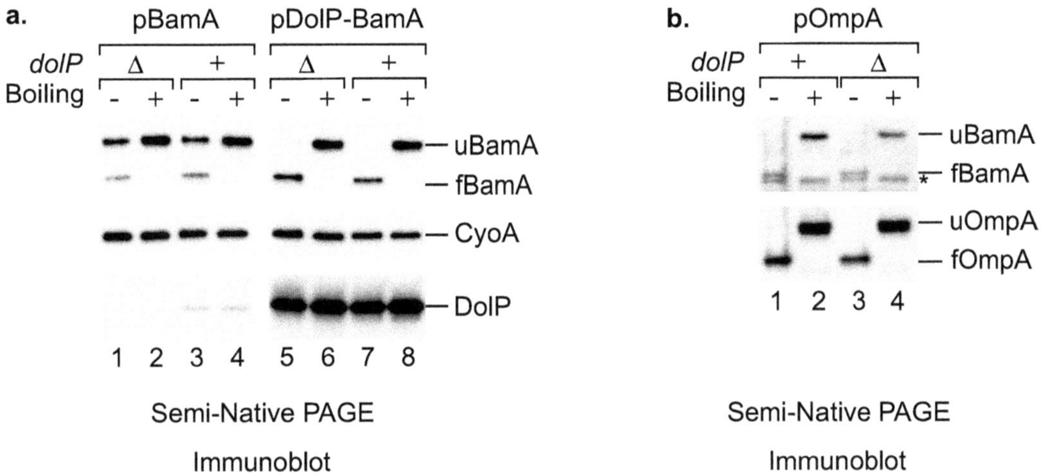

Fig. 3 Heat modifiability gel shift assays of BamA over-expressed alone (lanes 1–4) or together with DolP lipoprotein (lanes 5–8) in wild type or in Δ*dolP E. coli* strains (**a**) compared to the one of OmpA overexpressed in the same strains (**b**). Each protein is detected by immunoblot with specific antibodies. Migration of unfolded (u-) and folded (f-) β-barrel proteins is indicated on the side. * indicates a non-specific cross-reaction. This Figure is adapted from Ranava et al. [15], which was published under the Creative Commons Attribution 4.0 International Public License (CC-BY-4.0)

complex composed of the β-barrel protein BamA and the 4 lipoproteins BamB, BamC, BamD, and BamE^{His} (*see* **Note 4**).

3. 10× BN sample loading dye: 500 mM ε-amino n-caproic acid, 100 mM Bis-Tris-HCl, pH 7.0, 5% Coomassie Blue G250.
4. High Molecular Weight calibration kit for native electrophoresis (Amersham™) composed of thyroglobulin (669 kDa), ferritin (443 kDa), catalase (230 kDa), lactate dehydrogenase (140 kDa), and albumin (66 kDa).

2.1.3 BN Gel Staining

1. Coomassie staining solutions: 0.05% (w/v) Coomassie Blue R250, 50% (v/v) ethanol, 10% (v/v) acetic acid.
2. Coomassie destaining solution: 20% (v/v) ethanol, 10% (v/v) acetic acid.
3. Gel imaging system.

2.2 Heat Modifiability Assay

2.2.1 Bacterial Strains and Expression Plasmids

1. Competent wild type (e.g., BW25113) and a Δ*dolP* derivative *E. coli* strains.
2. LB (lysogeny broth) agar plates containing 100 μg/mL ampicillin.
3. BamA overproduction plasmid, pBamA (*see* **Note 5**).
4. OmpA overproduction plasmid, pOmpA, (*see* **Note 5**).
5. DolP and BamA overproduction plasmid, pDolP-BamA (*see* **Note 5**).

2.2.2 Bacterial Growth and Protein Expression

1. Sterile liquid LB media.
2. 100× ampicillin stock solution: 100 mg/mL ampicillin in ultra-pure water. Sterilize by filtration and store at −20 °C.
3. IPTG stock solution: 1 M isopropyl β-D-1-thiogalactopyranoside (IPTG). Sterilize by filtration and store at −20 °C.
4. Spectrophotometer.
5. Refrigerated benchtop microcentrifuge.
6. 4× Laemmli sample buffer: 277.8 mM Tris-HCl, pH 6.8, 44.4% (v/v) glycerol, 4.4% (w/v) Lithium dodecyl sulfate (LDS), 0.02% (w/v) bromophenol blue, freshly supplemented with 1.4 M 2-β-mercaptoethanol (βME) (*see* **Note 6**).

2.2.3 Semi-native PAGE

1. 4–20% acrylamide gradient gel (TGX, Bio-Rad) (*see* **Notes 7** and **8**).
2. Heated metal block for Eppendorf tubes at 98 °C.
3. SDS-PAGE running buffer: 25 mM Tris-HCl, pH 8.3, 192 mM glycine, 0.1% (w/v) SDS. Store at 4 °C (*see* **Note 9**).
4. Vertical gel electrophoresis system.
5. Ice bucket sufficiently large to host the vertical gel electrophoresis system.
6. Prestained protein ladder.

2.2.4 Proteins Immuno-Detection

1. Polyvinylidene fluoride (PVDF) transfer membrane, 0.45 μm pore.
2. Whatman blotting paper.
3. Semi-dry blotting apparatus.
4. Blotting buffer: 25 mM Tris, 192 mM glycine, 0.02% (w/v) SDS, 20% (v/v) ethanol.
5. Tris-buffered saline (TBS) solution: 20 mM Tris-HCl, pH 7.5, 125 mM NaCl.
6. Blocking buffer: 5% (w/v) skimmed milk, 0.05% v/v Tween in TBS.
7. Specific anti-OmpA, anti-BamA antibodies.
8. Anti-rabbit IgG antibodies conjugated to horseradish peroxidase (HRP) (*see* **Note 10**).
9. Rotary shaker.
10. Enhanced chemiluminescence (ECL) kit.
11. Chemiluminescence detection camera (e.g., ChemiDoc MP Imaging System, Bio-Rad).

3 Methods

3.1 BN-PAGE Assay

In this protocol, we illustrate how to assemble BN gels (Fig. 1) and the electrophoresis procedure to resolve and detect protein complexes of interest. Our example shows that purified BAM and DolP (Fig. 2a) interact to form a larger complex in detergent micelles (Fig. 2b).

3.1.1 Preparation of BN Polyacrylamide Gradient Gels

1. Wash the glass plates with deionized water, then wipe their surfaces with absolute ethanol and let them dry.
2. Assemble the glass plates with the spacer in the gel caster (*see* **Note 11**).
3. Place the gradient maker, with a stir bar in each chamber, on a magnetic stirrer.
4. Set up the connection of the gradient maker to the peristaltic pump with tubing and insert a 22G needle at the extremity.
5. Verify the tubing and adjust the flow rate of the pump (~2.5 mL/min) with water. Empty completely the mixer and the tubing, and close all valves.
6. For 160 × 180 mm glass plates, prepare 9 mL of both Light and Heavy solutions:
   - Light solution: 5% acrylamide, 1× BN gel buffer.
   - Heavy solution: 13% acrylamide, 1× BN gel buffer, 20% glycerol (*see* **Note 12**).
7. Position the needle on the top of the chamber formed between the glass plates (Fig. 1). Pour the two solutions in each compartment of the gradient maker with the heavy solution closest to the pump. The heavy solution is poured first, and the progressively lighter acrylamide solution layers are deposited on top (*see* **Note 13**).
8. Switch on the magnetic stirrer, add 0.02% APS and 0.02% TEMED in each solution, open the valves at the bases of both chambers, and switch on the pump. The mixing and flowing of heavy and light acrylamide solutions will generate a 5–13% gradient of polyacrylamide solution (resolving gel).
9. Gently add a thin layer of isopropanol on top of the gel solution and wait for the latter to polymerize (*see* **Note 14**).
10. Once the resolving gel is polymerized, remove the isopropanol and rinse the gel surface with water. Carefully remove the remaining drops of water. Pour the stacking solution: 4% acrylamide solution, 1× BN gel buffer, 0.02% APS, and 0.02% TEMED. Insert immediately the 1-mm comb. Ensure that the stacking solution forms a gel layer of ~1 cm below the comb. Let the stacking gel solution polymerize at room temperature. Store the gel at 4 °C before use (*see* **Notes 15** and **16**).

3.1.2 Preparation of Protein Sample for Migration on BN-PAGE

1. Prepare three reaction tubes. In one, add 5 μg of purified DolP diluted in protein dilution buffer to a final volume of 45 μL. In the second tube, dilute 5 μg of purified BAM in the same buffer and volume. In the third reaction, the BAM-DolP complex is generated by mixing approximately 5 μg of purified DolP and 5 μg of purified BAM complex (roughly a 5-molar excess of DolP with respect to BAM complex) and incubating on ice for at least 30 min. Dilute the mix in protein dilution buffer to a final volume of 45 μL (*see* **Note 17**).
2. Supplement all samples with 5 μL of 10× BN sample loading dye.
3. Keep samples on ice until the BN-PAGE apparatus is ready for loading.

3.1.3 Migration of Proteins by BN-PAGE

1. Prechill the BN-PAGE apparatus anode chamber before sample loading and electrophoresis.
2. Gently assemble the gel with the cathode buffer chamber.
3. Position gently the gel and the cathode chamber into the anode chamber containing cold anode buffer. Carefully fill the cathode chamber with blue cathode buffer and start electrophoresis.
4. Once the Coomassie Blue front of the samples reaches the resolving gel, replace the blue cathode buffer with clear cathode buffer (which does not contain Coomassie Blue).
5. Stop the electrophoresis just before the blue front of the samples runs out of the gel into the anode buffer. Protein complexes can be revealed by gel staining or immunoblotting (*see* **Note 18**).

3.1.4 Staining of BN Gels

1. Stain the gel by incubating it with Coomassie staining solution for 1 h.
2. Destain the gel with the destaining solution until the appearance of well-contrasted protein bands.
3. Rinse the gel in water.
4. Scan the gel using a gel imaging system.

3.2 Heat Modifiability Assay

In this protocol example, we show that upon overproduction, a large portion of BamA does not fold correctly. In fact, using semi-native PAGE, non-heated BamA generates two protein bands. A thicker band migrates similar to denatured BamA in heated samples whereas a thinner protein band migrates faster. The latter corresponds to folded BamA as it is abolished upon heating (Fig. 3a, compare lanes 1–4). This behavior seems to be specific of overproduced BamA. In fact, another OMP, such as OmpA, can efficiently fold when overproduced using an identical expression system (Fig. 3b). Concomitant overproduction of DolP together with BamA favors efficient assembly of BamA as a properly folded β-barrel structure, which is resistant to denaturation by SDS in the absence of heating (Fig. 3a, lanes 5–8).

3.2.1 Overexpression of OmpA and BamA either Alone or with DolP and Preparation of Total Protein Extracts

1. Transform wild-type and Δ*dolP* *E. coli* cells with the overproduction plasmids pOmpA, pBamA, or pDolP-BamA.
2. Select transformants by plating on LB agar plates containing 100 μg/mL ampicillin.
3. Use a single colony to inoculate an overnight liquid culture in LB containing 100 μg/mL ampicillin.
4. Dilute overnight cultures of *E. coli* BW25113 or Δ*dolP* transformed strains 1:100 in 10 mL LB containing 100 μg/mL ampicillin. Place at 37 °C with shaking at 180 rpm.
5. When the cultures reach an OD_{600} of 0.4, add 0.2 mM IPTG to induce protein expression during 1 h 30 min. Measure the final OD_{600} of the culture.
6. Pellet 0.2 optical density units of cells from each culture using a microcentrifuge set at 4 °C, for 3 min at top speed. The bacterial pellets can be stored at −20 °C if not used immediately.
7. Lyse bacterial cells by resuspending each pellet in 100 μL Laemmli sample buffer freshly supplemented with βME. Prepare two 10 μL-aliquots in separate micro-centrifuge tubes (corresponding to 0.02 optical units of the original culture). Each aliquot will be loaded on the gel after heating treatment. Keep samples on ice.

3.2.2 Loading and Migration on Semi-native PAGE

The semi-native PAGE system lacks SDS in the gel solution, but contains SDS (or LDS) in the running and in the loading (Laemmli) buffers. The electrophoresis is performed at low temperature (4 °C or on ice) and at low current to minimize heating of the gel during the electrophoresis.

1. Assemble the chilled polyacrylamide gel with cold running buffer into the electrophoresis chamber. Place the chamber into an ice bucket (*see* **Notes 19** and **20**).
2. Set up a heated block at 98 °C.
3. Take the two 10 μL-aliquots from each cell lysate. Place one aliquot at room temperature and the other at 98 °C for heat denaturation.
4. For each cell lysate, load nonheated and heated samples in adjacent lanes of 4–20% TGX mini-gel and run it at 150 to 170 V until bromophenol blue reaches the bottom of the gel (*see* **Note 21**).

3.2.3 Protein Transfer to a PVDF Membrane

1. During the electrophoresis, prepare the PVDF membrane and six pieces of blotting paper of a size similar to that of the gel.
2. Activate the PVDF membrane by rinsing it in technical-grade ethanol.
3. Soak the activated membrane and the blotting paper in blotting buffer.
4. After electrophoresis, rinse the gel in pure water and incubate it for 2 to 5 min in blotting buffer (*see* **Note 22**).
5. Assemble the transfer stack on the positive electrode of the semi-dry blotting chamber. Place from bottom to top: three wet filter papers, the PVDF membrane, the acrylamide gel, other three wet filter papers. Close the chamber with the negative electrode (*see* **Note 23**).
6. Run the transfer at constant current of 250 mA for 120–150 min.

3.2.4 Immunodetection

1. Disassemble the blotting system and recover the PVDF membrane (*see* **Notes 24** and **25**).
2. Wash the PVDF membrane with TBS and then saturate it by incubation with blocking buffer for 1 h at room temperature on an orbital shaker.
3. Incubate the PVDF membrane with primary antibodies (*see* **Notes 26** and **27**).
4. Wash the PVDF membrane four times in TBS for 10 min.
5. Incubate the PVDF membrane with secondary anti-rabbit IgG antibodies conjugates to HRP for 1 h at room temperature, following manufacturer's instructions.
6. Wash the membrane four times in TBS for 10 min.
7. Incubate the membrane with ECL detection kit containing the HRP substrate.
8. Image the ECL signal using a chemiluminescence detector imager.

4 Notes

1. "T" represents the total concentration (expressed as a w/v percentage) of acrylamide and bis-acrylamide, whereas "C" represents the percentage of cross-linker over the total amount of acrylamide and bis-acrylamide. Because of the high toxicity of acrylamide and bis-acrylamide, wear a lab coat, gloves, and a mask before weighing the powders under a fume hood. For preparation, dissolve 96 g of acrylamide and 3 g of bis-acrylamide powder in less than 100 mL ultra-pure water.

Let the acrylamide completely dissolve at room temperature for several hours to be sure that all powder particles are dissolved. Complete with pure water to 200 mL. After filtering (0.45 μm), store at room temperature in a bottle protected from light (e.g., by covering the bottle with aluminum foil).

2. Coomassie Blue G is soluble in water, differently from Coomassie Blue R, which is only soluble in organic solvents. The binding of this dye to the proteins confers negative charges promoting their migration through the gel, towards the anode.
3. To avoid precipitation of membrane proteins, ensure that the concentration of the mild detergent of choice is kept slightly above its critical micelle concentration (CMC).
4. Both DolP and the BAM complex were purified by Ni^{2+} affinity chromatography [15, 17].
5. The overproduction plasmids are derived from the pTrc vector that harbor an ampicillin-resistance cassette and an isopropyl β-D-1-thiogalactopyranoside (IPTG) inducible promoter to induce protein expression. This promoter is followed by ORFs encoding the indicated proteins (with endogenous signal peptides) alone or in operons [15].
6. Sodium dodecyl sulfate (SDS) can be used instead of LDS.
7. Any acrylamide gel, without SDS, can be used. The acrylamide concentration has to be chosen based on the size of the protein of interest.
8. Gel can be supplemented with 0.5% (v/v) 2,2,2-Trichloroethanol (TCE) during polymerization, which reacts with tryptophan residues in proteins allowing protein detection by fluorescence upon UV excitation. This step can be performed before immunoblot or before Coomassie Blue staining of the gel.
9. It is possible to use other gels such as acrylamide Bis-Tris gels.
10. Overproduced proteins can be detected up UV excitation if the TCE dye is incorporated into the gel.
11. As an alternative to gel casters, plates can be sealed with tape and maintained together with clamps positioned in correspondence with the lateral spacers.
12. The gradient of acrylamide can be adjusted depending on the size of the protein complexes of interest that the experimenter will need to separate. For the light solution consider a concentration equal to or higher than 4%, which is also the concentration of the stacking gel.
13. The needle can also be placed to reach the bottom of the chamber formed between the glass plates (the needle must be sufficiently long) and inject layers of acrylamide from the

bottom. In this case, the light acrylamide solution should be poured in the chamber closest to the pump, so that progressively heavier acrylamide layers are deposited at the bottom of the glass plates, while the lighter layers move up to fill the chamber. In all cases, the needle has to be removed very carefully to avoid disturbing the gradient.

14. Water can be used instead of isopropanol. Pour it very slowly to avoid mixing with the acrylamide solution.
15. It is possible to prepare the heavy and the light gel solutions in advance and to keep them at 4 °C (for less than a month) ready for use.
16. To optimally resolve high-molecular-weight complexes that would migrate on the very top part of the resolving gel, consider pouring the staking gel solution immediately after pouring the resolving gel solution, obtaining a continuously polymerized gel.
17. It is important that all samples have roughly the same volume and the same composition (buffer, salt, and detergent) to compare their migration profiles in different lanes. If the BN-PAGE is followed by immunoblot for protein detection, it will be possible to load even lower amounts of proteins.
18. To reveal protein complexes, consider two possibilities: (1) to proceed with Coomassie Blue staining as illustrated in this example protocol; (2) when samples contain low amounts of proteins that would be inefficiently stained, consider performing an immunoblot.
19. Ensure that the polyacrylamide gel is chilled prior to sample loading.
20. Electrophoresis can also be conducted in a cold (4 °C) room.
21. The running time approximately doubles compared to an electrophoresis conducted at 200 V and room temperature.
22. When TCE is added to the gel, proteins can be visualized by fluorescence upon exposure to UV light. This step is useful to verify the amount of proteins in each lane.
23. Ensure there is no air bubble between the layers of the sandwich.
24. The PVDF membrane can be briefly incubated with Coomassie staining solution and subsequently washed with the destaining solution. This step allows to verify the efficiency of protein transfer onto the membrane. The membrane can be cut in portions corresponding to different molecular weight ranges for incubation with different antibodies.
25. At this point, you can let the membrane dry and eventually store it before proceeding with immunodetection.

26. The rabbit primary antibody used here are against BamA, DolP, and OmpA, all diluted in blocking solution 1:1000.
27. The incubation with primary antibodies can be conducted at room temperature for 1 h or at 4 °C overnight.

Acknowledgments

This work was sponsored by the French CNRS, the ATIP grant and the Agence Nationale de la Recherche (ANR-23-CE11-0025) to R. Ieva, and an FRM 4th-year PhD extension fellowship to L. Orenday-Tapia.

References

1. Narita SI, Tokuda H (2017) Bacterial lipoproteins; biogenesis, sorting and quality control. Biochim Biophys Acta Mol Cell Biol Lipids 1862:1414–1423
2. Schulz GE (2000) beta-Barrel membrane proteins. Curr Opin Struct Biol 10:443–447
3. Fairman JW, Noinaj N, Buchanan SK (2011) The structural biology of β-barrel membrane proteins: a summary of recent reports. Curr Opin Struct Biol 21:523–531
4. Inouye M, Yee ML (1973) Homogeneity of envelope proteins of Escherichia coli separated by gel electrophoresis in sodium dodecyl sulfate. J Bacteriol 113:304–312
5. Nakamura K, Mizushima S (1976) Effects of heating in dodecyl sulfate solution on the conformation and electrophoretic mobility of isolated major outer membrane proteins from Escherichia coli K-12. J Biochem 80:1411–1422
6. Ranava D, Caumont-Sarcos A, Albenne C, Ieva R (2018) Bacterial machineries for the assembly of membrane-embedded β-barrel proteins. FEMS Microbiol Lett 365. https://doi.org/10.1093/femsle/fny087
7. Hagan CL, Kim S, Kahne D (2010) Reconstitution of outer membrane protein assembly from purified components. Science (New York, NY) 328:890–892
8. Noinaj N, Gumbart JC, Buchanan SK (2017) The β-barrel assembly machinery in motion. Nat Rev Microbiol 15:197–204
9. Cho SH, Szewczyk J, Pesavento C et al (2014) Detecting envelope stress by monitoring β-barrel assembly. Cell 159:1652–1664
10. Narita S, Masui C, Suzuki T et al (2013) Protease homolog BepA (YfgC) promotes assembly and degradation of β-barrel membrane proteins in Escherichia coli. Proc Natl Acad Sci U S A 110:E3612–E3621
11. Soltes GR, Martin NR, Park E et al (2017) Distinctive roles for periplasmic proteases in the maintenance of essential outer membrane protein assembly. J Bacteriol 199:e00418-17
12. Alvira S, Watkins DW, Troman L et al (2020) Inter-membrane association of the Sec and BAM translocons for bacterial outer-membrane biogenesis. elife 9:9
13. Iqbal H, Kenedy MR, Lybecker M, Akins DR (2016) The TamB ortholog of Borrelia burgdorferi interacts with the β-barrel assembly machine (BAM) complex protein BamA. Mol Microbiol 102:757–774
14. Schiffrin B, Machin JM, Karamanos TK et al (2022) Dynamic interplay between the periplasmic chaperone SurA and the BAM complex in outer membrane protein folding. Commun Biol 5:560
15. Ranava D, Yang Y, Orenday-Tapia L et al (2021) Lipoprotein DolP supports proper folding of BamA in the bacterial outer membrane promoting fitness upon envelope stress. elife 10:e67817
16. Rhodius VA, Suh WC, Nonaka G et al (2006) Conserved and variable functions of the sigmaE stress response in related genomes. PLoS Biol 4:e2
17. Roman-Hernandez G, Peterson JH, Bernstein HD (2014) Reconstitution of bacterial autotransporter assembly using purified components. elife 3:e04234

Chapter 10

Affinity Purification of Membrane β-Barrel Proteins via Biotin-Tagged Peptidiscs

Zhiyu Zhao, John William Young, and Franck Duong van hoa

Abstract

β-barrel membrane proteins play a crucial role in bacterial pathogenesis and antibiotic resistance, making them a prime focus for the development of new antibiotics and therapeutics. However, their inherent hydrophobic nature and limited presence pose challenges for their high-throughput characterization using conventional methods. In this context, we present a simple but efficacious approach using peptidisc, a membrane mimetic, to overcome the low abundance and hydrophobicity of these proteins. Our methodology, illustrated here using *Escherichia coli* (*E. coli*) as a model organism, covers the entire process from outer membrane fraction preparation to data analysis. This detailed protocol outlines the purification of a diverse collection of β-barrel membrane proteins, rendering them water-soluble and readily amenable to mass spectrometry and downstream drug screening strategies.

Key words β-barrel proteins, Outer membrane proteome, Peptidiscs, Membrane proteome purification, Affinity purification

1 Introduction

β-barrel membrane proteins play numerous essential roles in cell biology, such as nutrient transport, membrane biogenesis, cell communication, and receptor activities [1–3]. They are critical to the membrane structural integrity and are exclusively conserved in the outer membranes of Gram-negative bacteria, as well as in mitochondria and chloroplasts of eukaryotic cells [4, 5]. In Gram-negative bacteria, this group of proteins is involved in bacterial pathogenesis [6, 7], and mutations that occur on them can result in antibiotic resistance [8–11]. Therefore, they are attractive targets for therapeutic development, including new antibiotics, vaccines, and antibodies [8].

Numerous studies have used mass spectrometry (MS)-based techniques to survey the outer membrane proteome [12–16]. However, most have applied a traditional in-gel digestion

Raffaele Ieva (ed.), *Transmembrane β-Barrel Proteins: Methods and Protocols*, Methods in Molecular Biology, vol. 2778, https://doi.org/10.1007/978-1-0716-3734-0_10,

strategy, which requires extensive sample handling, incompatible with high-throughput proteome identification [17, 18]. The in-solution digestion strategy is commonly applied to proteome characterization, but the technique is challenging to perform on membrane proteins due to their intrinsic insolubility in an aqueous solution. Detergents, chaotropic agents, and organic solvents can be used to increase the solubility of membrane proteins. However, they also pose technical limitations. For example, detergents are mostly incompatible with LC/MS-based identification [19–21], and chaotropic or organic reagents can lead to protein sample loss [17]. In this chapter, we apply the peptidisc membrane mimetic to facilitate a high-throughput β-barrel membrane proteomics analysis.

Peptidisc is composed of multiple copies of amphipathic ApoA1-derived peptides, which wrap around membrane proteins to stabilize them in a water-soluble state [22–26]. When functionalized with an affinity tag, such as a His_6- or biotin-tag, the membrane proteome reconstituted in the peptidisc (membrane protein library) can be affinity purified [24, 27]. Using *E. coli* as a model organism, we present below a complete protocol to identify β-barrel membrane proteins via MS analysis. Starting from a crude membrane preparation, we apply a dual-detergent strategy to deplete the inner membrane components before solubilization of the outer membrane fraction. Following protein reconstitution in biotinylated peptidiscs and purification over a streptavidin-mutein matrix, soluble contaminants are washed away. We describe how to prepare the sample for MS analysis and use the MS analytical software and databases to annotate the proteins (Fig. 1). The β-barrel proteins share similar basic structures across bacteria, mitochondria, and chloroplasts. With the success of identifying a peptidisc library rich in β-barrel proteins from *E.coli* [26], a similar protocol may be very useful for facilitating proteomics-based analysis of β-barrel proteins in other bacterial strains, as well as in mitochondria and chloroplasts.

2 Materials

2.1 Laboratory Equipment

1. Microfluidizer Processor.
2. Beckman Coulter Optima (TLX) Ultracentrifuge, with the TLA-55 rotor.
3. Beckman Coulter Optima (LE-80K) Ultracentrifuge, with Type 45-Ti rotor.
4. Beckman Coulter Optima (Avanti J-E) centrifuge with rotor JLA 10.500.
5. Table top centrifuges for Eppendorf and conical tubes.

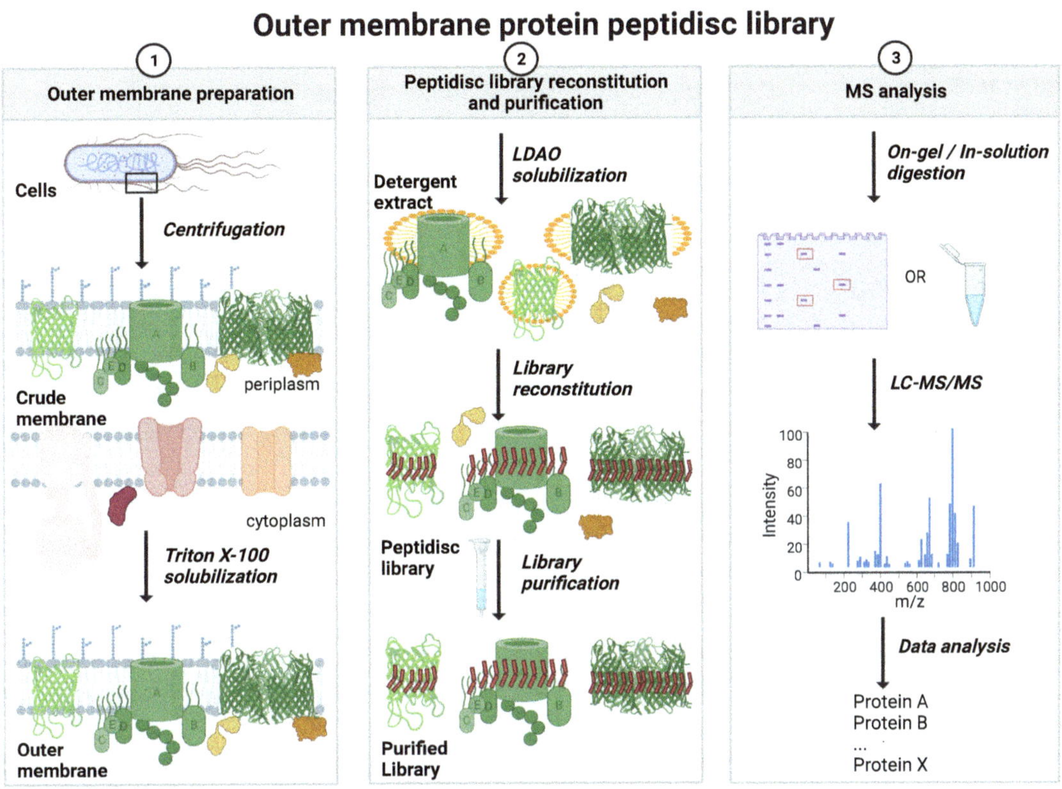

Fig. 1 Outer membrane peptidisc library workflow. The workflow contains three parts: (1) outer membrane protein isolation, (2) outer membrane protein peptidisc library reconstitution and purification, and (3) protein identification through mass spectrometry

6. Dounce homogenizer.
7. SpeedVac.
8. ThermoStat plus Heating block.

2.2 Crude Membrane Preparation

1. *Escherichia coli* K12 strain W3110.
2. Luria-Bertani (LB) medium: 10 g/L tryptone, 5 g/L yeast extract, 10 g/L NaCl, supplemented with 15 g/L for LB agar plate.
3. Phenylmethylsulfonyl fluoride (PMSF): 1 M PMSF stock solution in 100% isopropanol.
4. TSG buffer: 50 mM Tris/HCl, pH 7.8, 100 mM NaCl, 10% glycerol.

2.3 Membrane Solubilization

1. Triton X-100: 10% (v/v) stock solution in distilled H_2O.
2. Lauryldimethylamine oxide (LDAO): 10% (v/v) stock solution in distilled H_2O.
3. TSG buffer: 50 mM Tris/HCl, pH 7.8, 100 mM NaCl, 10% (v/v) glycerol.

2.4 Peptidisc Library Reconstitution and Purification

1. Biotinylated peptidisc peptides, referred to as "Bio-peptidisc" (Peptidisc Biotech).
2. 100 kDa-cutoff Amicon centrifugal filters.
3. Streptavidin Mutein Matrix.
4. TS buffer: 50 mM Tris, pH 7.8, 100 mM NaCl.
5. TS buffer supplemented with 5 mM Biotin.

2.5 Sodium Dodecyl Sulfate-Polyacrylamide Gel Electrophoresis (SDS-PAGE)

1. Resolving gel buffer: 1 M Tris-HCl, pH 8.8, 40% acrylamide mixture (37.5:1 (w/w) acrylamide/bis-acrylamide), 10% (w/v) SDS.
2. Stacking gel buffer: 1 M Tris-HCl, pH 6.8, 40% acrylamide mixture (37.5:1 (w/w) acrylamide/bis-acrylamide), 10% (w/v) SDS.
3. Ammonium persulfate: 10% (w/v) solution in distilled H_2O, store at 4 °C for short-term use; store at −20 °C for long-term storage.
4. TEMED: store at 4 °C.
5. SDS-PAGE sample buffer (5×): 10% (w/v) SDS, 400 mM Tris-HCl, pH 6.8, 50% (v/v) glycerol, 0.32 mg/mL bromophenol blue.
6. SDS-PAGE running buffer (8×): 115 g/L glycine, 24 g/L Tris, 8 g/L SDS.
7. Coomassie staining buffer: 40% (v/v) ethanol, 0.2% (w/v) Coomassie Brilliant Blue G-250, 10% (v/v) acetic acid in distilled H_2O.
8. Destaining buffer: 20% (v/v) ethanol, 10% acetic acid (v/v) in distilled H_2O.

2.6 Mass Spectrometry Sample Preparation

1. MS-Grade trypsin is dissolved in 50 mM NH_4HCO_3 to obtain 1 mg/mL stock. Store the stock solution at −20 °C until use.
2. Urea.
3. 1 M Dithiothreitol (DTT): freshly made before use.
4. 500 mM iodoacetamide (IAA): freshly made before use and kept in the dark.
5. Gel wash solution: 100 mM NH_4HCO_3, 50% (v/v) acetonitrile, prepared in HPLC-Grade H_2O.
6. C18 solid phase extraction disc (Empore).
7. Polygoprep 300-20 C18 slurry.
8. Stage tip wash buffer: 0.1% (v/v) formic acid.
9. Stage tip elution buffer: 80% (v/v) acetonitrile, 0.1% (v/v) formic acid.

2.7 Protein Identification

1. Thermo Scientific Q Exactive Mass Spectrometer outfitted with a nanospray source and EASY-nLC 1000 nano-LC system (Thermo Fisher).
2. Buffer A: 0.1% v/v formic acid in HPLC-grade H_2O.
3. Buffer B: 80% (v/v) acetonitrile, 0.1% (v/v) formic acid in HPLC-Grade H_2O.

3 Methods

3.1 Cell Growth and Crude Membrane Preparation

1. Revive the bacteria, here, W3110 strain, on LB agar plate at 37 °C for 16 h.
2. Inoculate a single colony into 10 mL LB and grow the bacteria at 37 °C for 16–18 h.
3. Inoculate the starter bacterial culture into 1 L fresh LB with 1/100 dilution factor. Grow cells at 37 °C until OD_{600} reaches 1.0.
4. Harvest cells by centrifugation at 6000 × *g* for 10 min (Beckman Coulter Optima Avanti J-E centrifuge with rotor JLA 10.500).
5. Resuspend the cell pellet in Buffer TSG to a final volume of 20 mL, dounce the cells, and add 1 mM PMSF. Keep on ice.
6. Break cells by 3 passes through a Microfluidizer at 12,000 psi (*see* **Note 1**) and collect the lysate in a conical tube.
7. Remove unbroken cells via centrifugation at 3000 × *g* for 10 min at 4 °C using a centrifuge for conical tubes.
8. Pellet the crude membranes by ultracentrifugation (LE-80K) at 100,000 × *g* for 45 min using a Type 45-Ti rotor.
9. Resuspend the crude membrane pellet in ~4 mL TSG Buffer (*see* **Note 2**).
10. Aliquot the crude membranes. Use the membrane fraction aliquots immediately or store them at −80 °C until use.

3.2 Outer Membrane Solubilization

1. Solubilize the crude membranes (e.g., 4 mg in 2 mL) with 1% Triton X-100 at room temperature for 1 h to deplete the inner membrane fraction (*see* **Note 3**).
2. Pellet the outer membrane fraction via ultracentrifugation (Beckman Coulter Optima Ultracentrifuge) at 186,000 × *g* for 45 min using a TLA-55 rotor.
3. Resuspend the outer membrane pellet in 1 mL TSG buffer and solubilize with 1% LDAO at 4 °C overnight with gentle rocking (*see* **Note 4**).

4. Pellet the LDAO-insoluble material via ultracentrifugation (Beckman Coulter Optima Ultracentrifuge) at 186,000 × *g* for 20 min using a TLA-55 rotor. Keep the LDAO-solubilized material.

3.3 Peptidisc Library Reconstitution and Purification

1. Prepare peptidisc peptides: dissolve 20 mg biotinylated peptidiscs in distilled water, and adjust pH to 7.8 using 1 M Tris-HCl, pH 7.8. The final peptidisc peptide concentration is around 5 mg/mL (*see* **Note 5**).
2. Mix LDAO-solubilized material (e.g., 700 μL at 2 mg/mL) with four-fold mass excess of Bio-tagged peptidisc peptides (e.g., 1120 μL at 5 mg/mL) (*see* **Note 6**).
3. Load the LDAO-solubilized membrane proteins and peptidisc peptides mixture over a 15 mL 100 kDa-cutoff Amicon centrifugal filter.
4. Dilute the mixture with TS buffer up to 15 mL and concentrate to ~500 μL via centrifugation (centrifuge for conical tubes) at 3000 × *g* at 4 °C (*see* **Note 7**).
5. Repeat **step 4** one more time to obtain reconstituted peptidisc "Starting Library" (final volume ~ 500 μL at 2 mg/mL). Save an aliquot of the Starting Library for SDS-PAGE analysis.
6. Load the Starting Library (~900 μg) onto 200 μL Streptavidin Mutein Matrix (*see* **Note 8**).
7. Incubate the Starting Library with purification resins for 2 h at 4 °C with gentle shaking.
8. Pellet the resins via centrifugation (table-top) at 800 × *g* for 1 min and save the flow through (FT) for SDS-PAGE analysis.
9. Wash the resins using TS buffer (1 mL each time for 5 min). Wash the resins 3 times. Save an aliquot for SDS-PAGE analysis.
10. Elute the "Purified Library" in 120 μL TS buffer supplemented with 5 mM biotin. Repeat the elution procedure. Combine the elution fractions and save an aliquot for SDS-PAGE analysis.

3.4 SDS-PAGE and In-Gel Digestion

1. Solubilize samples (detergent-extract, Starting Library, Flow through, wash, and Purified Library) in SDS sample buffer and analyze them by SDS-PAGE (4% stacking gel, 15% resolving gel). A representative analysis is shown in Fig. 2.
2. Stain gels with Coomassie staining buffer.
3. Distain gels with the destaining buffer.
4. Excise the protein bands of interest with care (*see* **Note 9**). Transfer each gel slice to a clean 1.5 mL Eppendorf tube.
5. To further destain the protein slices, wash the gel slices with the gel wash solution for 30 min. Repeat this procedure until the Coomassie blue stain is barely seen on the gel slices.

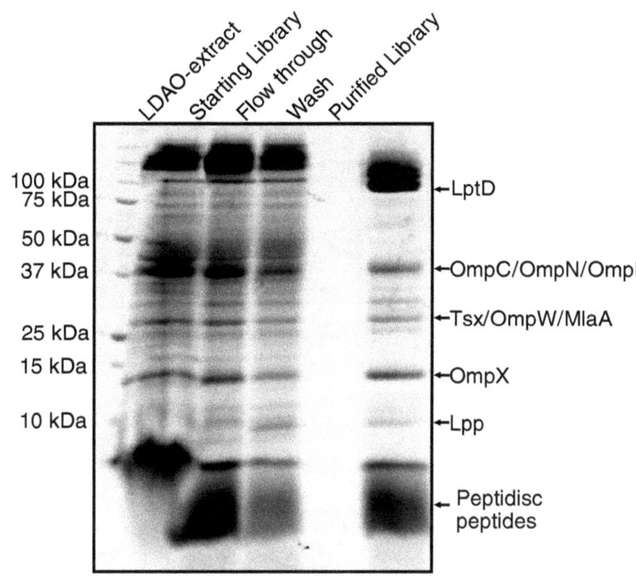

Fig. 2 SDS-PAGE analysis. LDAO-extract, Starting Library, Flow through, wash, and Purified Library are analyzed on 15% SDS-PAGE. Proteins from five major bands were extracted following the in-gel digestion protocol and analyzed by MS. The corresponding outer membrane proteins are labeled

6. Dehydrate the gel slices with 500 μL 100% acetonitrile for 10 min.
7. Remove acetonitrile and dry the gel slices via SpeedVac.
8. Add 100 μL of 10 mM DTT to each gel slice and incubate at 55 °C in a heating block for 45 min. Remove DTT solution by aspiration after incubation (*see* **Note 10**).
9. Add 100 μL of 55 mM IAA to each gel slice and incubate at room temperature for 20 min in the dark (*see* **Note 11**). Remove IAA solution by aspiration after incubation.
10. Wash each gel slice with 50 mM NH_4HCO_3 twice.
11. Digest proteins from each gel slice with 0.1 μg trypsin at 37 °C overnight. Cover the gel slice with trypsin solution (*see* **Note 12**).
12. On the following day, collect the supernatant, which contains digested peptides, and transfer it to a new clean tube (*see* **Note 13**).
13. Add 50 μL 5% formic acid to each gel slice and incubate for 15 min; collect the supernatant and combine it with the supernatant collected in **step 12** of this section.
14. Add 50 μL 100% acetonitrile to each gel slice and incubate for 15 min; collect the supernatant and combine it with the supernatant collected in **step 12** of this section.

15. Quench the trypsin reaction with 10% (v/v) formic acid until the solution reaches pH 3 (*see* **Note 14**).
16. Dry the peptides using a SpeedVac.
17. Resuspend the peptides in 100 μL 0.1% (v/v) formic acid. To desalt the peptides, proceed to Subheading 3.6.

3.5 In-Solution Digestion

1. Unfold the proteins captured in peptidisc (total ~ 80 μg proteins in 80 μL) with 6 M urea and incubate with urea at room temperature for 30 min (*see* **Note 15**).
2. Add 1 M DTT to a final concentration of 10 mM into samples and incubate at room temperature for 1 h.
3. Add 500 mM IAA to a final concentration of 20 mM into samples, and incubate at room temperature for 30 min in the dark.
4. Quench the IAA reaction by adding 1 M DTT to a final concentration of 10 mM, and incubate at room temperature for another 30 min.
5. Dilute the sample six-fold with TS buffer.
6. Add 1.5 μg of trypsin and incubate at room temperature overnight.
7. Quench the trypsin reaction with 10% (v/v) formic acid until the solution reaches pH 3. To desalt the peptides, proceed to Subheading 3.6.

3.6 Peptides Desalting Using C18 Stage Tips

1. Prepare Stage tips: punch out a small disk of C18 Empore filter using a 17 G flat-tipped syringe and eject the disk into a P200 pipette tip (*see* **Note 16**). For high-capacity tips, add C18 slurry (*see* **Note 17**).
2. Activate the Stage tip with 75 μL 100% methanol by forcing the organic solvent through the tip, or via centrifugation (2000 × *g*, 30 s).
3. Equilibrate the Stage tip by washing it with 200 μL wash buffer. Remove the wash buffer by centrifugation (2000 × *g*, 1 min) (*see* **Note 18**).
4. Load the peptide samples (obtained from Subheadings 3.4 or 3.5) on the Stage tip, each time 200 μL and centrifuge (2000 × *g*, 1 min).
5. After all the samples bind onto the Stage tip, wash the Stage tip 3 times with wash buffer.
6. Elute desalted peptides from the Stage tip with 65–75 μL of elution buffer into a new 1.5 mL Eppendorf tube.
7. Dry the peptides via SpeedVac. Store the lyophilized samples at −20 °C.

3.7 Protein Identification (See Note 19)

1. Resuspend the lyophilized peptides in 0.1% (v/v) formic acid and load onto a 75 μm × 50 cm PepMax RSLC EASY-Spray column filled with 2 μM C18 beads (Thermo Fisher) at 900 bar and 60 °C.
2. Elute the peptides over 60 min at a rate of 250 nL/min using a gradient (0–35% Buffer B over 50 min; 35–100% Buffer B over 2 min; 100% Buffer B for 8 min).
3. Perform MS1 acquisition with a scan range of 400–1600 *m/z* with resolution set to 70,000, maximum injection time of 100 ms, and AGC target set to 3×10^6.
4. Perform MS2 acquisition with a scan range of 200–2000 *m/z* with resolution set to 17,500, maximum injection time of 50 ms, and AGC target set to 1×10^5.
5. Set the isolation window to 2 *m/z*. Perform the dynamic exclusion with a duration of 10 s.

3.8 Data Analysis

1. Upload the raw MS data to MaxQuant v1.6.17.0 [28] (*see* **Note 20**).
2. Set parameters: peptide mass accuracy is set to <5 ppm; fragment mass accuracy is set to 0.006 Da; Enzyme specificity is set to trypsin, with a maximum of two missed cleavages allowed. Set cysteine carbamidomethyl as a fixed modification and methionine oxidation, asparagine, glutamine, and N-acetyl peptide deamidation as variable modifications. The rest parameters are set by default.
3. Open the MaxQuant output file "proteinGroups.txt." Convert the .txt file to .xlsx file.
4. Analyze and annotate identified proteins using UniprotKB database (https://www.uniprot.org).
5. Go to panel "ID mapping" in Uniprot. Copy and paste protein accession IDs from the data .xlsx file in the dialog box, then click "Map xxx IDs" (xxx: the total input ID number).
6. After ID mapping is completed, go to "Customize columns" and select relevant columns. Here, we use "Gene Ontology (GO)" as an example, then click "save."
7. Download the annotated data: go to "download" and select "Excel" format (*see* **Note 21**).
8. Annotate proteins as "membrane proteins" if they have a "membrane" Go-term. Annotate proteins as "outer membrane proteins" if they have an "outer membrane" Go-term.

4 Notes

1. Cool down the Microfluidizer system before cell lysis by adding half ice and water in the "lysis unit" (where cell lysis takes place) container. Equilibrate the lysis unit with TSG buffer. After cell lysis is done, clean the lysis unit with water and 20% ethanol.
2. Crude membranes can be hard to resuspend by pipetting. Use the pestle of a douncer homogenizer to help the resuspension and vortex. After resuspension, transfer the crude membrane into the douncer homogenizer and dounce it until it gets homogenized.
3. Inner membrane proteins can be solubilized with the detergent Triton X-100. After ultracentrifugation, the solubilized inner membrane fraction is in the supernatant, which can be discarded.
4. If the pellet is hard to resuspend, vortex vigorously or sonicate briefly in a bath sonicator.
5. Bio-tagged peptidisc peptides precipitate at higher pH (>pH 6.5), so 0.2% detergent n-dodecyl-β-d-maltoside (DDM) is typically added to increase their solubility in solution.
6. Peptidisc peptides are supplemented with an excess amount—three to four folds higher than the detergent-solubilized material in mass ratio—to keep the membrane protein in a water-soluble environment.
7. Add TS buffer in a step-wise manner: add 500 μL each time, repeat ~10 times. Then add 1 mL each time, repeat this procedure, and fill the filter up to 15 mL. We do this to avoid abrupt dilution of detergents, which may cause protein precipitation/aggregation. It is ideal for replacing detergent with peptidiscs gradually.
8. The purification resins are equilibrated with TS buffer before use.
9. Rinse the gel with water and excise the protein bands with care. Do this in a clean fume hood and use a clean scalpel to avoid keratin contamination.
10. Be careful not to break the gel slice into pieces.
11. The samples can be left in a drawer.
12. The enzyme-to-protein sample ratio is typically between 1:50 and 1:100. About 0.1 μg of trypsin can digest 5–10 μg of protein samples.
13. The samples can evaporate when incubated at 37 °C. Use parafilm to seal the tube.

14. Use pH test strips.
15. The sample volume is around 80 μL. Add 36 mg of urea powder directly into the sample solution. The final volume of the sample is around 100 μL.
16. If a 17 G flat-tipped syringe is not available, prepare a blunt-tip pipetting needle.
17. C18 slurry preparation: mix Polygoprep C18 powder with 50% acetonitrile. Force the C18 slurry into Stage tips. Remove the organic solvent, and the final C18 column volume is around 8–10 μL. Do not over-pack; otherwise, it will be hard to pass liquid through the column.
18. Avoid drying the Stage tip any time before eluting the peptides; otherwise, it may cause sample loss.
19. All our MS samples are analyzed at Toronto SickKids SPARC BioCentre.
20. MaxQuant search was set to be against the *E.coli* K12 source organism database and common contaminations.
21. Not all the IDs can be retrieved from UniprotKB database. For example, some proteins may have different accession IDs (A0A0xxxxxx and Pxxxxx). One of the ID may no longer be used; therefore, its information cannot be found in the database. It is possible to have 1000 ID input, but with 998 annotated.

Acknowledgments

We thank Dr. Craig Simpson and Dr. Leanne Wybenga-Groot at SPARC Molecular Analysis (SickKids Hospital, University of Toronto) for LC-MS/MS sample analyses. This work is supported by a grant from the NSERC. Figure 1 is created with BioRender.com.

References

1. Yamashita S, Buchanan SK (2010) Solute and ion transport: outer membrane pores and receptors. EcoSal Plus 4. https://doi.org/10.1128/ecosalplus.3.3.1
2. Wimley WC (2003) The versatile beta-barrel membrane protein. Curr Opin Struct Biol 13: 404–411. https://doi.org/10.1016/s0959-440x(03)00099-x
3. Fairman JW, Noinaj N, Buchanan SK (2011) The structural biology of β-barrel membrane proteins: a summary of recent reports. Curr Opin Struct Biol 21:523–531. https://doi.org/10.1016/j.sbi.2011.05.005
4. Tamm LK, Hong H, Liang B (2004) Folding and assembly of beta-barrel membrane proteins. Biochim Biophys Acta 1666:250–263. https://doi.org/10.1016/j.bbamem.2004.06.011
5. Diederichs KA, Buchanan SK, Botos I (2021) Building better barrels - β-barrel biogenesis and insertion in bacteria and mitochondria. J Mol Biol 433:166894. https://doi.org/10.1016/j.jmb.2021.166894
6. Krishnan S, Prasadarao NV (2012) Outer membrane protein A and OprF: versatile roles in Gram-negative bacterial infections. FEBS J 279:919–931. https://doi.org/10.1111/j.1742-4658.2012.08482.x
7. Hejair HMA, Zhu Y, Ma J et al (2017) Functional role of ompF and ompC porins in

pathogenesis of avian pathogenic Escherichia coli. Microb Pathog 107:29–37. https://doi.org/10.1016/j.micpath.2017.02.033

8. Hermansen S, Linke D, Leo JC (2022) Transmembrane β-barrel proteins of bacteria: from structure to function. Adv Protein Chem Struct Biol 128:113–161. https://doi.org/10.1016/bs.apcsb.2021.07.002
9. Roterman I, Stapor K, Fabian P, Konieczny L (2021) The functional significance of hydrophobic residue distribution in bacterial beta-barrel transmembrane proteins. Membranes (Basel) 11. https://doi.org/10.3390/membranes11080580
10. Bartsch A, Ives CM, Kattner C et al (2021) An antibiotic-resistance conferring mutation in a neisserial porin: structure, ion flux, and ampicillin binding. Biochim Biophys Acta Biomembr 1863:183601. https://doi.org/10.1016/j.bbamem.2021.183601
11. Choi U, Lee C-R (2019) Distinct roles of outer membrane porins in antibiotic resistance and membrane integrity in Escherichia coli. Front Microbiol 10:953. https://doi.org/10.3389/fmicb.2019.00953
12. Nally JE, Whitelegge JP, Bassilian S et al (2007) Characterization of the outer membrane proteome of Leptospira interrogans expressed during acute lethal infection. Infect Immun 75:766–773. https://doi.org/10.1128/IAI.00741-06
13. Lee E-Y, Bang JY, Park GW et al (2007) Global proteomic profiling of native outer membrane vesicles derived from Escherichia coli. Proteomics 7:3143–3153. https://doi.org/10.1002/pmic.200700196
14. Distler AM, Kerner J, Hoppel CL (2008) Proteomics of mitochondrial inner and outer membranes. Proteomics 8:4066–4082. https://doi.org/10.1002/pmic.200800102
15. Zhu T, Lei Z, Qu S et al (2022) Comparison of the outer membrane proteomes between clinical carbapenem-resistant and -susceptible Acinetobacter baumannii. Lett Appl Microbiol 74:873–882. https://doi.org/10.1111/lam.13672
16. Liu J, Hsieh C-L, Gelincik O et al (2019) Proteomic characterization of outer membrane vesicles from gut mucosa-derived fusobacterium nucleatum. J Proteome 195:125–137. https://doi.org/10.1016/j.jprot.2018.12.029
17. Prados-Rosales RC, Aragoneses-Cazorla G, Estevez H et al (2019) Strategies for membrane protein analysis by mass spectrometry. Adv Exp Med Biol 1140:289–298. https://doi.org/10.1007/978-3-030-15950-4_16
18. Swart C, Martínez-Jaime S, Gorka M et al (2018) Hit-Gel: streamlining in-gel protein digestion for high-throughput proteomics experiments. Sci Rep 8:8582. https://doi.org/10.1038/s41598-018-26639-3
19. Pauwels J, Fijałkowska D, Eyckerman S, Gevaert K (2021) Mass spectrometry and the cellular surfaceome. Mass Spectrom Rev. https://doi.org/10.1002/mas.21690
20. Berridge G, Chalk R, D'Avanzo N et al (2011) High-performance liquid chromatography separation and intact mass analysis of detergent-solubilized integral membrane proteins. Anal Biochem 410:272–280. https://doi.org/10.1016/j.ab.2010.11.008
21. Macklin A, Khan S, Kislinger T (2020) Recent advances in mass spectrometry based clinical proteomics: applications to cancer research. Clin Proteomics 17:17. https://doi.org/10.1186/s12014-020-09283-w
22. Carlson ML, Young JW, Zhao Z et al (2018) The Peptidisc, a simple method for stabilizing membrane proteins in detergent-free solution. elife 7. https://doi.org/10.7554/eLife.34085
23. Carlson ML, Stacey RG, Young JW et al (2019) Profiling the Escherichia coli membrane protein interactome captured in Peptidisc libraries. elife 8. https://doi.org/10.7554/eLife.46615
24. Young JW, Wason IS, Zhao Z et al (2020) His-tagged peptidiscs enable affinity purification of the membrane proteome for downstream mass spectrometry analysis. J Proteome Res 19:2553–2562. https://doi.org/10.1021/acs.jproteome.0c00022
25. Young JW, Wason IS, Zhao Z et al (2022) Development of a method combining peptidiscs and proteomics to identify, stabilize, and purify a detergent-sensitive membrane protein assembly. J Proteome Res 21:1748–1758. https://doi.org/10.1021/acs.jproteome.2c00129
26. Young JW, Zhao Z, Wason IS, Duong van Hoa F (2022) A dual detergent strategy to capture a bacterial outer membrane proteome in peptidiscs for characterization by mass spectrometry and binding assays. J Proteome Res. https://doi.org/10.1021/acs.jproteome.2c00560
27. Saville JW, Troman L, Duong Van Hoa F (2019) PeptiQuick, a one-step incorporation of membrane proteins into biotinylated peptidiscs for streamlined protein binding assays. J Vis Exp. https://doi.org/10.3791/60661
28. Cox J, Mann M (2008) MaxQuant enables high peptide identification rates, individualized p.p.b.-range mass accuracies and proteome-wide protein quantification. Nat Biotechnol 26:1367–1372. https://doi.org/10.1038/nbt.1511

Chapter 11

Monitoring the Interaction of the Peptidoglycan with the Bacterial β-Barrel Assembly Machinery

Federico Corona and Waldemar Vollmer

Abstract

Gram-negative bacteria coordinate the biosynthesis of their different cell envelope components. Growth of the outer membrane (OM) requires the essential β-barrel assembly machine (BAM), which inserts OM proteins (OMPs) into the OM. The underlying peptidoglycan (PG) sacculus grows by the insertion of nascent glycan chains. We have previously identified interactions between BAM and PG in *E. coli* and showed that these interactions coordinate OM biogenesis with PG growth. BAM responds to the maturation state of the PG, and this mechanism activates preferentially BAM complexes at sites of active PG synthesis. Here we present protocols to purify soluble Bam proteins and full-length BamABCDE, isolate PG and soluble PG fragments, and study BAM-PG interactions with the isolated components. We also describe the protocol to detect interactions between Bam proteins and PG in cells.

Key words BAM complex, Peptidoglycan, Protein-peptidoglycan interactions, Peptidoglycan pull-down assay, Microscale thermophoresis, Chemical cross-linking

1 Introduction

Gram-negative bacteria have a multi-layered cell envelope with a lipopolysaccharide (LPS)-containing outer membrane (OM), which acts as permeability barrier, protecting the cell against many toxic molecules [1–4]. The cytoplasmic membrane and outer membrane (OM) encase the periplasm, an aqueous compartment that hosts the thin layer of peptidoglycan (PG) called sacculus [1–3, 5, 6]. The OM is tightly linked to the PG by some of the most abundant proteins of the cell, and an intact OM-PG superstructure is crucial to maintain the shape of the cell and counteract the cytoplasmic osmotic pressure (turgor) [5, 7, 8].

PG is made of linear glycan strands containing alternating *N*-acetylglucosamine (Glc*N*Ac) and *N*-acetylmuramic acid (Mur*N*Ac), connected by short peptides [5, 6, 9]. In the cell, glycan strands are arranged parallel to the cytoplasmic membrane and

Raffaele Ieva (ed.), *Transmembrane β-Barrel Proteins: Methods and Protocols*, Methods in Molecular Biology, vol. 2778, https://doi.org/10.1007/978-1-0716-3734-0_11,

(on average) perpendicular to the long axis of the cell [5, 10]. In *E. coli*, the glycan chains have an average length between 21 and 35 disaccharide units [11, 12]. The peptides are attached to Mur*N*Ac, and in *E. coli* about half of all peptides are connected with a peptide from an adjacent strand, resulting in the mesh-like architecture of the sacculus [5, 6].

Biogenesis of the PG layer in growing and dividing bacteria is a highly dynamic process, in which the synthesis, remodelling, and hydrolysis of PG are intimately connected with the cell cycle and ultimately controlled by the elongasome and divisome machineries. In *E. coli*, newly synthesised glycan strands carry pentapeptides with the sequence L-Ala-D-iGlu-*meso*-Dap-D-Ala-D-Ala (*meso*-Dap, *meso*-diaminopimelic acid). The inserted PG matures within minutes through removal of the terminal D-Ala residue (position 5) by DD-carboxypeptidase, which produce tetrapeptides, the most abundant peptides in mature PG [6, 11].

PG homeostasis is interconnected with OM dynamics, but the mechanisms underpinning PG-OM coordination have long remained unknown. In *E. coli*, the OM is tethered to PG by Lpp and other highly abundant proteins such as OmpA and Pal [13–16]. Moreover, several PG synthases and hydrolases require activation by OM-anchored lipoproteins. The lipoproteins LpoA and LpoB activate the major bi-functional PG synthases, PBP1A and PBP1B, respectively [17–20]. The OM-anchored NlpD and ActS activate septum-splitting PG amidases during cell division [21, 22]. The OM-anchored NlpI acts as an adaptor for PG DD-endopeptidases active during cell elongation [23] and targets one of these, MepS, to the protease Prc [24]. Severe OM assembly stress can affect the PG layer: The inhibition of LPS export to the OM causes defects in the PG layer, which the cell repairs through the action of LD-transpeptidases and other PG enzymes in order to survive [25].

While the above examples highlight the dependence of PG biogenesis on an intact OM, recent work revealed that the opposite is also true, that is, OM biogenesis is responsive to PG growth [26]. The OM is enriched in membrane-spanning β-barrel proteins (OMPs) [2, 4, 27] which are assembled into the OM by the essential β-Barrel Assembly Machinery (BAM) [2, 27–30]. In *E. coli*, BAM is a heteropentameric complex composed of the essential β-barrel BamA and four associated lipoproteins, BamB-E. The five Bam proteins engage in an extensive network of protein-protein interactions, in which the N-terminal periplasmic arm of BamA, containing five POlypeptide TRansport Associated (POTRA) domains, acts as a scaffold for the four lipoproteins [31–34]. LptD, the essential β-barrel component of the LPS export machinery Lpt, is a substrate of BAM [27, 35, 36]; hence, the insertase activity of BAM is a key orchestrator of OM growth, controlling both OMP and LPS insertion in the OM.

Extensive biochemical and structural work has revealed the mechanism of BAM-mediated OMP folding [37–39] and recent evidence points to a possible continuous, trans-envelope supercomplex that transports the OMP substrates from the cytoplasmic membrane to the BAM complex [40–42]. Together with the Kleanthous group, we recently showed that BAM is primarily active at sites near active PG synthesis in *E. coli* and other Gram-negative bacteria despite its uniform distribution across the cell envelope [26]. Interestingly, the maturation state of PG controls the spatial distribution of BAM activity in the cell envelope: BAM is able to bind mature, tetrapeptide-rich PG (abundant at the cell periphery and poles), and this BAM-PG interaction suppresses the foldase activity of BAM at these areas with low PG synthesis activity. By contrast, BAM does not interact with and is not inhibited by nascent, pentapeptide-rich PG present at PG growth sites. This elegant mechanism directs BAM-mediated OMP insertion to areas of active PG growth, such as midcell [26].

In this book chapter, we describe the biochemical tools we used to study BAM-PG interactions. We performed in vitro BAM-PG binding assays with the purified individual Bam proteins, or the full BAM complex, and purified PG from *E. coli*. Here, we report the detailed protocols for the purification of monomeric, soluble versions of *E. coli* Bam proteins, including three different fragments from the N-terminal periplasmic domain of BamA, encompassing POTRA domains 1–2 (BamA P1,2), 3–4 (BamA P3,4), and 4–5 (BamA P4,5), respectively, and soluble versions of BamB, BamC, and BamE. We also describe the purification of the full-length *E. coli* BamABCDE complex and the standard procedure for the isolation of PG sacculi from *E. coli* cells. We provide the protocols for two different in vitro assays to detect interactions between Bam proteins and purified PG. The PG pull-down assay [17, 43] relies on the ability of isolated PG sacculi to pellet upon centrifugation. Proteins of interest are incubated with sacculi to allow binding, then PG is pelleted by centrifugation, washed, and pelleted again. PG-binding proteins co-sediment with sacculi and are then released from PG after stirring the pellet in 2% SDS (Fig. 1a). All fractions (supernatant, wash fraction, and pellet fraction) are collected and analysed by SDS-PAGE to identify protein-containing fractions. PG-binding proteins are present in the final pellet fraction, while proteins that do not interact with PG remain in the supernatant or wash fractions (Fig. 1b).

The second assay is designed to study interactions between purified Bam proteins and soluble PG fragments. Here we report the purification procedure for the DD-endopeptidase MepM and its use as a tool to generate soluble PG fragments in the form PG glycan strands containing tetraptides (Tetra$_n$). We also describe microscale thermophoresis (MST) experiments to test for

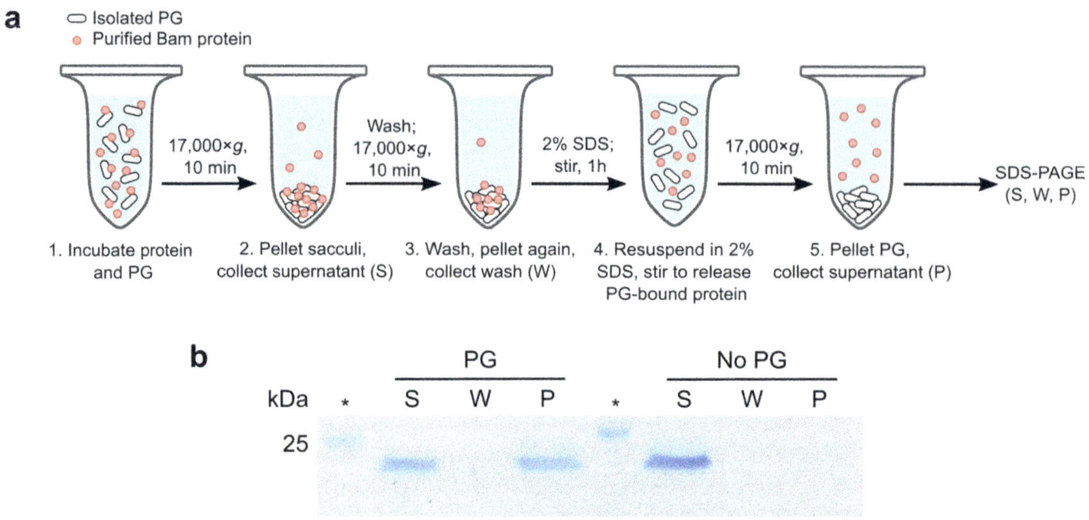

Fig. 1 Principle of the PG pull-down assay. (**a**) Experimental workflow to detect protein-PG interactions in vitro, based on the ability of purified PG sacculi to sediment by centrifugation. PG-binding proteins co-sediment with PG and can be found retained in the pellet fraction, P. (**b**) PG binding by BamA P3,4 analysed by PG pull-down, SDS-PAGE, and Coomassie Blue staining (S, supernatant; W, wash; P, pellet). Retention of protein in the pellet fraction, P, indicates binding to PG. Asterisks indicate molecular weight markers

interactions between $Tetra_n$ and purified Bam proteins (Fig. 2). This method can be adapted to other types of PG fragments by using PG hydrolases that cleave different bonds in PG.

We also describe a method to detect the interaction between Bam proteins and PG in *E. coli* cells through chemical cross-linking [44]. In this procedure, bacterial cells are treated with the soluble cross-linking agent 3,3′-dithiobis(sulfosuccinimidylpropionate) (DTSSP), which covalently cross-links PG-binding proteins to PG (Fig. 3a). PG-bound proteins can be detected upon PG isolation and treatment with a reducing agent such as β-mercaptoethanol, which cleaves the disulphide bridge in DTSSP, releasing cross-linked proteins from PG. Proteins are then detected by Western Blot with specific antibodies (Fig. 3b).

All protocols to study BAM-PG interactions described in this chapter can be adapted to investigate interactions between PG and other putative PG binding proteins.

2 Materials

All solutions are prepared with ultrapure water (18 MΩ cm at 25 °C) or ddH_2O (purified deionised water) and are filter-sterilised. Growth medium is autoclaved. Use analytical grade reagents for all solutions.

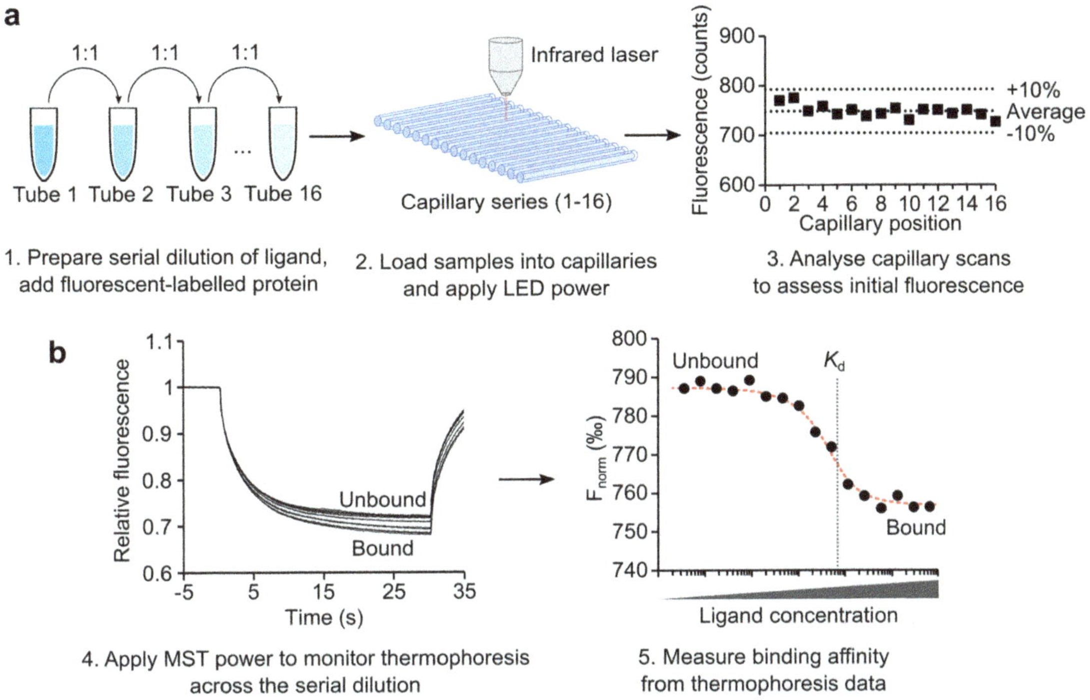

Fig. 2 Microscale thermophoresis experiment to analyse the interaction between a Bam protein and Tetra$_n$. (**a**) Preparation of the serial dilution of ligand and assessment of initial fluorescence. A stock solution of Tetra$_n$ is serially diluted two-fold into MST buffer and the fluorescent-labelled protein is added. Samples are loaded into capillaries and initial fluorescence is assessed across the series to ensure all samples have fluorescent emission within ±10% from the average of the whole series. (**b**) Microscale thermophoresis analysis of Bam-Tetra$_n$ interaction. MST power is applied to each sample across the serial dilution and thermophoresis is monitored. Normalised fluorescence in the steady-state for each sample is plotted against ligand concentration. Binding affinity is calculated from the binding curve

2.1 Protein Purification

1. *Escherichia coli* BL21(DE3) competent cells.
2. Antibiotic stock solutions: kanamycin (50 mg/mL) and ampicillin (100 mg/mL). Store at −20 °C.
3. LB medium: 10 g/L tryptone, 5 g/L yeast extract, and 10 g/L NaCl. Adjust pH to 7.2.
4. LB agar (1.5% w/v) plates supplemented with kanamycin (30 μg/mL) or ampicillin (100 μg/mL).
5. Static incubator at 37 °C.
6. Erlenmeyer flasks (100 mL, 2 L, and 5 L).
7. Orbital shaker.
8. Spectrophotometer.
9. IPTG (isopropyl 1-thio-β-D-galactopyranoside), 1 M stock solution. Store at −20 °C.
10. Ice bucket.

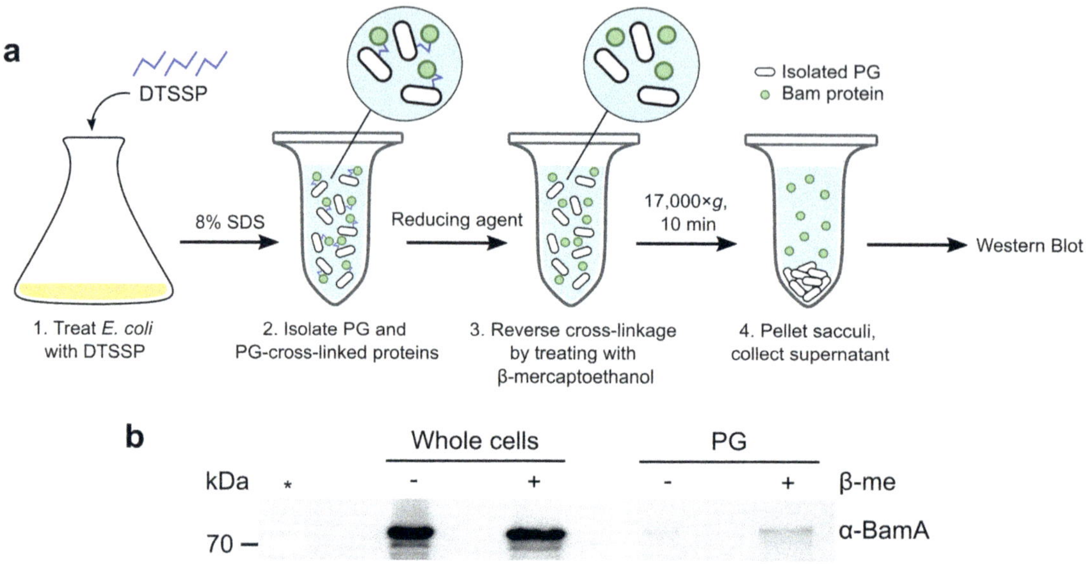

Fig. 3 Detection of BAM-PG interactions by in vivo cross-linking. (**a**) Experimental workflow to detect protein-PG interactions in *E. coli* cells. DTSSP cross-links PG binding proteins to the sacculus. PG is then isolated and treated with a reducing agent, which disrupts the disulphide bridge in DTSSP, releasing PG-bound proteins, which are detected by Western Blot analysis. (**b**) PG binding by BamA detected by in vivo cross-linking and Western Blot analysis. Whole cell lysates are also shown as controls. The asterisk indicates the molecular weight marker

11. Beckman-Coulter centrifuge compatible with JLA-8.1000 rotor.
12. 1 L centrifuge bottles.
13. DNase I from bovine pancreas, freeze-dried.
14. Protease inhibitor cocktail, 5 mg/mL in DMSO, or EDTA-free protease inhibitor tablets.
15. PMSF (phenylmethanesulfonylfluoride), 100 mM in ethanol.
16. Emulsiflex cell homogeniser or probe sonicator.
17. Refrigerated centrifuge for 50 mL tubes.
18. Falcon tubes, 50 mL.
19. Beckman-Coulter centrifuge compatible with SW 32 Ti rotor.
20. Syringe filters, 0.45 μm.
21. Plasmids pQE70-*bamA* P1,2 [45], pET26b-*bamA* P3,4, pET26b-*bamA* P4,5, pET22b-*bamB*, pET22b-*bamE* [26], or pET16b-*bamC* [46].
22. sBam extraction buffer: 50 mM sodium phosphate pH 7.5, 300 mM NaCl, and 10 mM imidazole.
23. Plasmid pJH114 [47].
24. BamABCDE extraction buffer: 20 mM Tris/HCl, pH 8.0.

25. BamABCDE solubilisation buffer: 50 mM Tris/HCl pH 8.0, 150 mM NaCl, 1% DDM.
26. Plasmid pMN86 [48].
27. MepM extraction buffer: 25 mM Tris/HCl pH 7.0, 300 mM NaCl, 10 mM $MgCl_2$, 20 mM imidazole, and 10% glycerol.

2.2 Ni-NTA Affinity Chromatography

1. 5 mL His-Trap HP (Cytiva) or Ni-NTA agarose beads (Qiagen).
2. Tube roller.
3. Disposable gravity flow polypropylene columns or ÄKTA FPLC (Fast Protein Liquid Chromatography) system (Cytiva).
4. Heating block.
5. Sodium dodecyl sulphate-polyacrylamide gel electrophoresis (SDS-PAGE) apparatus and buffers.
6. Polyacrylamide gels (15%).
7. 2× SDS-PAGE loading buffer: 100 mM Tris/HCl pH 6.8, 4% (w/v) SDS, 20% (v/v) glycerol, 0.2% (w/v) bromophenol blue, and 10% (v/v) β-mercaptoethanol.
8. Coomassie Brilliant Blue staining and destaining solution.
9. sBam elution buffer: 50 mM sodium phosphate pH 7.5, 300 mM NaCl, and 500 mM imidazole.
10. BamABCDE wash buffer: 50 mM Tris/HCl pH 8.0, 150 mM NaCl, 0.05% (w/v) n-Dodecyl-β-D-Maltoside (DDM), and 50 mM imidazole.
11. BamABCDE elution buffer: 50 mM Tris/HCl pH 8.0, 150 mM NaCl, 0.05% (w/v) DDM, 500 mM imidazole.
12. MepM elution buffer: 25 mM Tris/HCl pH 7.0, 300 mM NaCl, 10 mM $MgCl_2$, 400 mM imidazole, and 10% (v/v) glycerol.

2.3 Size-Exclusion Chromatography

1. Spectra/Por dialysis membrane (Carl Roth), MWCO (molecular weight cut-off) 12–14 kDa.
2. Amicon Ultra or Vivaspin centrifugal ultrafiltration units, MWCO 10, 30 or 100 kDa.
3. ÄKTA Fast protein liquid chromatography (FPLC) system.
4. Superdex 200 (10/300) or Superdex 75 (16/600) size-exclusion chromatography columns (Cytiva).
5. Heating block.
6. SDS-PAGE apparatus and buffers.
7. Polyacrylamide gels (15%).
8. 2× SDS-PAGE loading buffer (*see* Subheading 2.2, **item 7**).
9. Coomassie Blue.

10. Glycerol, analytical grade, >99%.
11. Bicinchoninic acid assay (BCA) kit or spectrophotometer.
12. sBam SEC buffer: 50 mM sodium phosphate pH 7.5 and 300 mM NaCl. Filter and degas.
13. BamABCDE SEC buffer: 50 mM Tris/HCl pH 8.0, 150 mM NaCl, and 0.05% (w/v) DDM. Filter and degas.
14. MepM storage buffer: 25 mM Tris/HCl pH 7.0, 300 mM NaCl, 10 mM $MgCl_2$, and 10% (v/v) glycerol. Filter and degas.

2.4 Preparation of PG Sacculi and Soluble PG Fragments

1. LB medium: 10 g/L tryptone, 5 g/L yeast extract, and 10 g/L NaCl. Adjust pH to 7.2.
2. LB agar (1.5% w/v) plates.
3. Static incubator at 37 °C.
4. Sterile plastic culture tubes.
5. Erlenmayer flasks, 250 mL and 2 L.
6. Orbital shaker.
7. Spectrophotometer.
8. Ice bucket.
9. Beckman-Coulter centrifuge compatible with JLA-8.1000 rotor.
10. 1 L centrifuge bottles.
11. Sterile ultrapure water.
12. Glass pipettes, 10 mL.
13. Glass Pasteur pipettes and rubber bulbs.
14. SDS (sodium dodecyl sulphate) stock solutions: 8% and 4% (w/v) HPLC-grade SDS.
15. Magnetic hotplate stirrer.
16. Magnetic stir bars.
17. Aluminium foil.
18. Beckman-Coulter centrifuge compatible with SW 32 Ti rotor.
19. Glass stir rods, 5 mm.
20. Benchtop vortex-mixer.
21. 10 mM Tris/HCl, 10 mM NaCl, pH 7.0.
22. α-amylase stock solution: 10 mM Tris/HCl, pH 7.0, 10 mM NaCl, and 10 mg/mL α-amylase.
23. Pronase E stock solution: 10 mM Tris/HCl, pH 7.0, 10 mM NaCl, and 10 mg/mL pronase E.
24. Heating block.
25. 0.02% (w/v) NaN_3 stock solution.

26. Microcentrifuge tubes, 2 mL.
27. Sterile ultrapure water.
28. 0.7 M sodium phosphate, pH 7.0.
29. 0.5% (w/v) methylene blue.
30. Chloroform.
31. Benchtop vortex-mixer.

2.5 Generation of Tetra$_n$ Chains

1. Microcentrifuge tubes, 2 mL.
2. Purified MepM DD-endopeptidase.
3. Purified PG sacculi from *E. coli* BW25113Δ6LDT [49].
4. Benchtop vortex-mixer.
5. 4× MepM activity buffer: 100 mM Tris/HCl pH 7.5, 600 mM NaCl, and 0.2% Triton X-100.
6. Eppendorf Thermomixer.
7. Heating block.
8. Benchtop microcentrifuge.
9. Spectra/Por dialysis membrane (Carl Roth), MWCO 3–5 kDa.
10. Tetra$_n$ dialysis buffer: 50 mM sodium phosphate and 150 mM NaCl, pH 7.0.

2.6 PG Pull-Down Assay

1. Microcentrifuge tubes, 2 mL.
2. Benchtop microcentrifuge.
3. 10× PG binding buffer: 100 mM Tris/maleate, 500 mM NaCl, 100 mM $MgCl_2$, pH 7.5.
4. Triton X-100, 10% (w/v) stock solution.
5. 2% (w/v) SDS stock solution.
6. Magnetic stir bars, 3 × 6 mm and larger.
7. Floating microcentrifuge tube rack.
8. Magnetic stir plate.
9. Heating block.
10. 2× SDS-PAGE loading buffer (*see* Subheading 2.2, **item** 7).
11. SDS-PAGE apparatus and buffers.
12. Polyacrylamide gels (15%).
13. Coomassie Blue.

2.7 Microscale Thermophoresis Experiments

1. Red-NHS labelling kit, glass capillaries and Monolith NT.115 (NanoTemper Technologies).
2. Benchtop microcentrifuge.
3. MST buffer: 50 mM sodium phosphate, pH 7.0, 150 mM NaCl, 10% glycerol, and 0.05% (w/v) Triton X-100.

4. Spectrophotometer.
5. PCR tubes, 0.2 mL.
6. Tetra$_n$ chains stock or digestion control.
7. Ice bucket.
8. Aluminium foil.
9. MO.Control and MO.Affinity Analysis software (NanoTemper Technologies).

2.8 SDS-Denaturation Test for Microscale Thermophoresis

1. PCR tubes, 0.2 mL.
2. MST buffer: 50 mM sodium phosphate, pH 7.0, 150 mM NaCl, 10% (v/v) glycerol, 0.05% (w/v) Triton X-100.
3. Glass capillaries and Monolith NT.115 (NanoTemper Technologies).
4. Benchtop microcentrifuge.
5. 2× SD-mix: 40 mM DTT, 4% (w/v) SDS.
6. Thermal cycler.

2.9 In Vivo Cross-linking of Bam Proteins to PG

1. LB medium: 10 g/L tryptone, 5 g/L yeast extract, and 10 g/L NaCl. Adjust pH to 7.2.
2. LB agar (1.5% w/v) plates.
3. Static incubator at 37 °C.
4. Sterile plastic culture tubes.
5. Erlenmayer flasks, 250 mL.
6. Orbital shaker.
7. Spectrophotometer.
8. Benchtop centrifuge for 50 mL Falcon tubes.
9. Falcon tubes, 50 mL.
10. Phosphate buffered saline (PBS).
11. Cross-linker stock solution: 5 mM sodium citrate, pH 5.0, and 10 mM DTSSP (3,3′-dithiobis(sulfosuccinimidylpropionate)).
12. 1 M Tris/HCl, pH 8.0.
13. 2× SDS-PAGE loading buffer (*see* Subheading 2.2, **item** 7).
14. β-Mercaptoethanol.
15. 2% and 8% (w/v) SDS stock solutions.
16. Glass Pasteur pipettes and rubber bulbs.
17. Magnetic hotplate stirrer.
18. Magnetic stir bars.
19. Aluminium foil.
20. Beckman-Coulter centrifuge compatible with SW 32 Ti rotor.

21. SDS-PAGE apparatus and buffers.
22. Polyacrylamide gels (15%).
23. Western Blot transfer apparatus and buffers.
24. Nitrocellulose or PVDF membranes.
25. Anti-Bam antibodies and HRP (horseradish peroxidase)-conjugated secondary antibodies.
26. Chemiluminescent HRP substrate reagents and imaging system.

3 Methods

3.1 Protein Purification

3.1.1 Purification of Bam Proteins

1. Transform plasmid(s) of interest in *E. coli* BL21(DE3) (*see* **Note 1**). Plate on LB agar containing 100 μg/mL ampicillin (BamB, BamC, BamE, and BamA P1,2) or 30 μg/mL kanamycin (BamA P3,4 and P4,5) and incubate overnight at 37 °C.
2. Pick single transformant colonies and inoculate in LB supplemented with the correct antibiotic. Grow overnight at 37 °C with orbital shaking (175 rpm).
3. Dilute overnight cultures 1:100 in 2 L of LB supplemented with the correct antibiotic. Grow aerobically at 37 °C with orbital shaking (175 rpm) up to OD_{600} ~ 0.4.
4. Shift cultures to 18 °C. After 30 min, induce protein overproduction by supplementing 1 mM IPTG. Incubate at 18 °C with orbital shaking (175 rpm) overnight.
5. Harvest cells by centrifugation at 6,200 × *g*, 4 °C for 15 min (*see* **Note 2**).
6. Resuspend cell pellet in 50 mL of cold sBam extraction buffer, supplemented with EDTA-free protease inhibitor.
7. Lyse cells in an Emulsiflex C3 cell disruptor.
8. Centrifuge at 75,000 × *g*, 4 °C for 45 min to pellet insoluble material.
9. Filter the supernatant through a 0.45 μm syringe filter.
10. Apply the supernatant to a 5 mL HisTrap HP column, pre-equilibrated in sBam extraction buffer, at 1 ml/min to immobilise His-tagged Bam proteins.
11. Wash the column with 5 column volumes of sBam extraction buffer at 1 mL/min.
12. Elute His-tagged Bam proteins at 1 mL/min with 5 column volumes of sBam elution buffer, collecting 2 mL-fractions.

13. Mix 10 μL of each collected elution fraction with 10 μL of 2× SDS-PAGE loading buffer and boil at 100 °C for 10 min. Load 10 μL of each sample on a 15% polyacrylamide gel and check purity and yield of eluted fractions by SDS-PAGE and Coomassie Blue staining.
14. Pool fractions of interest together and concentrate the sample to <5 mL in an Amicon Ultra/Vivaspin ultrafiltration unit, choosing an appropriate MWCO. Apply the sample to a Superdex 75 (16/600) column for size exclusion chromatography, pre-equilibrated in filtered and degassed sBam storage buffer at 1 mL/min. Collect 2 mL-fractions in correspondence of the major elution peak and analyse by SDS-PAGE as described in **step 16**.
15. Pool fractions with highest purity and yield together. Assess protein concentration spectrophotometrically and store at −80 °C in 100 μL-aliquots.

3.1.2 Purification of BamABCDE

1. Transform plasmid pJH114 (54) (*see* **Note 3**) in *E. coli* BL21 (DE3) competent cells. Plate on LB agar containing 100 μg/mL ampicillin and incubate overnight at 37 °C.
2. Pick one transformant colony and inoculate in 50 mL of LB supplemented with 100 μg/mL ampicillin. Grow overnight at 37 °C with orbital shaking (175 rpm).
3. Dilute overnight culture 1:100 in 2 L of LB supplemented with 100 μg/mL ampicillin. Grow aerobically at 37 °C with orbital shaking (175 rpm) up to OD_{600} ~ 0.5–0.6.
4. Induce protein overproduction by supplementing the culture with 0.4 mM IPTG. Incubate at 37 °C with orbital shaking (175 rpm) for 90 min, then arrest cell growth by incubating flasks on ice for 10 min.
5. Harvest cells by centrifugation at 6,200 × *g*, 4 °C for 15 min (*see* **Note 2**).
6. Resuspend cell pellet in 50 mL of cold BamABCDE extraction buffer, supplemented with EDTA-free protease inhibitor cocktail, 200 μM PMSF and 10 μg/mL DNase I from bovine pancreas.
7. Transfer cell suspension to a clean glass beaker. Place the beaker on ice. Lyse cells by sonication (*see* **Note 4**).
8. Transfer cell lysate to 50 mL Falcon tubes and spin down cell debris at 3,220 × *g*, 4 °C for 15 min.
9. Transfer supernatant to ultracentrifuge tubes (*see* **Note 5**). Carefully balance the tubes using cold BamABCDE extraction buffer. Harvest cell membranes by ultracentrifugation at 130,000 × *g*, 4 °C for 1 h.

10. Solubilise cell membranes in 20 mL of cold BamABCDE solubilisation buffer by carefully pipetting up and down. Incubate on ice for 1 h.
11. Transfer samples to ultracentrifuge tubes, carefully balanced using cold BamABCDE extraction buffer. Remove insoluble material by ultracentrifugation at 130,000 × *g*, 4 °C for 1 h.
12. Apply supernatant to 5 mL (bed volume) of Ni-NTA beads, pre-equilibrated in BamABCDE extraction buffer. Incubate on a tube roller overnight at 4 °C.
13. Apply beads to a gravity flow column at 4 °C and collect the flow-through.
14. Wash beads with 5 bed volumes of BamABCDE wash buffer and collect the wash fraction.
15. Elute BamABCDE in 2 bed volumes of BamABCDE elution buffer, collecting 1 mL fractions.
16. Mix 10 μL of each collected elution fraction with 10 μL of 2× SDS-PAGE loading buffer and boil at 100 °C for 10 min. Load 10 μL of each sample on a 15% polyacrylamide gel and check purity and yield of eluted fractions by SDS-PAGE and Coomassie Blue staining.
17. Pool fractions with the best purity and yield together and concentrate the sample to <500 μL in an Amicon Ultra/Vivaspin ultrafiltration unit (100 kDa MWCO). Apply the sample to a Superdex 200 (10/300) column for size exclusion chromatography, pre-equilibrated in filtered and degassed BamABCDE SEC buffer (*see* **Note 6**). Collect 0.5 mL elution fractions and analyse by SDS-PAGE as described in **step 16**.
18. Measure protein concentration in aliquots with highest purity and yield using BCA Protein Assay Kit according to the manufacturer's protocol.
19. Store purified BamABCDE in 10% glycerol at −80 °C in 100 μL-aliquots.

3.1.3 Purification of MepM DD-endopeptidase to Generate Soluble PG Fragments for Bam-PG Interaction Studies

1. Transform plasmid pMN86 (50) (*see* **Note 7**) into *E. coli* BL21 (DE3) competent cells. Plate on LB agar containing 100 μg/mL ampicillin and incubate overnight at 37 °C.
2. Pick one transformant colony and inoculate in 50 mL of LB supplemented with 100 μg/mL ampicillin. Grow overnight at 37 °C with orbital shaking (175 rpm).
3. Dilute overnight culture 1:100 in 3 L of LB supplemented with 100 μg/mL ampicillin. Grow aerobically at 37 °C with orbital shaking (175 rpm) up to OD_{600} ~ 0.6.

4. Shift culture to 25 °C and induce protein overproduction by supplementing 50 μM IPTG. Incubate at 25 °C with orbital shaking (175 rpm) for 2 h, then arrest cell growth by incubating flasks on ice for 10 min.
5. Harvest cells by centrifugation at 6,200 × *g*, 4 °C for 15 min (*see* **Note 2**).
6. Resuspend cell pellet in 80 mL of cold MepM extraction buffer, supplemented with EDTA-free protease inhibitor cocktail, 200 μM PMSF and 10 μg/mL DNase I from bovine pancreas.
7. Transfer cell suspension to a clean glass beaker. Place the beaker on ice. Lyse cells by sonication (*see* **Note 4**).
8. Transfer cell lysate to 50 mL Falcon tubes and spin down cell debris at 3,220 × *g*, 4 °C for 15 min.
9. Transfer supernatant to ultracentrifuge tubes (*see* **Note 5**). Carefully balance the tubes using cold MepM extraction buffer. Pellet cell membranes by ultracentrifugation at 130,000 × *g*, 4 °C for 1 h.
10. Collect the supernatant and apply to a 5 mL HisTrap HP column, pre-equilibrated in MepM extraction buffer, at 1 mL/min to immobilise His-tagged MepM.
11. Wash the column with 5 column volumes of MepM extraction buffer at 1 mL/min.
12. Elute His-tagged MepM at 1 mL/min with 5 column volumes of MepM elution buffer, collecting 4 mL-fractions.
13. Mix 10 μL of each collected elution fraction with 10 μL of 2× SDS-PAGE loading buffer and boil at 100 °C for 10 min. Load 10 μL of each sample on a 15% polyacrylamide gel and check purity and yield of eluted fractions by SDS-PAGE and Coomassie Blue staining.
14. Pool fractions of interest together and dialyse overnight in a dialysis membrane of appropriate MWCO (12–14 kDa is recommended) against 3 L of MepM storage buffer. Perform dialysis at 4 °C overnight with gentle stirring on a magnetic stir plate.
15. Concentrate the sample to <5 mL in an Amicon Ultra/Vivaspin ultrafiltration unit (30 kDa MWCO). Apply the sample to a Superdex 75 (16/600) column for size exclusion chromatography, pre-equilibrated in filtered and degassed MepM storage buffer at 1 mL/min. Collect 2 mL fractions in correspondence of the major elution peak and analyse by SDS-PAGE as described in **step 16**.
16. Pool fractions with highest purity and yield together. Assess protein concentration spectrophotometrically and store at −80 °C in 100 μL-aliquots.

3.2 Preparation of PG Sacculi and Soluble PG Fragments for Protein-PG Interaction Assays

3.2.1 PG Isolation from *E. coli*

1. Streak *E. coli* strain of interest on LB agar. Incubate the plate overnight at 37 °C.
2. Inoculate 5 mL of LB with one single colony from the plate. Grow overnight with shaking at 37 °C.
3. Dilute overnight culture 1:100 into 4 L of fresh, pre-warmed LB medium, split across four 2 L Erlenmayer flasks (*see* **Note 8**). Grow aerobically at 37 °C with orbital shaking (175 rpm) up to OD_{600} ~ 0.5–0.6.
4. Incubate flasks on an ice-water mixture for 10 min to arrest cell growth.
5. Harvest cells by centrifugation at 6,200 × *g*, 4 °C for 15 min (*see* **Note 2**).
6. Resuspend cell pellets in 40 mL of ice-cold, sterile Milli-Q water (*see* **Note 9**).
7. Add 40 mL of HPLC-grade 8% SDS and a magnetic stirrer bar to a 250 mL Erlenmayer flask. Boil SDS solution on a magnetic hotplate stirrer, keeping the flask closed with aluminium foil to prevent evaporation (*see* **Note 10**).
8. Using a glass Pasteur pipette and a rubber bulb, add the cell suspension dropwise to the boiling SDS solution while stirring vigorously (*see* **Note 11**). After the whole cell suspension has been added to the boiling SDS, keep boiling the sample while stirring for 30 more minutes (*see* **Note 12**). Cool down samples and store at room temperature (*see* **Note 13**).
9. Transfer the sample to ultracentrifuge tubes (*see* **Note 5**). Carefully balance the tubes using SDS-free water at room temperature.
10. Ultracentrifuge samples to pellet PG sacculi at 130,000 × *g*, 28 °C for 1 h.
11. Discard supernatant. Carefully resuspend pellet in 1 mL of SDS-free water pipetting up and down (*see* **Note 14**). Add 5 mL of water, then wash the sample into the ultracentrifuge tube with a 5 mm glass stir rod while vortexing vigorously for 1 min. Add 5 mL more and repeat until maximum volume is reached.
12. Repeat **steps 9** and **10** for at least 3 times (*see* **Note 15**). Monitor the presence of residual SDS in the sample by performing a Hayashi's test (*see* Subheading 3.2.2). If SDS is still present, repeat **steps 9** and **10** until samples are SDS-free.
13. Resuspend pellets in total 9 mL of 10 mM Tris/HCl, 10 mM NaCl, pH 7.0. Add 1.5 mg of α-amylase (150 μL of a 10 mg/mL stock solution) and incubate at 37 °C for 2 h with shaking to degrade high molecular weight glycogen trapped into sacculi.

14. Add 2 mg of pre-activated pronase E (200 μL of a 10 mg/mL stock solution, *see* **Note 16**) and incubate at 60 °C for 1 h to remove covalently bound Lpp lipoprotein from PG.
15. Mix samples 1:1 (v/v) with 4% HPLC-grade SDS and boil for 15 min at 100 °C to inactivate enzymatic reactions. Cool down at room temperature.
16. Repeat **steps 8–11** until the sample is SDS-free (*see* **Note 17**).
17. Resuspend pellet in 1 mL of 0.02% NaN_3 (*see* **Note 18**) and store at 4 °C.

3.2.2 Hayashi's Test for the Detection of SDS During PG Isolation

1. After pelleting sacculi by ultracentrifugation, transfer 335 μL of supernatant into a 2 mL microcentrifuge tube. Transfer 335 μL of SDS-free water to a parallel control tube.
2. Add 170 μL of 0.7 M sodium phosphate, pH 7.0, and 7 μL of 0.5% methylene blue. Mix by pipetting up and down.
3. Add 1 mL of chloroform and vortex vigorously for 20–30 s, then let phases separate.
4. Assess the presence of SDS in the sample visually: The chloroform phase will be blue/cyan in the presence of SDS, and transparent/pink in SDS-free samples.

3.2.3 Generation of $Tetra_n$ Chains as Ligands for the Bam Proteins in Microscale Thermophoresis Experiments

1. Isolate PG sacculi from *E. coli* BW25113Δ6LDT (*see* **Note 19**) as described in Subheading 3.1.
2. Digest sacculi with 3 μM purified MepM DD-endopeptidase (*see* Subheading 3.1.3) in MepM activity buffer. Assemble reactions in 1 mL of volume into 2 mL microcentrifuge tubes as follows: 250 μL of 4× MepM activity buffer, the required volume of purified MepM to obtain 3 μM final concentration, and sacculi suspension up to 1 mL (*see* **Note 21**). Prepare in parallel a mock digestion mixture containing equal volumes of activity buffer and MepM but no PG sacculi. Incubate mixtures at 37 °C for ~18 h with shaking at 900 rpm on an Eppendorf Thermomixer.
3. Inactivate enzymatic reactions by boiling digestion mixtures at 100 °C for 10 min.
4. Collect soluble $Tetra_n$ and mock digestions by centrifugation at 17,000 × *g*, room temperature for 15 min. Transfer samples to a clean microcentrifuge tube.
5. Dialyse $Tetra_n$ and mock digestions in 3–5 kDa molecular weight cut-off dialysis membranes against 2 L of $Tetra_n$ dialysis buffer (*see* **Note 20**). Perform dialysis at room temperature with gentle stirring for 24 h.
6. Collect dialysed $Tetra_n$ and mock digestions into clean microcentrifuge tubes. Store at −20 °C.

3.3 BAM-PG Interaction Assays

3.3.1 PG Pull-Down Assay

1. Transfer 100 μL of purified sacculi suspension at 10 mg/mL (~1 mg of purified PG) to a 2 mL microcentrifuge tube (*see* **Note 21**). Pellet sacculi at 17,000 × *g*, room temperature for 10 min and remove supernatant. Prepare a parallel mock sample containing no sacculi to be used as a negative control.
2. Substitute the volume of the removed supernatant with 10 μL of 10× PG binding buffer, water and the required volume of purified protein (in the case of Bam proteins, the volumes that correspond to 25 μg of purified proteins, *see* **step 3** and **Note 22**). The total volume of the sample (including the PG pellet) has to be 100 μL (*see* **Note 23**). When performing pull-down experiments with a membrane spanning complex, such as BamABCDE, supplement 10× PG binding buffer with 0.5% Triton X-100 and use it throughout the whole procedure.
3. Add 25 μg of purified Bam protein of interest (*see* **Note 24**) to sacculi suspensions or negative control sample and incubate on ice for 30 min.
4. Pellet samples at 17,000 × *g*, room temperature for 10 min, and collect supernatant (S) into a clean 1.5 ml microcentrifuge tube. Keep collected supernatant on ice.
5. Wash pellets by carefully resuspending in 200 μL of 1× PG binding buffer (*see* **Note 25**).
6. Pellet samples at 17,000 × *g*, room temperature for 10 min, and collect supernatant (wash fraction, W) into a clean 1.5 mL microcentrifuge tube. Keep collected wash fraction on ice.
7. Carefully resuspend pellets in 100 μL of 2% SDS, avoiding foam formation. Add a 3 × 6 mm magnetic stir bar into each tube (*see* **Note 26**) and place tubes on a floating microcentrifuge tube rack. Place the rack on a magnetic stir plate and stir samples vigorously for 1 h.
8. Remove the magnetic stir bars from the samples by sliding a larger stir bar from outside the tube along the tube wall.
9. Pellet samples at 17,000 × *g*, room temperature for 10 min and collect supernatant (pellet fraction, P) into a clean 1.5 mL microcentrifuge tube. Keep this fraction at room temperature.
10. Mix all collected fractions (S, W, P) 1:1 (v/v) with 2× SDS-PAGE loading buffer containing fresh 10% β-mercaptoethanol. Boil at 100 °C for 10 min. Briefly spin down tubes after boiling to remove condensation from the lid.
11. Analyse fractions collected throughout the experiment by SDS-PAGE (Fig. 1b). Load 20 μL of each sample on a 12–15% polyacrylamide gel in this order: S, W, P. After SDS-PAGE, stain the gel in Coomassie Brilliant Blue, then destain to remove the background. Binding of Bam proteins to PG will be indicated by the detection of the protein in the P fraction in the presence of sacculi.

3.3.2 Microscale Thermophoresis Experiments to Detect the Interaction Between Purified Bam Proteins and Tetra$_n$ PG Chains

1. Label purified Bam protein of interest with Red-NHS (NanoTemper Technologies) in MST buffer according to the manufacturer's protocol (*see* **Note 24**). Measure protein concentration and estimate efficiency of labelling as described by the manufacturer. Store labelled protein in 100 μL-aliquots at −80 °C (*see* **Note 27**).
2. Dilute the labelled protein in MST buffer to 200, 100, 50, and 25 nM. Perform initial fluorescence test runs (capillary scans) on a NanoTemper Monolith NT.115 in both standard- and premium-coated MST capillaries by testing fluorescent emission for each concentration at different LED powers (*see* **Note 28**). A fluorescent emission between 250 and 2500 counts is desired for the sample containing 50 nM protein and the shape of the fluorescence peak must be sharp and symmetrical (*see* **Note 29**).
3. To test interactions between the labelled protein and Tetra$_n$, prepare a serial dilution of dialysed Tetra$_n$ or mock MepM digestion (negative control) in MST buffer up to 16 samples in 0.2 mL PCR tubes on ice as follows: place 20 μL of dialysed Tetra$_n$ stock or mock digestion in tube n. 1 (*see* **Note 30**), and aliquot 10 μL of MST buffer in tubes 2–16. Transfer 10 μL from tube n. 1 to tube n. 2 and mix thoroughly by pipetting up and down. Transfer then 10 μL from tube n. 2 to tube n. 3, mix thoroughly, and proceed until the end of the series (Fig. 2a). Discard 10 μL from tube n. 16 so that it contains only 10 μL.
4. Dilute the labelled Bam protein of interest to 100 nM in MST buffer in a microcentrifuge tube, wrapped in aluminium foil to protect it from the light. Keep on ice.
5. Add 10 μL of labelled protein solution to each of the 16 tubes containing the serial dilution of Tetra$_n$. Cover the samples with aluminium foil to protect them from the light and incubate for 5 min on ice, then for 5 min at room temperature.
6. Place a glass capillary into each of the tube. Let each capillary fill with sample (*see* **Note 31**).
7. Place capillaries on the capillary tray of the Monolith NT.115 and start capillary scan analysis (Fig. 2a), setting the LED power to the same value used in **step 2** (*see* **Note 32**).
8. Perform measurement setting MST power to 50% (Fig. 2b; *see* **Note 33**). An amplitude (i.e. the difference between bound and unbound state) above five response units and a baseline noise three times lower than the amplitude are desired (*see* **Note 34**).
9. Repeat experiments in triplicate. If concentration-dependent changes in initial fluorescence were observed in **step 7** (*see* **Note 32**), refer to Subheading 3.3.3 for validation.

3.3.3 SDS-Denaturation Test to Validate Ligand-Dependent Changes in Initial Fluorescence During Microscale Thermophoresis Experiments

1. Prepare serial dilutions of dialysed Tetra$_n$ and assemble reactions in the presence of labelled Bam proteins of interest as described in Subheading 3.3.2, **steps 3–5**.
2. Keep sample 1–3 (unbound fraction) and sample 14–16 (bound fraction). Discard other samples.
3. Place MST capillaries into the tubes and analyse initial fluorescence as described in Subheading 3.3.2, **steps 6–7**, to assess difference in initial fluorescence between bound and unbound state.
4. Centrifuge samples at 15,000 × *g*, room temperature for 15 min, then transfer 10 μL of each sample into clean 0.2 mL PCR tubes.
5. Mix samples 1:1 (v/v) with 2× SD-mix and incubate at 95 °C for 10 min on a thermal cycler.
6. Place MST capillaries into the tubes and analyse initial fluorescence again. Compare differences in initial fluorescence between bound and unbound state measured in **step 3** with differences between bound and unbound state after sample denaturation. If no differences are observed after boiling the samples, the initial changes are confirmed to be ligand-dependent.
7. Repeat experiments in triplicate.

3.3.4 In Vivo Cross-linking of Bam Proteins to PG

1. Streak *E. coli* strain of interest on LB agar. Incubate the plate overnight at 37 °C.
2. Inoculate 5 mL of LB with one single colony from the plate. Grow overnight with shaking at 37 °C.
3. Dilute overnight culture 1:100 into 50 mL of fresh, pre-warmed LB medium into a 250 mL Erlenmayer flask. Grow aerobically at 37 °C with orbital shaking (175 rpm) up to OD_{600} ~ 0.5.
4. Pellet cells at 3,220 × *g*, room temperature for 10 min (*see* **Note 35**). Wash pellet by resuspending in 50 mL of PBS. Repeat twice.
5. Adjust OD_{600} to 2.0 in PBS and add freshly prepared 10 mM DTSSP stock to a final concentration of 0.5 mM. Incubate at room temperature for 10 min.
6. Add 1 M Tris/HCl, pH 8.0 to a final concentration of 50 mM to quench the cross-linking reaction. Incubate at room temperature for 15 min.
7. Take two 300 μL-control samples for Western Blot analysis of whole cell lysates. Spin down cell suspensions at 15,000 × *g*, room temperature for 10 min. Resuspend samples in 100 μL of 1× SDS-PAGE loading buffer, supplemented with 10%

β-mercaptoethanol for one of the sample and no β-mercaptoethanol for the other sample. Boil at 100 °C for 10 min.

8. Add the rest of the cell suspension to an equal volume of boiling 8% SDS and boil with vigorous stirring to isolate PG sacculi as described in Subheading 3.2.1, **steps 6** and 7.
9. Ultracentrifuge samples to pellet PG sacculi at 130,000 × *g*, 28 °C for 1 h.
10. Discard supernatant. Wash pellet in 25 mL of 2% SDS and ultracentrifuge again. Repeat once.
11. Resuspend pellet in 100 μL of 2% SDS and split sample in half. Mix one-half 1:1 (v/v) with 2× SDS-PAGE loading buffer supplemented with 10% β-mercaptoethanol and the other half with 2× SDS-PAGE loading buffer containing no β-mercaptoethanol. Boil samples for 10 min at 100 °C.
12. To detect PG-bound Bam proteins, briefly centrifuge samples at 15,000 × *g*, room temperature for 10 min to pellet insoluble sacculi. Load 10 μL of supernatant on 15% polyacrylamide gels and perform SDS-PAGE, including the whole cell lysate controls from **step** 7. Transfer proteins to nitrocellulose or PVDF and detect proteins of interest with specific antibodies. PG-bound Bam proteins will be detected in the isolated PG fraction treated with β-mercaptoethanol, which reverses cross-linking by reducing the intramolecular disulphide bridge of DTSSP (Fig. 3b).

4 Notes

1. BamA P1,2 (residues 21–174) is expressed from a pQE70 vector with a C-terminal 4 × His-tag [45]. BamA P3,4 (residues 175–345) and BamA P4,5 (residues 266–422) are expressed from pET26b(+) with a C-terminal 6 × His-tag. BamB and BamE, lacking their periplasmic export sequences and with the cysteine residue at the N-terminus of the mature protein mutated to serine to remove their N-terminal acylation sites, are expressed from pET22b(+) with a C-terminal 6 × His-tag [26]. BamC, lacking its periplasmic export sequence and N-terminal acylation site, is expressed from pET16b with an N-terminal 6 × His-tag [46].
2. The use of a Beckman JLA-8.1000 rotor and 1 L centrifuge bottles is recommended.
3. Plasmid pJH114 encodes all five Bam subunits with an 8 × His-Tag at the C-terminus of BamE [47].
4. Our standard procedure is carried out using 20 s power on/30 s power off cycles at 10% power increase per cycle, from 10% to 60%.

5. A Beckman SW 32 Ti rotor is recommended. Fill up ultracentrifuge tubes to at least 4/5 of maximum volume to prevent them from collapsing during ultracentrifugation.
6. BamABCDE will elute as a single peak between ~7 and 10 mL.
7. Plasmid pMN86 encodes MepM lacking its N-terminal periplasmic export sequence, and membrane-anchoring α-helix, such as MepM, is overproduced in the cytoplasm in soluble form [48].
8. Volumes can be scaled up or down at convenience.
9. The use of glass pipettes is recommended.
10. Avoid spillage from boiling over by adjusting the boiling temperature and the stirring speed. To prevent excessive evaporation and subsequent thickening of the sample while adding the cell suspension to the boiling SDS, lift the aluminium foil cover on the flask just enough for the Pasteur pipette to enter the flask.
11. If excessive foaming occurs, add a few drops of cold Milli-Q water from time to time to avoid spillages.
12. The sample should look clear after this step.
13. Avoid storage at 4 °C to prevent SDS precipitation.
14. Avoiding the formation of clumps and excessive foaming from residual SDS in the sample.
15. Different *E. coli* strains can require a different number of washing steps. After each step, the pellet should become progressively thinner.
16. Pronase E is pre-activated by preparing a 10 mg/mL stock solution in 10 mM Tris/HCl pH 7.0, 10 mM NaCl and incubating for 2 h at 60 °C [11].
17. After α-amylase and pronase E treatment, the PG pellet should look translucent and might be more difficult to visualise.
18. This will result in a PG suspension at a concentration of ~10 mg/mL. Many impurities from the preparation protocol might be present in the sample at this stage, but their presence should not affect downstream applications.
19. *E. coli* BW25113Δ6LDT lacks all LD-transpeptidases that catalyse the formation of LD-cross-links [49]; thus, its PG contains almost entirely tetrapeptide stems.
20. Tris/HCl cannot be used in microscale thermophoresis experiments as the fluorescent dye will react with primary amine groups on Tris buffer molecules. Sample volumes of mixtures to be dialysed are usually below 1 mL; therefore, extra care is recommended to avoid losing sample. Pre-wet the dialysis membrane in dialysis buffer before transferring the sample

into the membrane and carefully close the ends of the dialysis membrane with clamps. Leave the membrane loose to prevent it from opening during dialysis. Also, consider that dialysis will not remove detergent from the digestion mixture.

21. Vortex the sacculi suspension thoroughly before pipetting.
22. This amount corresponds to ~1.5 nanomoles of BamA P1,2, ~1.3 nanomoles of BamA P3,4, ~1.4 nanomoles of BamA P4,5, ~0.6 nanomoles of BamB, ~0.7 nanomoles of BamC, ~2.4 nanomoles of BamE, ~0.4 nanomoles of BamCD, ~0.3 nanomoles of BamCDE, and ~0.1 nanomoles of BamABCDE.
23. Measure with a micropipette how much supernatant was removed after the first pelleting step and substitute the removed volume with 10 μL of 10× PG binding buffer, the volume of purified protein to be tested for PG binding and water up to total 100 μL including the volume of sacculi. For example, if after pelleting 100 μL of sacculi suspension at 10 mg/mL, the amount of removed supernatant is 75 μL and the amount of protein to be added will be 2 μL, add to the pellet 63 μL of water, 10 μL of 10× PG binding buffer, and 2 μL proteins to obtain a total sample volume of 100 μL. Carefully resuspend the pellet pipetting up and down after the components have been added.
24. Centrifuging protein of interest (17,000 × *g*, 4 °C, 10 min) to remove aggregates prior to their use in protein-PG interaction assays is recommended.
25. To obtain 1× PG binding buffer, simply dilute 10× PG binding buffer ten-fold.
26. Carefully clean the magnetic stir bars with ethanol prior to their use to avoid contamination.
27. Use labelled proteins in MST experiments within ~3 months.
28. In our experience, the use of standard-coated capillaries in protein-Tetra$_n$ MST experiments provided the best results. It is recommended to test fluorescent emission starting with 40% LED power.
29. Irregular peaks usually indicate protein aggregation or adsorption to capillary walls.
30. Consider that the sample in tube n. 1 will be diluted 1:1 (v/v) with the labelled protein in the next steps; therefore, the concentration of Tetra$_n$ in tube n. 1 during the experiment will be half of the concentration present in tube n. 1 at this step.
31. Do not let the solution reach the upper end of the capillary to avoid spillage while handling. Leave the liquid column ~2–3 mm away from the end of the capillary.

32. Some Bam proteins (e.g. BamE) exhibit concentration-dependent differences in initial fluorescence in the presence of a serial dilution of $Tetra_n$. If this happens and if the measurements exhibit variations in initial fluorescence greater than ±10% of the average fluorescence of all 16 samples prior to the application of the temperature gradient, binding curve fitting must be performed on the initial fluorescence measurements and validated with SDS-denaturation tests (*see* Subheading 3.3.3).
33. Alternatively, two consecutive measurements at 40% MST power and 60% MST power are recommended. Adjust accordingly to increase the amplitude of the thermophoretic signal.
34. For more details on how to interpret and analyse MST data, consult the User Starting guide for Monolith NT.115 from NanoTemper Technologies.
35. Avoid low temperatures to prevent cells from entering cold shock, which can alter the properties of the cell envelope.

Acknowledgements

This work was funded by the European Union's Horizon 2020 research and innovation programme under the Marie Skłodowska-Curie grant agreement No 721484 (International Training Network Train2Target).

References

1. Nikaido H (2003) Molecular basis of bacterial outer membrane permeability revisited. Microbiol Mol Biol Rev 67:593–656
2. Silhavy TJ, Kahne D, Walker S (2010) The bacterial cell envelope. Cold Spring Harb Perspect Biol 2:a000414
3. Dufresne K, Paradis-Bleau C (2015) Biology and assembly of the bacterial envelope. Adv Exp Med Biol 883:41–76
4. May KL, Silhavy TJ (2017) Making a membrane on the other side of the wall. Biochim Biophys Acta Mol Cell Biol Lipids 1862:1386–1393
5. Höltje JV (1998) Growth of the stress-bearing and shape-maintaining murein sacculus of *Escherichia coli*. Microbiol Mol Biol Rev 62: 181–203
6. Vollmer W, Blanot D, De Pedro MA (2008) Peptidoglycan structure and architecture. FEMS Microbiol Rev 32:149–167
7. Nanninga N (1998) Morphogenesis of *Escherichia coli*. Microbiol Mol Biol Rev 62:110–125
8. Rojas ER, Billings G, Odermatt PD et al (2018) The outer membrane is an essential load-bearing element in Gram-negative bacteria. Nature 559:617–621
9. Schleifer KH, Kandler O (1972) Peptidoglycan types of bacterial cell walls and their taxonomic implications. Bacteriol Rev 36:407–477
10. Vollmer W, Höltje JV (2004) The architecture of the murein (peptidoglycan) in gram-negative bacteria: vertical scaffold or horizontal layer(s)? J Bacteriol 186:5978–5987
11. Glauner B (1988) Separation and quantification of muropeptides with high-performance liquid chromatography. Anal Biochem 172: 451–464
12. Harz H, Burgdorf K, Höltje JV (1990) Isolation and separation of the glycan strands from murein of *Escherichia coli* by reversed-phase high-performance liquid chromatography. Anal Biochem 190:120–128
13. Li GW, Burkhardt D, Gross C et al (2014) Quantifying absolute protein synthesis rates

reveals principles underlying allocation of cellular resources. Cell 157:624–635
14. Braun V, Hantke K (2019) Lipoproteins: structure, function, biosynthesis. Subcell Biochem 92:39–77
15. Magnet S, Bellais S, Dubost L et al (2007) Identification of the L,D-transpeptidases responsible for attachment of the Braun lipoprotein to *Escherichia coli* peptidoglycan. J Bacteriol 189:3927–3931
16. Egan AJF (2018) Bacterial outer membrane constriction. Mol Microbiol 107:676–687
17. Typas A, Banzhaf M, Den Berg V, Van Saparoea B et al (2010) Regulation of peptidoglycan synthesis by outer-membrane proteins. Cell 143:1097–1109
18. Lupoli TJ, Lebar MD, Markovski M et al (2014) Lipoprotein activators stimulate *Escherichia coli* penicillin-binding proteins by different mechanisms. J Am Chem Soc 136:52–55
19. Egan AJF, Jean NL, Koumoutsi A et al (2014) Outer-membrane lipoprotein LpoB spans the periplasm to stimulate the peptidoglycan synthase PBP1B. PNAS 111:8197–8202
20. Egan AJF, Maya-Martinez R, Ayala I et al (2018) Induced conformational changes activate the peptidoglycan synthase PBP1B. Mol Microbiol 110:335–356
21. Uehara T, Parzych KR, Dinh T et al (2010) Daughter cell separation is controlled by cytokinetic ring-activated cell wall hydrolysis. EMBO J 29:1412–1422
22. Gurnani Serrano CK, Winkle M, Martorana AM et al (2021) ActS activates peptidoglycan amidases during outer membrane stress in *Escherichia coli*. Mol Microbiol. https://doi.org/10.1111/mmi.14712
23. Banzhaf M, Yau HC, Verheul J et al (2020) Outer membrane lipoprotein NlpI scaffolds peptidoglycan hydrolases within multi- enzyme complexes in *Escherichia coli*. EMBO J 39:1–20
24. Su MY, Som N, Wu CY et al (2017) Structural basis of adaptor-mediated protein degradation by the tail-specific PDZ-protease Prc. Nat Commun 8:1516
25. Morè N, Martorana AM, Biboy J et al (2019) Peptidoglycan remodeling enables *Escherichia coli* to survive severe outer membrane assembly defect. MBio 10:e02729–e02718
26. Mamou G, Corona F, Cohen-Khait R et al (2022) Peptidoglycan maturation controls outer membrane protein assembly. Nature 606:953–959
27. Ricci DP, Silhavy TJ (2012) The Bam machine: a molecular cooper. Biochim Biophys Acta Biomembr 1818:1067–1084
28. Rollauer SE, Sooreshjani MA, Noinaj N et al (2015) Outer membrane protein biogenesis in Gram-negative bacteria. Philos Trans R Soc Lond Ser B Biol Sci 370:20150023
29. Konovalova A, Kahne DE, Silhavy TJ (2017) Outer membrane biogenesis. Annu Rev Microbiol 71:539–556
30. Tomasek D, Kahne D (2021) The assembly of β-barrel outer membrane proteins. Curr Opin Microbiol 60:16–23
31. Bakelar J, Buchanan SK, Noinaj N (2016) The structure of the β-barrel assembly machinery complex. Science 351:180–186
32. Gu Y, Li H, Dong H et al (2016) Structural basis of outer membrane protein insertion by the BAM complex. Nature 531:64–69
33. Han L, Zheng J, Wang Y et al (2016) Structure of the BAM complex and its implications for biogenesis of outer-membrane proteins. Nat Struct Mol Biol 23:192–196
34. Schiffrin B, Brockwell DJ, Radford SE (2017) Outer membrane protein folding from an energy landscape perspective. BMC Biol 15: 1–16
35. Braun M, Silhavy TJ (2002) Imp/OstA is required for cell envelope biogenesis in *Escherichia coli*. Mol Microbiol 45:1289–1302
36. Lee J, Xue M, Wzorek JS et al (2016) Characterization of a stalled complex on the β-barrel assembly machine. PNAS 113:8717–8722
37. Doyle MT, Bernstein HD (2019) Bacterial outer membrane proteins assemble via asymmetric interactions with the BamA β-barrel. Nat Commun 10:1–13
38. Tomasek D, Rawson S, Lee J et al (2020) Structure of a nascent membrane protein as it folds on the BAM complex. Nature 583:473–478
39. Lee J, Tomasek D, Santos TM et al (2019) Formation of a β-barrel membrane protein is catalyzed by the interior surface of the assembly machine protein BamA. elife 8:e49787
40. Wang Y, Wang R, Jin F et al (2016) A supercomplex spanning the inner and outer membranes mediates the biogenesis of β- barrel outer membrane proteins in bacteria. J Biol Chem 291:16720–16729
41. Alvira S, Watkins DW, Troman L et al (2020) Inter-membrane association of the Sec and BAM translocons for bacterial outer-membrane biogenesis. elife 9:e60669
42. Watkins DW, Williams SL, Collinson I (2022) A bacterial secretosome for regulated envelope biogenesis and quality control? Microbiology (Reading). https://doi.org/10.1099/mic.0.001255

43. Ursinus A, Van Den Ent F, Brechtel S et al (2004) Murein (peptidoglycan) binding property of the essential cell division protein FtsN from *Escherichia coli*. J Bacteriol 186:6728–6737
44. Li G, Peter Howard S (2017) *In vivo* and *in vitro* protein-peptidoglycan interactions. Methods Mol Biol 1615:143–149
45. Knowles TJ, Jeeves M, Bobat S et al (2008) Fold and function of polypeptide transport-associated domains responsible for delivering unfolded proteins to membranes. Mol Microbiol 68:1216–1227
46. Knowles TJ, Sridhar P, Rajesh S et al (2010) Secondary structure and ^{1}H, ^{13}C and ^{15}N resonance assignments of BamE, a component of the outer membrane protein assembly machinery in *Escherichia coli*. Biomol NMR Assign 4:179–181
47. Roman-Hernandez G, Peterson JH, Bernstein HD (2014) Reconstitution of bacterial autotransporter assembly using purified components. elife 3:e04234
48. Singh SK, Saisree L, Amrutha RN et al (2012) Three redundant murein endopeptidases catalyse an essential cleavage step in peptidoglycan synthesis of *Escherichia coli* K12. Mol Microbiol 86:1036–1051
49. Kuru E, Lambert C, Rittichier J et al (2017) Fluorescent D-amino-acids reveal bi-cellular cell wall modifications important for *Bdellovibrio bacteriovorus* predation. Nat Microbiol 2: 1648–1657

Chapter 12

MitoLuc: A Luminescence-Based Assay to Study Real-Time Protein Import into Mitochondria

Holly C. Ford and Ian Collinson

Abstract

All but a few mitochondrial proteins are translated into the cytosol and imported in via complicated and varied pathways. These processes occur over short time frames and, as such, are difficult to monitor with classical approaches such as Western blotting or autoradiography that require sample collection at discrete time points. The development of an assay based on a split version of the small luciferase—Mitoluc—has allowed us to monitor the import of proteins into mitochondria in high resolution and real time (Pereira et al., J Mol Biol 431:1689–1699, 2019). Luminescence measurements are acquired using a plate reader in the order of seconds. This allows scores of experiments to be conducted in parallel in a single multi-well plate and permits kinetic analysis yielding information about import mechanisms (Ford et al., Elife 11: e75426, 2022).

Key words Mitochondria, Luminescence, NanoLuc, MitoLuc, Protein import, Split-luciferase

1 Introduction

The intact NanoLuc enzyme is a 19.1 kDa 10-stranded β-barrel that catalyzes the oxidation of its substrate furimazine to produce furimamide, CO_2, and high-intensity luminescence with glow-type kinetics [1]. It is thermally stable [2], active over a broad pH range, and ATP-independent [3]. It was engineered several years ago for use in protein-fragment complementation assays. The C terminus was truncated, removing the tenth beta strand, and the remaining protein mutagenized to improve expression, stability, and activity, hereafter called 11S. Variants of the 13 amino acid peptide removed from the C terminus were screened for affinity for 11S and subsequent activity of the reconstituted enzyme [2]. Pep86, a shortened (11 amino acid) version of this peptide with high affinity for 11S, was selected for use in the NanoLuc protein translocation assay—designed to interrogate both bacterial protein secretion and mitochondrial protein import [4].

Raffaele Ieva (ed.), *Transmembrane β-Barrel Proteins: Methods and Protocols*, Methods in Molecular Biology, vol. 2778,
https://doi.org/10.1007/978-1-0716-3734-0_12,

The latter (MitoLuc) translocation assay was developed to monitor the arrival of preproteins into the mitochondrial matrix following import through the translocases of the outer and inner mitochondrial membranes (TOM and TIM complexes, respectively). Typically, a N-terminal mitochondrial targeting sequence (MTS) drives the import of preproteins through a membrane-embedded β-barrel channel in the TOM complex and the α-helical TIM complex. To monitor this reaction, yeast cells or mammalian cells in culture (both methods described below) were genetically engineered to express a mitochondrial matrix-targeted 11S. Precursor proteins of interest were purified as fusions with pep86, usually at the C-terminus but also at other locations as required for specific analyses [5]. Precusor-pep86 fusion proteins are added to either isolated mitochondria (from yeast cells) [4, 5] or permeabilized mammalian cells [6] and imported into the matrix, and upon arrival, pep86 binds to 11S with high affinity. In the presence of furimazine and O_2, the reconstituted enzyme produces luminescence that can be detected with a luminometer or plate reader. We describe detailed methods for performing the MitoLuc assay in isolated yeast mitochondria and the adaptation for measuring mitochondrial protein import in whole mammalian cells. Typical data generated, data processing, and interpretation for the yeast and mammalian systems are described in the papers listed above. To date, these assays have been utilized to elucidate mechanisms of protein transport. They are also being, and have been, deployed for the analysis of import failure, including for the characterization of human health problems associated with import defects. Thus, we anticipate their broader application such as in screening for small molecules that modulate the import process towards the development of drugs that can offset the consequences of import failure associated with disease and aging.

2 Materials

Chemicals were purchased from Sigma, UK unless stated otherwise, and were of high purity. Aqueous solutions were prepared using ultrapure, deionized water, and for nonaqueous preparations, pure ethanol or DMSO was used instead. For culture of cells, media and solutions were sterilized by filtering through a 0.22 μm membrane or autoclaving.

2.1 Producing Yeast That Constitutively Express Mitochondrial Matrix-Targeted 11S

1. pYES2/CT vector.
2. DNA encoding mt-11S (ATP synthase subunit alpha MTS residues 1–37 and 11S), codon-optimized for yeast expression and ordered as a gene strand or in a plasmid with suitable restriction sites for cloning into pYES2/CT.

3. Chemically competent YPH499 yeast cells.
4. 10× Tris-EDTA: 100 mM Tris-HCl, 10 mM EDTA, pH 7.5, filter-sterilized.
5. 1 M stock solution of lithium acetate, filter-sterilized.
6. 50% (w/v) solution of PEG 3000 (filter-sterilized).
7. 9–12 mg/mL single-stranded DNA sodium salt from salmon testes.
8. Centrifuge.
9. Benchtop orbital shaker incubator.
10. 2× YPD medium: 20 g/L yeast extract, 40 g/L bactopeptone, 4% (w/v) glucose.
11. Selective agar plates: agar plates prepared with 1.93 g/L synthetic complete (SC) minus uracil dropout mixture, 6.91 g/L yeast nitrogen base (without amino acids), 15 g/L agar, autoclaved, and then supplemented with 2% (w/v) glucose (from a sterile 40% (w/v) solution).
12. Static incubator.
13. Preculture medium: 1.93 g/L synthetic complete (SC) minus uracil dropout mixture, 6.91 g/L yeast nitrogen base (without amino acids), autoclaved, and then supplemented with sterile glucose solution for 2% (w/v) as a final concentration.
14. Freezing medium: 1.93 g/L synthetic complete (SC) minus uracil dropout mixture, 6.91 g/L yeast nitrogen base (without amino acids), 30% (v/v) glycerol. Autoclave to sterilize.

2.2 Isolation of Mitochondria from Yeast Cells with Matrix-Localized 11S

1. YPH499-mt-11S yeast cells produced in Subheading 3.1.
2. Preculture medium: 1.93 g/L synthetic complete (SC) minus uracil dropout mixture, 6.91 g/L yeast nitrogen base (without amino acids), autoclaved, and then supplemented with sterile glucose solution for 2% (w/v) as a final concentration.
3. 10,000 units/mL penicillin and 10 mg/mL streptomycin mixture (100× working concentration).
4. 1 L unbaffled Erlenmeyer flask and 2 L baffled Erlenmeyer flasks
5. Aerobic culture medium: 1.93 g/L synthetic complete (SC) minus uracil dropout mixture, 6.91 g/L yeast nitrogen base (without amino acids), 3% (v/v) glycerol (autoclaved to sterilize).
6. Shaking incubator.
7. Cuvette spectrophotometer for measuring OD_{600} of yeast cultures and colorimetric BCA assay.

8. 20% (w/v) solution of galactose, filter sterilized.
9. DTT buffer: 100 mM Tris-SO_4, 10 mM DTT, pH 9.4. *See* **Note 1** for preparation suggestions.
10. SM buffer: 250 mM sucrose, 10 mM MOPS-KOH, pH 7.2. Stored in aliquots at −20 °C.
11. Zymolyase buffer: 1.2 M sorbitol, 20 mM potassium phosphate, pH 7.4. *See* **Note 2** for preparation suggestions.
12. 100 mM PMSF stock solution in MeOH. Store at −20 °C.
13. Homogenization buffer: 0.6 M sorbitol, 10 mM Tris pH 7.4, 0.5% (w/v) BSA (fatty-acid free), 1 mM PMSF added immediately before use.
14. Centrifuge and centrifuge bottles/tubes.
15. Glass-teflon elvehjem potter homogenizer and overhead stirrer.
16. Zymolyase® -20T (from *Arthrobacter luteus*).
17. Bicinchoninic acid (BCA) protein quantification kit.
18. Bovine serum albumin (BSA).

2.3 Expression and Purification of GST-Dark, GST-rPFO (Recombinant Perfringolysin O), and pep86-Precursor Fusion Proteins

1. Protein expression vectors: pBAD, pRSFDuet, pGEX-1.
2. Codon-optimized DNA sequences encoding proteins for expression with restriction sites for cloning into expression vectors.
3. Chemically competent *E. coli* BL21(DE3) cells.
4. LB bacterial growth media.
5. 2× YT bacterial growth media
6. Antibiotics for plasmid transformation selection.
7. Orbital shaking incubator.
8. Cuvette spectrophotometer for measuring OD_{600}.
9. 20% (w/v) arabinose solution (filter-sterilized).
10. 1 M IPTG solution (filter-sterilized).
11. Centrifuge.
12. Ultracentrifuge.
13. Beckman Ti70 or Ti45 rotor.
14. TK buffer: 20 mM Tris-HCl, 50 mM KCl, pH 7.5.
15. Reduced glutathione.
16. Cell disrupter (Constant Systems Ltd.).
17. DNase (bovine origin).
18. FPLC protein purification system (AKTA™).

19. 5 mL HisTrap FF column (Nickel affinity column).
20. Ion exchange columns: 5 mL HiTrap Q HP and 5 mL HiTrap SP HP.
21. GSTrap column affinity column: 5 mL GSTrap FF.
22. Centrifugal ultrafiltration spin columns for protein concentration (with appropriate molecular weight cut-off values).
23. Microvolume spectrophotometer.
24. Binding buffer: 20 mM Tris-HCl, 50 mM KCl, 6 M urea, pH adjusted to more than 1 pH unit away from the proteins predicted isoelectric point (*see* **Note 3** for suggested buffer pH and column choice).
25. Tube roller or an orbital shaking incubator set to ambient temperature and 100 rpm.
26. Nickel elution buffer: 20 mM Tris-HCl, 50 mM KCl, 6 M urea, 1 M imidazole, pH to match that of binding buffer.
27. Ion-exchange elution buffer: 20 mM Tris-HCl, 1 M KCl, 6 M urea, pH to match that of binding buffer.
28. SDS-PAGE system, including polyacrylamide gels, running buffer, power supply, and Coomassie stain and destain solutions.
29. Liquid nitrogen.

2.4 NanoLuc-Based Import Assay with Isolated Yeast Mitochondria

1. Low binding, round bottom, white 96 well plate.
2. Nano-Glo™ luciferase assay substrate (furimazine) from Promega.
3. Prionex solution (10% (w/v), aqueous).
4. 0.2 M ATP, adjusted to pH 7.2 with Tris. Store at −20 °C in single-use aliquots.
5. 0.2 M creatine kinase. Store at −20 °C in single-use aliquots.
6. 0.2 M phosphocreatine. Store at −20 °C in single-use aliquots.
7. NADH, reduced disodium salt hydrate.
8. Microplate reader compatible with luminescence with automatic injectors.
9. 2× import buffer: 500 mM sucrose, 160 mM KCl, 2 mM K_2HPO_4/KH_2PO_4, 10 mM $MgCl_2$, 20 mM MOPS-KOH, adjust pH to 7.2 with KOH. Store in 1 mL aliquots at −20 °C.
10. Purified preprotein (*see* Subheading 3.3).
11. Purified GST-dark protein (*see* Subheading 3.3).
12. Mitochondrial poisons, import inhibitors or test compounds at 100× working concentration (such as 100 μM valinomycin,

400 μM oligomycin, 800 μM antimycin A, 3 mM MitoBlock-10, 2 mM MitoBlock-20, 1 mM MitoBlock-6, 1 mM dequalinium chloride (MitoBlock-12), 2 mM MKT-077, 2 mM VER-155008, 2 mM SW02).

13. mt-11S yeast mitochondria (*see* Subheading 3.2).

2.5 NanoLuc-Based Import Assay in Mammalian Cells (MitoLuc Import Assay)

1. pLVX-eqFP670-P2A-Cox8a-11S plasmid. *See* description in **Note 4**.
2. Laminar flow cell culture cabinet.
3. Cell culture flasks.
4. 6 well plates.
5. White flat-bottom 96 well plates.
6. CO_2 incubator for mammalian cell culture.
7. Lipofectamine™ 3000 reagent, Opti-MEM™, and P3000™ reagent for transfection (ThermoFisher).
8. Culture medium, trypsin solution, and HBSS (calcium and magnesium free).
9. HBSS imaging buffer: 1× HBSS (Thermo Scientific), 5 mM D-(+)-glucose, 10 mM HEPES, 1 mM MgCl2, 1 mM CaCl2, pH 7.4.
10. MitoLuc assay buffer: 225 mM mannitol, 10 mM HEPES, 2.5 mM $MgCl_2$, 40 mM KCl, 2.5 mM KH_2PO_4, 0.5 mM EGTA, adjust to pH 7.4 with KOH.
11. Nano-Glo™ luciferase assay substrate, furimazine (Promega).
12. Prionex solution (10% (w/v), aqueous).
13. 0.2 M ATP, adjusted to pH 7.2 with Tris. Store at −20 °C in single-use aliquots.
14. 0.2 M creatine kinase. Store at −20 °C in single-use aliquots.
15. 0.2 M phosphocreatine. Stored at −20 °C in single-use aliquots.
16. 300 mM succinate in H_2O, filter-sterilized.
17. 2.5 mM Rotenone in EtOH.
18. Purified preprotein (*see* Subheading 3.3).
19. Purified GST-dark protein and GST-rPFO protein (*see* Subheading 3.3).
20. Binding buffer (*see* Subheading 2.3, **step 24**).
21. Mitochondrial poisons (such as oligomycin, antimycin A, and valinomycin) and import inhibitors or test compounds.
22. Microplate reader compatible with luminescence and fluorescence with automatic injectors.

3 Methods

3.1 Generating a Yeast Cell Line That Constitutively Expresses Mitochondrial Matrix-Targeted 11S

1. Optimize DNA encoding the amino acid sequence for 11S [2] for expression in yeast and at the 5′ add DNA encoding the mitochondrial targeting sequence of yeast ATP synthase subunit alpha (amino acid residues 1–37). Clone this construct into the yeast expression vector pYES2/CT downstream of the GAL1 promotor, with a stop codon immediately after the 11S sequence, thus avoiding the V5 tag and 6xHis tag contained in the parent vector. This vector is hereafter referred to as pYES—mt-11S.
2. Once pYES—mt-11S is constructed or obtained, gather materials for transformation of competent YPH499 cells.
3. Prepare a mixture of 1× TE and 100 mM lithium acetate (50 μL 10× TE, 50 μL 1 M lithium acetate, and 400 μL sterile H_2O is plenty for a single transformation) and chill on ice.
4. Prepare another mixture, as above (**step 3**) but instead of H_2O use 50% PEG 3000 (1× TE, 100 mM lithium acetate, 40% PEG 3000). Chill on ice.
5. Heat 10 μL DNA sodium salt from salmon testes at 98 °C for 10 min and then place on ice for at least 5 min.
6. Thaw a 50 μL aliquot of competent YPH499 yeast cells on ice and then harvest cells from storage medium by centrifuging at 3000 × *g* for 1 min and removing the supernatant.
7. Resuspend the cells in 50 μL in ice-cold 1× TE, 100 mM lithium acetate (prepared in **step 3** of this section).
8. To this cell suspension, add 1 μg pYES—mt-11S, 5 μL DNA sodium salt from salmon testes (prepared **step 5**), 300 μL of 1× TE, 100 mM lithium acetate, and 40% PEG 3000 (prepared in **step 4** of this section).
9. Incubate the above at 30 °C, shaking on a benchtop orbital shaker incubator at 800 rpm for 30 min, then increase the temperature to 42 °C for 15 min.
10. Harvest the yeast cells from the transformation mixture by centrifuging at 3000 × *g* for 1 min, then add 500 μL 2× YPD and incubate for 2–3 h at 30 °C, shaking at 200 rpm.
11. Harvest cells by centrifugation (as before) and spread all of them onto a selective agar plate.
12. This plate should be incubated at either 30 °C for 2 days, or 24 °C for 3 days (whichever is more convenient).
13. Single colonies should be picked from this plate, streaked onto fresh selective agar, and grown as described above (**step 12** of this section). The colonies from these plates can be used to inoculate a liquid culture in synthetic growth preculture medium lacking uracil, supplemented with 2% glucose.

14. To keep a stock of the YPH499-mt-11S yeast cell line, resuspend a colony in 500 μL freezing medium and store in an Eppendorf tube at −80 °C. This can be streaked out onto fresh selective agar, grown as described in **step 12** of this section and used to start liquid cultures.

3.2 Isolation of Mitochondria from YPH499-mt-11S Yeast Cells

1. Pick a colony from a freshly grown plate of YPH499-mt-11S yeast cells and use this to inoculate 100 mL preculture medium supplemented with penicillin and streptomycin in a 1 L flask. Grow at 30 °C, 200 rpm in an orbital shaker incubator overnight.

   All subsequent steps can be carried out in triplicate in order to obtain three independent preparations of mitochondria, which can be used for performing experiments in biological triplicate (*see* **Note 5**).
2. Add 20 mL of the yeast overnight preculture to a 2 L baffled flask containing 1 L of aerobic culture medium supplemented with penicillin and streptomycin. Grow this culture at 30 °C, 180 rpm until OD_{600} reaches 0.4–0.6 (mid-log phase). This takes approximately 8–10 h.
3. Induce expression of mt-11S by adding 50 mL of 20% (w/v) galactose (for 1% final concentration) and continue growth at 30 °C, 180 rpm for 16 h.
4. Prepare 20 mL DTT buffer and 350 mL zymolyase buffer and warm both to 30 °C. Prepare 100 mL homogenization buffer and keep on ice. This will be more than enough for a 1 L culture but scale up if working with more than 1 L culture.
5. Thaw an aliquot of SM buffer.
6. Measure OD_{600} of the yeast culture. The optimal OD_{600} at this point is between 1.2 and 1.8, but between 0.9 and 3 is acceptable.
7. Harvest cells by centrifugation at 4000 × *g* for 10 min at room temperature.
8. After removing the supernatant, resuspend the cell pellet in about 200 mL deionized H_2O and transfer the cell suspension to a pre-weighed centrifuge bottle.
9. Harvest the cells, this time at 1670 × *g* for 5 min at room temperature, discard the supernatant, and reweigh the bottle to calculate the weight of the cell pellet.
10. Resuspend the pellet in DTT buffer at a concentration of 0.5 g/mL.
11. Incubate the suspension in a shaker incubator at 30 °C, at 100 rpm for 15 min.
12. Harvest the cells at 1670 × *g* for 5 min at room temperature and discard the supernatant.

13. Resuspend the cell pellet in about 100 mL zymolyase buffer then harvest cells again, discarding the supernatant, as in the previous step.
14. Resuspend the cell pellet in zymolyase buffer at a concentration equivalent to 6.5 mL per gram of original cell pellet (calculated in **step 9** of this section).
15. Add Zymolyase® -20T for a concentration equivalent to 4.5 mg per gram of original cell pellet. It may be easier to dissolve the Zymolyase® -20T in a small volume (~1 mL) of zymolyase buffer before adding it to the cell suspension.
16. Incubate in a shaker incubator at 30 °C, at 100 rpm for 30 min.
17. Pellet the cells by centrifugation at $1670 \times g$ for 5 min at room temperature.
18. Wash the pellet in about 150 mL zymolyase buffer, then centrifuge again, this time at $2970 \times g$ for 5 min at room temperature.
19. From this point onwards, continue the mitochondrial isolation at 4 °C, with the sample, buffers, homogenizer, and centrifuge tubes on ice. Chill the centrifuge and rotor to 4 °C.
20. Add PMSF to homogenization buffer to a final concentration of 1 mM, then resuspend the pellet in 6.5 mL of this buffer per gram of original cell pellet.
21. Homogenize the sample using a glass-teflon elvehjem potter homogenizer with an overhead stirrer at 500 rpm (15 strokes is sufficient) and transfer the homogenate to centrifuge tubes.
22. Centrifuge the sample at low speed ($1480 \times g$) for 5 min to pellet cell debris.
23. Transfer the homogenate (supernatant) to a fresh centrifuge tube, then resuspend the pellet in homogenization buffer, homogenize the sample, and centrifuge as in the previous step.
24. Add the secondary homogenate to the first (this can be divided between centrifuge tubes as necessary), and centrifuge at high speed ($17{,}370 \times g$) for 15 min.
25. Discard supernatants, then combine pellets in 1 mL SM buffer with a p1000 pipette, before diluting to approximately 15 mL in SM buffer.
26. Centrifuge the sample at low speed ($1480 \times g$) for 5 min to pellet cell debris. Mitochondria are now in the supernatant.
27. Using a serological pipette, and taking care not to disturb the pellet, which will be quite soft, transfer the supernatant to a clean centrifuge tube.
28. Centrifuge at high speed ($17{,}370 \times g$) for 15 min to pellet mitochondria, then discard the supernatant.

29. Thoroughly resuspend the pellet of mitochondria in SM buffer to 250–300 μL. This brings them to an appropriate concentration for estimation of protein concentration using a BCA assay kit if performed as described in **step 30**.
30. Use the kit according to manufacturer's instructions with a pre-prepared standard curve using BSA. Prepare 50 μL samples in cuvettes: a blank (10 μL SM buffer and 40 μL H_2O) and triplicate mitochondrial samples (1 μL mitochondrial suspension, 9 μL SM buffer and 40 μL H_2O).
31. Measure the volume of the mitochondrial sample using a pipette, and adjust the concentration to 10 mg/mL by adding SM buffer.
32. Snap freeze 0.5 mg aliquots in liquid nitrogen and store at −80 °C.

3.3 Expression and Purification of GST-Dark, GST-rPFO (Recombinant Perfringolysin O), and pep86-Precursor Fusion Proteins

1. Design a gene, codon-optimized for expression in *E. coli* with mitochondrial preprotein in frame with pep86 and a 6xHis tag. It may be necessary to insert GSG linker regions on either side of pep86.
2. Using standard cloning methods, insert the gene designed in **step 1** into a protein expression plasmid, such as pBAD or pRSFDuet, at a multiple cloning site downstream of the promoter region.
3. Clone DNA encoding glutathione-S-transferase (GST), fused C-terminally to the "dark" peptide (a peptide that binds 11S with high affinity but doesn't reconstitute luminescence activity) into the pGEX-1 vector [4].
4. Clone DNA encoding glutathione-S-transferase (GST), TEV protease cleavage sequence, and rPFO into pGEX-1 [6].
5. Transform a protein expression strain of *E. coli*, such as BL21 (DE3), with the plasmids created in **steps 1–4** of this section.
6. Inoculate 100 mL LB with a single colony from a plate of freshly transformed expression cells, include appropriate antibiotic for selection, and grow overnight at 37 °C, 200 rpm.
7. Add 1–10 mL overnight culture to 1 L 2xYT medium with selection antibiotic, and grow at 37 °C, 200 rpm until OD600 reaches 0.6. This can be scaled up as required.
8. Induce protein expression by adding arabinose (0.2% w/v) if using pBAD or IPTG (1 mM) if using pRSFDuet or pGEX-1 (GST-dark and GST-rPFO), and continue to grow under the same conditions for a further 3 h.
9. From this point onward, keep samples, buffers, centrifuge bottles, and rotors on ice or at 4 °C.
10. Harvest cells by centrifugation (4000 × *g* for 20 min is sufficient).

11. Wash the cell pellet in ice-cold TK buffer, then centrifuge again and remove the supernatant. At this stage, pellets can be stored at −20 °C. When resuming the purification, thaw pellets and proceed to **step 12**.
12. Disrupt cells by passing three times through a cell disruptor at 25 kpsi. It can be beneficial to add a pinch of DNase (bovine origin) and homogenize the sample before disrupting cells, as the sample can become viscous.
13. Centrifuge the sample at 38,000 rpm in a Beckman Ti70 or Ti45 rotor for 45 min.

    The following steps are for purification of GST-dark and GST-rPFO proteins. Continue from **step 23** for purification of pep86-tagged preproteins.
14. For purification of GST-dark protein or GST-rPFO from 6 L culture, attach two 5 mL GSTrap FF columns to a chromatographic protein purification system and equilibrate with 5 column volumes (50 mL) TK buffer.
15. Load the supernatant from **step 13** onto the columns at 1 mL/min, and collect the flow through.
16. Wash the columns with TK buffer until the UV absorbance returns to baseline level.
17. Elute bound material by running TK buffer supplemented with 20 mM glutathione through the columns at 4 mL/min. Collect 5 mL fractions.
18. Once UV absorbance returns to baseline, wash columns with 5 column volumes of TK buffer.
19. The binding capacity of the two columns will have been saturated in **step 15** of this section, so repeat **steps 15–18**, this time loading the flow through from the previous load step. This can be done a number of times but the protein yield will diminish each time (3–4 times is usually sufficient).
20. Combine the fractions with high UV absorbance. Two fractions from each elution will usually contain almost all the protein. Concentrate the combined fractions in a spin concentrator with molecular weight cut-offs of 10,000 Da for GST-dark and 50,000 Da for GST-rPFO.
21. Measure the protein concentration using a microvolume spectrophotometer using TK + 20 mM glutathione as a blank.
22. Once the protein concentration is between 6 and 9 mM, prepare 10 μL aliquots, snap freeze in liquid nitrogen, and store at −80 °C.

Preproteins are purified from inclusions bodies, from the pelleted material following centrifugation of disrupted cells (**step 13**).

23. Resuspended the cell pellet (containing inclusion bodies) in 5 mL binding buffer using a 5 mL glass homogenizer.
24. Add binding buffer for about 25 mL total volume and incubate on a roller at room temperature for 30 min.
25. Centrifuge the sample at 17,000 × *g* for 20 min. Keep the supernatant and discard the pellet.
26. Connect a 5 mL HisTrap FF column to a protein purification system and wash with 5 volumes of binding buffer.
27. Load the supernatant from **step 25** onto the column at 4 mL/min, and then wash the column with binding buffer until UV absorbance returns to baseline.
28. Wash the column with a column volume of 3% nickel elution buffer (97% binding buffer) for a 30 mM imidazole wash.
29. Elute bound material by washing the column with 33% nickel elution buffer (330 mM imidazole), and collect 5 mL fractions.
30. Combine fractions with UV absorbance.
31. Wash and disconnect the nickel column from the protein purification system (the column can be reused), and connect a 5 mL HiTrap Q HP or HiTrap SP HP column, for proteins carrying a negative or positive charge, respectively (*see* **Note 3**).
32. Wash the column with 5 volumes of binding buffer, then load the sample from **step 30** at 4 mL/min onto the column.
33. Wash the column with binding buffer until UV absorbance returns to baseline.
34. Elute bound material with a 0–100% gradient of ion-exchange elution buffer over 100 mL (at 5 mL/min this will take 20 min). Collect 5 mL fractions.
35. Assess the protein content and purity of each fraction by SDS-PAGE analysis.
36. Combine fractions that have a high ratio of protein of the correct molecular weight to contaminant proteins.
37. Concentrate this sample in a spin concentrator with appropriate molecular weight cut-off.
38. Remove excess KCl by diluting the concentrated sample in the binding buffer and concentrating again. Two dilutions of 1 in 10 is sufficient.
39. Measure protein concentration using a microvolume spectrophotometer and once >30 μM (*see* **Note 6**), snap freeze small aliquots in liquid nitrogen, and then store at −80 °C.

3.4 NanoLuc Assay to Measure Protein Import into Isolated Yeast Mitochondria

1. Prepare a fresh solution of 0.2 M NADH and keep on ice.
2. Thaw frozen aliquots of 0.2 M ATP, 10 mg/mL creatine kinase, 0.5 M phosphocreatine, 2× import buffer, 10% (w/v) prionex solution, GST-dark protein, and preprotein. Don't thaw mitochondria until immediately before use.
3. Prepare the reaction plate. The number of wells that can be measured in parallel will depend on the plate reader, specifically how fast it can perform injections and measurements. For kinetic experiments, we recommend measuring luminescence in each well at least every 6 s. Add drug of interest (such as a potential inhibitor of mitochondrial protein import) to wells at selected range of concentrations, no more than 1 μL/well. We recommend including a well with a known mitochondrial import inhibitor or inhibitor mixture. For this, 1 μL of a mixture of 125 μM valinomycin, 500 μM oligomycin, and 1 mM antimycin A are ideal. For alternative assay setups, *see* **Notes 7** and **8**.
4. Prepare two separate mixtures, one containing mitochondria, and the other containing preprotein. Mixture 1 contains supplements at 1.25× final concentration, and mixture 2 contains supplements at 5× concentration. Final concentrations are achieved by injecting 100 μL mixture 1 onto 25 μL mixture 2.
   (a) Mixture 1 (to be injected onto wells by the plate reader): 1× import buffer (from 2× stock), 12.5 μM GST-dark protein (for 10 μM final), prionex solution 0.1% (v/v), and 62.5 μg/mL mt-11S yeast mitochondria (for 50 μg/mL final). Add mitochondria immediately before use. Prepare enough for all wells at 100 μL per well, and the volume required to prime the plate reader injector.
   (b) Mixture 2 (to be added to wells 25 μL/well): 1× import buffer, prionex solution 0.1% (v/v), 10 mM NADH (for 2 mM final), 5 mM ATP (for 1 mM final), 0.5 mg/mL creatine kinase (for 0.1 mg/mL final), 25 mM phosphocreatine (for 5 mM final), 1.25× Nano-Glo™ luciferase assay substrate (for 0.25× final), and 5 μM purified pep86-preprotein (for 1 μM final). Prepare enough for all wells at 25 μL per well and 10% extra.
5. Turn on the plate reader and set it to the desired temperature (25 °C is recommended for yeast mitochondria), then prime the injector with mixture 1.
6. Program the plate reader method to inject 100 μL mixture 1 onto all wells containing mixture 2. Mixing is sufficient without shaking the plate. Luminescence can be measured without emission filters. Start reads immediately once the injection cycle is complete, with a minimum frequency of every 6 s. Thirty minutes of read cycles is usually sufficient to observe most import events and obtain kinetic information.

3.5 NanoLuc Assay to Measure Protein Import in Permeabilized Mammalian Cells (MitoLuc Assay)

1. Prepare buffers using sterile H_2O, and filter-sterilise in a cell culture cabinet. They can be aliquoted and stored at −20 °C, then kept sterile once open.
2. Seed cells at 3×10^5/well in a 6 well plate and grow for 48 h at 37 °C or until 60–80% confluent.
3. Transfect cells with pLVX-eqFP670-P2A-Cox8a-11S plasmid (0.5–2 μg) using Lipofectamine 3000 reagent according to manufacturer's instructions, with a ratio of DNA to lipofectamine at 1:3.
4. Grow cells at 37 °C for 48 h, then harvest cells with trypsin, and seed at 3×10^4/well in a sterile white, flat bottom, 96 well plate. Seed enough wells for each experimental condition in triplicate.
5. Grow cells for a further 16 h at 37 °C.
6. Thaw an aliquot of MitoLuc assay buffer and add prionex solution for 0.1% (v/v) and the following supplements for 1.25× final concentration: 12.5 μM GST-dark protein (for 10 μM final), 0.1% prionex solution, 0.125 mg/mL creatine kinase (for 0.1 mg/mL final), 6.25 mM phosphocreatine (for 5 mM final), 1.25 mM ATP (for 1 mM final), 3.75 mM GST-rPFO, 0.3125× Nano-Glo™ luciferase assay substrate (for 0.25× final), 6.25 mM succinate (for 5 mM final), and 1.25 μM rotenone (for 1 μM final). Prepare enough for 100 μL per well +10%. Wells will have preprotein added in 25 μL so will contain 125 μL final mixture.
7. Dilute preprotein stock to 30 μM in binding buffer. Prepare the equivalent to 0.2× volume of preprotein mixture (see below).
8. Prepare preprotein mixture for injecting into wells (25 μL/well), enough for all wells and injector priming: MitoLuc assay buffer, 0.1% (v/v) prionex solution, and 5 μM preprotein from 30 μM stock prepared above.
9. Remove medium from wells and wash cells twice with 200 μL HBSS (calcium and magnesium free).
10. Add 100 μL HBSS imaging buffer to each well and measure fluorescence on a plate reader with excitation set to 605 nm and emission 670 nm. This measures the cytosolic far-red fluorophore eqFP670 as a marker for Cox8a-11S expression (*see* Materials, Subheading 2.5), for normalizing luminescence data.
11. Remove buffer from all wells and add 100 μL MitoLuc assay buffer + supplements to each well.
12. Add drugs or import inhibitors to be tested to each well (we suggest 1 μL at 125× final concentration).
13. Gently mix the contents of each well with a multichannel pipette.

14. Turn on the plate reader and set to 37 °C.
15. Prime injector with preprotein mixture (prepared in **step 8**).
16. Measure the luminescence of wells with a minimum frequency of 6 s, without emission filters.

4 Notes

1. 1 M DTT can be prepared beforehand and stored as single-use aliquots in the dark at −20 °C. A 1 M solution of Tris-SO_4 at pH 9.4 can be prepared from Tris base and H_2SO_4, autoclaved, and stored at room temperature.
2. Zymolyase buffer is best prepared from a fresh 2.4 M solution of sorbitol and 1 M stock solutions of K_2HPO_4 and KH_2PO_4 at a ratio of 4:1.
3. Table of suggested buffers and columns for purification of preproteins, based on predicted pIs.

Predicted protein pI	Buffer pH	Ion-exchange column
<6.5	7.5	HiTrap Q HP
6.5–7.5	8.5	HiTrap Q HP
7.5–8.0	9.0	HiTrap Q HP
8.0–9.0	7.0	HiTrap SP HP
>9.0	7.5	HiTrap SP HP

4. In the pLVX-eqFP670-P2A-Cox8a-11S plasmid, downstream of the EF-1α promoter is a sequence encoding the constitutive far-red fluorophore eqFP670, followed by P2A sequence, and Cox8a-11S. The P2A sequence causes ribosomal skipping, resulting in 1:1 stoichiometry of eqFP670, which remains cytosolic, and Cox8a-11S, which is localized to the mitochondrial matrix.
5. For biological triplicates perform the experiment three times, each time with a different batch of mt-11S yeast mitochondria. These can be from mitochondrial isolations carried out in parallel with triplicate yeast cultures.
6. Up to 4.2 μL of preprotein can be added per well as the assay is unaffected by final urea concentrations of 200 mM of lower. As such preprotein stocks must be >30 μM.
7. Alternative assay setup 1: Titration of preprotein. In place of the drug of interest, add preprotein at a range of concentrations to wells and omit preprotein from mixture 2.

8. Alternative assay setup 2: Compare preproteins. In place of the drug of interest, add various preproteins to wells (1 μL 125 μM for 1 μM final concentration), and omit preprotein from mixture 2.

Acknowledgments

We would like to thank Drs. Hope Needs, Goncalo Pereira, James Lorriman, William Allen, and Prof. Mark Dillingham, who have all contributed to the development of the MitoLuc assay.

Author Contributions HCF wrote the manuscript and IC edited it.

Declarations The authors declare no competing interests. The funding agency and the University had no role in study design, data collection and analysis, decision to publish, or preparation of the manuscript.

Funding

Our work on mitochondrial import is funded by the Wellcome: Investigator Award to IC (104632/Z/14/Z).

References

1. Hall MP, Unch J, Binkowski BF et al (2012) Engineered luciferase reporter from a deep sea shrimp utilizing a novel imidazopyrazinone substrate. ACS Chem Biol 7(11):1848–1857
2. Dixon AS, Schwinn MK, Hall MP et al (2015) NanoLuc complementation reporter optimized for accurate measurement of protein interactions in cells. ACS Chem Biol 11:400–408
3. England CG, Ehlerding EB, Cai W (2016) NanoLuc: a small luciferase is brightening up the field of bioluminescence. Bioconjug Chem 27:1175–1187
4. Pereira GC, Allen WJ, Watkins DW et al (2019) A high-resolution luminescent assay for rapid and continuous monitoring of protein translocation across biological membranes. J Mol Biol 431:1689–1699
5. Ford HC, Allen WJ, Pereira GC et al (2022) Towards a molecular mechanism underlying mitochondrial protein import through the TOM and TIM23 complexes. elife 11:e75426
6. Needs HI, Lorriman JS, Pereira GC et al (2023) The MitoLuc assay system for accurate real-time monitoring of mitochondrial protein import within mammalian cells. J Mol Biol 435:168129

Chapter 13

Modular Assembly of Mitochondrial β-Barrel Proteins

Rituparna Bhowmik, Fabian den Brave, and Thomas Becker

Abstract

Mitochondrial β-barrel proteins fulfill crucial roles in the biogenesis and function of the cell organelle. They mediate the import and membrane insertion of proteins and transport of small metabolites and ions. All β-barrel proteins are made as precursors on cytosolic ribosomes and are imported into mitochondria. The β-barrel proteins fold and assemble with partner proteins in the outer membrane. The in vitro import of radiolabelled proteins into isolated mitochondria is a powerful tool to investigate the import of β-barrel proteins, the folding of the β-barrel proteins, and their assembly into protein complexes. Altogether, the in vitro import assay is a versatile and crucial assay to analyze the mechanisms of the biogenesis of mitochondrial β-barrel proteins.

Key words Mitochondria, β-barrel proteins, TOM complex, SAM complex, Protein import assay, Blue native electrophoresis

1 Introduction

Mitochondria originate from an endosymbiotic event that took place two billion years ago. A bacterium similar to α-proteobacteria was incorporated by a eukaryotic ancestor cell [1, 2]. During evolution most of the genetic information was lost or transferred to the host genome, leading to the development of nowadays mitochondria. Mitochondria carry out many crucial functions for the viability of the cell such as energy production, iron-sulfur cluster formation, lipid, heme, and amino acid biosynthesis and signaling in programmed cell death and innate immunity [3, 4]. They contain about 1000 proteins in baker's yeast *Saccharomyces cerevisiae* and about 1300 proteins in humans [5–7]. 99% of these proteins are synthesized as precursors on cytosolic ribosomes and have to be imported into mitochondria [8, 9].

Rituparna Bhowmik and Fabian den Brave contributed equally with all other contributors.

Raffaele Ieva (ed.), *Transmembrane β-Barrel Proteins: Methods and Protocols*, Methods in Molecular Biology, vol. 2778,
https://doi.org/10.1007/978-1-0716-3734-0_13,

The presence of a handful of β-barrel proteins in the outer mitochondrial membrane is a relic of the endosymbiotic origin of the organelle. β-barrel proteins can be found in the outer membrane of Gram-negative bacteria, mitochondria, and plant plastids, which are also of endosymbiotic origin [10–12]. These proteins carry out essential functions for the physiology and biogenesis of mitochondria. The β-barrel of Tom40 forms the protein-conducting channel of the translocase of the outer membrane (TOM complex), which imports about 1000 proteins from the cytosol into the mitochondria [13–19]. The β-barrel protein Sam50 (also termed Tob55, topogenesis of mitochondrial outer membrane β-barrel proteins) is the central subunit of the sorting and assembly machinery (SAM), which promotes membrane insertion and assembly of β-barrel proteins [20–25]. The voltage-dependent anion channel (VDAC) is the most abundant outer membrane protein and allows passage of small metabolites and ions via its β-barrel. There are three VDAC isoforms in humans and two isoforms in baker's yeast mitochondria [26–28]. Finally, the yeast-specific β-barrel protein Mdm10 mediates the assembly of the TOM complex and forms the membrane anchor for the endoplasmic reticulum (ER) mitochondria encounter structure (ERMES) that forms a molecular bridge between ER and mitochondria to facilitate lipid trafficking [24, 29–34]. There are some structural variations between the mitochondrial β-barrel proteins. Tom40, VDACs, and Mdm10 form a β-barrel with 19 β-strands, whereas the β-barrel of Sam50 consists of 16 β-strands [18, 19, 24, 35–39].

All β-barrel proteins are made on cytosolic ribosomes as precursor proteins and have to be imported into mitochondria. Cytosolic chaperones like Hsp70 and its co-chaperones guide the precursor proteins to the mitochondrial surface [40]. Receptors of the TOM complex recognize incoming β-barrel precursor proteins via a targeting signal that is formed by a β-hairpin motif [41]. The TOM complex transports these proteins across the outer membrane [21, 42]. Subsequently, the SAM complex folds and inserts these proteins into the target membrane (Fig. 1) [23, 43]. TOM and SAM complexes form a supercomplex to ensure an efficient transfer of the precursor proteins [43, 44]. The small TIM chaperones of the intermembrane space support the transfer of β-barrel proteins from the TOM to the SAM complex [43, 45–47]. The SAM complex binds to the precursor proteins via a conserved C-terminal motif within the last β-strand, termed β-signal [10]. This protein machinery consists of the central subunit Sam50 and the associated subunits Sam35 and Sam37. Sam50 comprises a 16-stranded β-barrel protein and a N-terminal polypeptide transport-associated (POTRA) domain. Sam50 is a

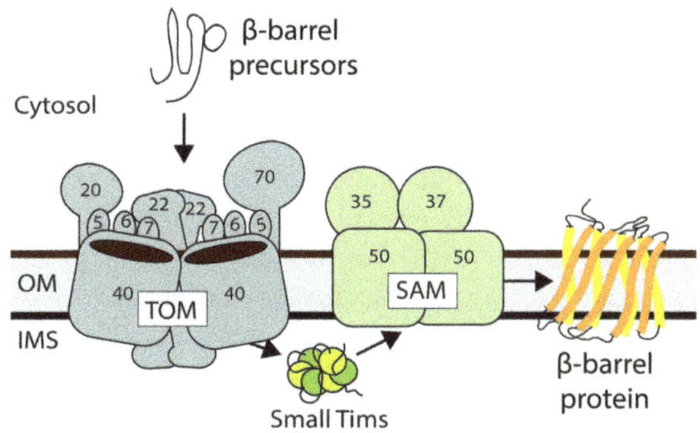

Fig. 1 Import of β-barrel proteins. Precursors of β-barrel proteins are synthesized in the cytosol and imported into the intermembrane space of mitochondria by the translocases of the outer membrane (TOM complex). TOM and SAM complexes form a supercomplex to ensure efficient transfer of the precursor protein. The small TIM chaperones facilitate the transfer of the precursor protein to the SAM complex. The SAM complex inserts and folds β-barrel precursor proteins into the outer mitochondrial membrane

homolog of the bacterial BamA, which is the central subunit of the β-barrel assembly machinery (BAM complex) that inserts β-barrel proteins into the outer membrane of Gram-negative bacteria [48].

The SAM complex mediates the folding of β-barrel proteins and their integration into the outer membrane by an elegant mechanism. In vitro protein import assays and recent cryo-electron microscopic structures led to a detailed view of how the SAM complex operates in the biogenesis of mitochondrial β-barrel proteins [23, 24, 39, 43, 49]. The β-barrel of Sam50 forms a lateral gate between its first and last β-strand to integrate the β-barrel precursor protein into the outer membrane [23]. Thereby, hairpin structures of two β-strands are inserted in a stepwise manner into the lateral gate, leading to the transient formation of a hybrid β-barrel of Sam50 and the growing precursor protein [23, 25]. Once the folding of the β-barrel is completed, the protein is released into the outer membrane. A β-barrel switching mechanism is important for the maturation of β-barrel proteins. The SAM complex contains two copies of Sam50. The growing β-barrel of the precursor proteins at the lateral gate displaces the second Sam50 subunit. To release the folded precursor protein into the outer membrane, another β-barrel protein such as Mdm10 or a second Sam50 has to bind to the SAM complex [24]. Binding of Mdm10 leads to the release of Tom40 from the SAM complex, whereas the association of the second Sam50 is sufficient to release VDAC from the SAM complex [24]. The SAM complex not only mediates the insertion and folding of β-barrel proteins but also forms a platform to assemble the Tom40 precursor with other subunits of the TOM complex [49–52].

Import of radiolabelled β-barrel precursor proteins into isolated yeast mitochondria is a key experiment that enabled the identification and characterization of the maturation pathway of mitochondrial β-barrel proteins. The usage of the blue native gel system allowed the analysis of how β-barrel proteins such as Tom40 are assembled in a stepwise manner with their partner proteins to form the mature complex [21, 42, 50–53]. The protein import assay was further developed to accumulate and characterize import intermediates and to monitor the formation of the β-barrel in isolated mitochondria [23, 43, 54]. Altogether, the protein import assay with its various applications is a central experiment to elucidate the steps of the biogenesis of mitochondrial β-barrel proteins.

2 Materials

2.1 In Vitro Transcription

1. In vitro transcription kit (e.g., MEGAscript SP6 kit (Thermo Fisher Scientific) or mMESSAGE mMACHINE Transcription kit (Thermo Fisher, Waltham, USA)).
2. RNA purification kit (e.g., MEGAclear Transcription Clean-Up kit (Thermo Fisher, Waltham)).

2.2 Cell-Free Protein Synthesis

1. Rabbit reticulocyte lysate system (e.g., TNT® Quick Coupled Transcription/Translation Kit (Promega, Madison, USA) or Flexi rabbit reticulocyte lysate Kit (Promega, Madison, USA)).
2. Plasmid DNA free of ethanol, calcium, RNase, and salt for coupled transcription/translation reactions or highly purified RNA for in vitro translation reaction.
3. 10 mCi/mL [^{35}S]methionine (e.g., Perkin Elmer, Waltham, USA).
4. 200 mM methionine.
5. 1.5 M sucrose.

2.3 Isolation of Mitochondria

1. DTT buffer: 100 mM Tris-H_2SO_4 pH 9.4, 10 mM dithiothreitol (DTT) should be freshly added before using the buffer from 1 M stock in dH_2O that can be stored at −20 °C.
2. Zymolyase buffer: 20 mM KP_i pH 7.4, 1.2 M sorbitol. A stock of 100 mL KP_i pH 7.4 is prepared by mixing 80.2 mL of 1 M K_2HPO_4 with 19.8 mL of 1 M KH_2PO_4.
3. Homogenization buffer: 600 mM sorbitol, 10 mM Tris-HCl pH 7.4, 1 mM EDTA, 1 mM PMSF, 0.2% (w/v) fatty acid–free bovine serum albumin (BSA). PMSF is freshly added before use. Stock solutions of 200 mM PMSF are prepared in 100% (v/v) isopropanol and can be stored for up to 6 months. The homogenization buffer should be pre-cooled at 4 °C (*see* **Note 1**).

4. SEM buffer: 250 mM sucrose, 10 mM MOPS-KOH pH 7.2, 1 mM EDTA.
5. Zymolyase 20T (e.g., from *Arthrobacter luteus*, Nacalai Tesque).

2.4 Protein Import Assay into Isolated Mitochondria

1. Import buffer: 250 mM sucrose (*see* **Note 2**), 2 mM KH_2PO_4, 5 mM methionine, 5 mM $MgCl_2$, 80 mM KCl, 3% (w/v) fatty acid–free BSA, 10 mM MOPS-KOH pH 7.2. Aliquots of import buffer are stored at −20 °C.
2. SEM buffer: 250 mM sucrose, 1 mM ETDA, 10 mM MOPS-KOH pH 7.2.
3. 200 mM ATP pH 7.2 prepared in H_2O, pH adjusted with KOH. Aliquots are stored at −20 °C.
4. 500 mM creatine phosphate in H_2O with aliquots stored at −20 °C.
5. 10 mg/mL creatine kinase in H_2O with aliquots stored at −20 °C.
6. 200 mM NADH in H_2O, prepared freshly.
7. Laemmli buffer: 60 mM Tris-HCl pH 6.8, 2% (w/v) sodium-dodecylsulfate (SDS), 10% (v/v) glycerol, 0.01% (w/v) bromophenol blue, 1% (v/v) β-mercaptoethanol.

2.5 Blue Native Electrophoresis

1. 10× Anode buffer: 500 mM Bis-Tris-HCl pH 7.0. Store at 4 °C.
2. 10× Cathode buffer: 500 mM Tricine, 150 mM Bis-Tris pH 7.0, 0.2% (w/v) Coomassie blue G-250. Store at 4 °C.
3. 3× Gel buffer: 200 mM ε-amino-n-caproic acid, 150 mM Bis-Tris-HCl pH 7.0.
4. Acrylamide solution: 49.5% (w/v) acrylamide, 3% (w/v) bis-acrylamide (*see* **Note 3**).
5. 10× BN-loading dye: 5% (w/v) Coomassie blue G-250, 500 mM ε-amino-n-caproic acid, 100 mM Bis-Tris pH 7.0.
6. Solubilization buffer: 0.4–1% (w/v) digitonin, 20 mM Tris-HCl pH 7.4, 2 mM EDTA pH 8, 50 mM NaCl, 10% (v/v) glycerol, 1 mM PMSF. Store at 4 °C without digitonin. PMSF and digitonin are added freshly before usage.
7. SEM buffer: 250 mM sucrose, 1 mM ETDA, 10 mM MOPS-KOH pH 7.2.
8. High molecular weight marker (Cytiva, HMW Native Marker Kit): Solve the marker in solubilization buffer (without digitonin) and add 10× BN-loading dye.

2.6 Proteinase K Treatment

1. 2.5 mg/mL proteinase K prepared freshly in SEM buffer.
2. 200 mM phenylmethylsulfonyl fluoride (PMSF) in 100% (v/v) isopropanol.

2.7 Carbonate Extraction

1. 100 mM Na_2CO_3 pH 11.5.
2. 10% (w/v) trichloroacetic acid (TCA).

2.8 Import of Chemical Amounts of β-Barrel Proteins

1. Wheat germ-based translation kit (e.g., RTS 1000 Wheat Germ CECF kit, Biotechrabbit, Philadelphia, USA).

3 Methods

3.1 In Vitro Transcription

Proteins are produced in a cell-free translation system using mRNA as template. RNA of high quality and sufficient concentration is required for in vitro translation. Therefore, genes of interest are frequently cloned into plasmids like the pGEM4 or pBluescript series under the control of a phage-derived promoter (e.g., SP6, T3, or T7). Alternatively, PCR products can be used as templates. The PCR template must contain the open reading frame, the information for a ribosomal binding site (Kozak sequence), and a strong promoter such as the SP6 promoter [55]. In both cases, the in vitro transcription is performed at 37 °C using an in vitro transcription kit following the manufacturer's recommendations. To avoid degradation of the RNA molecules, RNase-free tips and tubes should be used. Furthermore, RNase inhibitors may be added to the transcription reaction. Subsequently, mRNAs are purified and used for in vitro translation.

1. Use about 0.1–0.2 μg purified PCR-product or 1 μg purified plasmid-DNA for a standard transcription reaction of 20 μL volume.
2. Incubate the reaction at 37 °C for 2 h following the manufacturer's recommendations.
3. Purify the mRNA using a transcript purification kit following the manufacturer's recommendations.
4. Determine the concentration of the purified mRNA.

3.2 Cell-Free Protein Synthesis

Precursor proteins are typically labelled with [^{35}S]methionine in the in vitro translation to allow their detection by autoradiography after import into isolated mitochondria. Since [^{35}S] has higher energy and considerably shorter half-life compared to [^{3}H] and [^{14}C], it is suitable for easy detection, storage, and disposal, making [^{35}S] the preferable isotope for protein labeling [56].

For in vitro protein synthesis, crude cell extracts are obtained from cultured cells with a high rate of protein synthesis. Rabbit reticulocyte lysates are commonly used for in vitro translation to label proteins for protein import studies. Alternatively, yeast extracts or wheat germ lysates can be used [54, 57]. Yeast cell extracts have to be self-prepared in a multi-step procedure [57–59]. We will focus here on the description of how labeled proteins are produced in rabbit reticulocyte lysate. Reticulocyte lysates (without native DNA and mRNA) are supplemented with macromolecular components needed for translation, including an energy regeneration system (creatine phosphate and creatine phosphokinase), hemin to stabilize the translation initiation factor eIF2α and calf liver tRNAs to enable efficient translation [60].

Two different strategies can be used for in vitro translation systems in reticulocyte lysates: the coupled transcription/translation (TNT Quick Coupled Transcription/Translation) and separate transcription and translation reactions (Flexi Rabbit Reticulocyte Lysate). The TNT-Quick Coupled Transcription/Translation kit allows coupled transcription and translation reaction in the test tube and is powerful when plasmid-based templates like constructs in pGEM4z are used. Alternatively, high-quality mRNA can be used for the Flexi Rabbit Reticulocyte Lysate kit.

3.2.1 In Vitro Synthesis of Radiolabelled Proteins with the Coupled Transcription/Translation Reaction

1. The standard reaction mix:

Component	Amount (μL)
TNT SP6/T7 Quick Master Mix	40
10 mCi/mL [^{35}S]methionine	2
Template DNA (0.5–1 μg)	1–8
Nuclease-free H_2O to final volume of	50

2. Incubate at 30 °C for 60–90 min.
3. Add sucrose to a final concentration of 250 mM to the translation reaction to ensure isosmotic conditions in the import reaction. Add cold methionine (final concentration 25 mM) to block further labeling of proteins with [^{35}S]methionine (*see* **Note 4**).
4. Validate translation products by SDS-PAGE and autoradiography.

3.2.2 In Vitro Synthesis of Radiolabelled Proteins with the Translation Reaction Kit

1. The standard translation reaction mix (*see* **Note 5**):

Component	Amount (μL)
Flexi Rabbit Reticulocyte Lysate	33
10 mCi/mL [^{35}S]methionine	2
Amino acid mixture without methionine (1 mM)	1
MgOAc (25 mM)	1.4
KCl (2.5 M)	1.4
Ribonuclease Inhibitor (40 U/μL)	1
Template mRNA (2 μg)	1–10
Nuclease-free H_2O to final volume of	50

2. Incubate at 30 °C for 60–90 min.
3. Add sucrose to a final concentration of 250 mM to the translation reaction to ensure isosmotic conditions in the import reaction.
4. Add cold methionine (final concentration 25 mM) to block further labeling of proteins with [^{35}S]methionine (*see* **Note 4**).
5. Control translation products by SDS-PAGE followed by autoradiography.

3.3 Isolation of Mitochondria

The isolation of intact, import-competent mitochondria is a critical step for the in vitro protein import assay. Protocols for the purification of yeast and human mitochondria are well-established [55, 61]. We will focus here on the purification of yeast mitochondria. Yeast mitochondria are isolated from an exponentially growing culture (ideally between OD_{600} 1.0–1.5) in complete or selective media. After isolating the cells, the cell wall is removed by zymolyase treatment, followed by mechanical breakage of the cells. Mitochondria are then isolated by differential centrifugation.

1. Collect yeast cell pellets by centrifugation at 2500 × *g* at 20 °C for 8 min.
2. Carefully discard the supernatant and wash the pellet with dH_2O, centrifuge at 2500 × *g* at 20 °C for 8 min.
3. Determine the weight of the cell pellet.
4. Resuspend pellet in DTT buffer with 2 mL buffer per g cells.
5. Incubation for 20–30 min under constant shaking at 30 °C in DTT buffer. In case the used yeast mutant strains are temperature sensitive, this step can also be carried out at 24 °C for 30 min.

6. Centrifuge at 2500 × *g* at 20 °C for 5 min to collect the cells and discard the supernatant.
7. Resuspend cells in zymolyase buffer and centrifuge at 2500 × *g* at 20 °C for 5 min.
8. Critical step: Carefully resuspend cells in 7 mL/g cells zymolyase buffer with 3–4 mg zymolyase for every gram of cell pellet. Incubate under constant shaking at 30 °C for 30–45 min (or at 24 °C for 45 min for temperature-sensitive strains). Zymolyase degrades the cell wall of yeast, resulting in spheroplast formation. The spheroplast formation should be controlled (*see* **Note 5**).
9. Re-isolate spheroplasts, wash with zymolyase buffer, and collect them at 2500 × *g* at 20 °C for 5 min.
10. Discard supernatant, and resuspend pellet in 6.5 mL/g pellet in precooled (4 °C) homogenization buffer.

    All following steps have to be performed on ice (4 °C) to minimize the activity of proteases. In addition, the serine protease inhibitor PMSF is added to the homogenization buffer.
11. Critical step: Carefully homogenize the pellet in homogenization buffer with 15–18 strokes with a glass dounce homogenizer. Homogenization should be performed on ice, very slowly with a short time gap between each stroke to minimize bubble formation and nonspecific oxidation of cellular membranes.
12. Centrifuge at 2500 × *g* at 4 °C for 5 min to remove cell debris and nuclei.
13. Transfer supernatant to a fresh tube and enrich mitochondria by centrifugation at 17,000 × *g* at 4 °C for 15 min, and discard the supernatant.
14. Resuspend the pellet (containing mitochondria) in a small volume of SEM buffer, which depends on the size of the pellet.
15. Centrifuge at 2500 × *g* at 4 °C for 5 min to remove the remaining cell debris and nuclei.
16. Transfer the supernatant into a fresh tube and centrifuge at 17,000 × *g* at 4 °C for 15 min to isolate pure mitochondria.
17. Resuspend the mitochondrial pellet in a small volume of SEM buffer and measure the protein concentration using the Bradford assay.
18. Adjust concentration to 10 mg/mL with SEM buffer and prepare aliquots of about 50 μL.
19. Aliquots should be frozen in liquid nitrogen and stored at −80 °C until further use.

3.4 Protein Import Assay into Isolated Mitochondria

The in vitro import of proteins into purified mitochondria is a powerful assay to investigate β-barrel biogenesis. To perform these experiments, ^{35}S-labelled precursor proteins are imported into intact isolated mitochondria. Following this procedure, kinetics, and requirements of protein import can be analyzed [55, 62]. The protein import assay was initially established for precursor proteins that contain a cleavable presequence. Since the presequence is removed upon import by the mitochondrial processing peptidase (MPP), the import of such precursor proteins can be monitored by a size shift of the mature band on SDS-PAGE. β-barrel proteins do not contain any cleavable presequences. Their import into mitochondria can be studied by their protection against externally added proteinase K since imported β-barrel proteins into the outer membrane are hardly accessible to proteinase K [10, 42, 53, 63, 64]. Alternatively, membrane integration of imported β-barrel proteins can be analyzed by their resistance to carbonate extraction [43]. A commonly used approach to study the biogenesis of β-barrel proteins is the combination of the import assay with blue native electrophoresis [42, 53]. This approach allows the characterization of import intermediates such as binding of precursor proteins to the SAM complex and the stepwise assembly with partner proteins (Fig. 2). It is well-established for yeast mitochondria but was also used to monitor the import of Tom40 in human mitochondria [65]. Here, we describe the protein import procedure followed by analysis on blue native electrophoresis, which is commonly used to study the biogenesis of mitochondrial β-barrel proteins.

1. Prepare import reaction mix for each time point as follows:

Component	Amount (μL)
Import buffer	81
NADH 200 mM	1
ATP 200 mM	1
Creatine phosphate 500 mM	1
Creatine kinase 10 mg/mL	1
50 μg mitochondria (10 mg/mL protein concentration)	5
Total volume	90

2. Prewarm import samples at 25 °C for 2 min.
3. To start the import reaction, add 5–10 μL of reticulocyte lysate with radiolabelled precursor proteins to the import reaction mix.
4. Perform the import for different time periods. Collect samples after each time point and transfer them on ice. The import reaction will be stopped on ice.

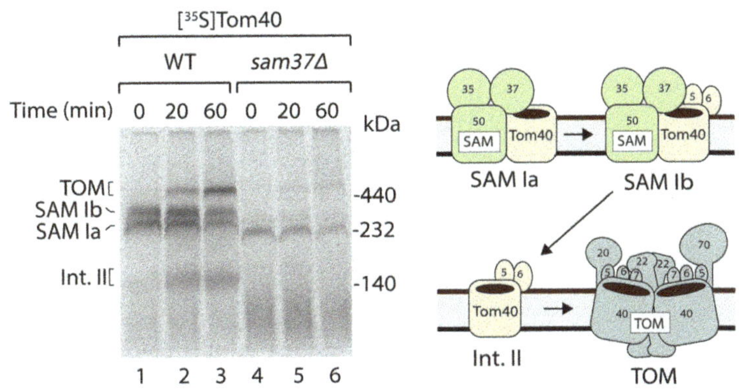

Fig. 2 Analysis of the assembly of Tom40 into the TOM complex. Import of radiolabelled Tom40 into isolated yeast mitochondria from wild-type (WT) and *sam37Δ* strains was performed for the indicated time periods. Subsequently, mitochondria were re-isolated and lysed with digitonin. The different assembly steps of the imported Tom40 precursor were analyzed by blue native electrophoresis and autoradiography. The Tom40 precursor binds first to the SAM complex (SAM-Ia), where the association of the small Tom proteins already occurs (SAM-Ib). The Tom40-small Toms module dissociates from the SAM complex (Int. II). Addition of Tom22 leads to the formation of the fully assembled TOM complex (TOM)

5. Centrifuge at 13,000 × *g* at 4 °C for 10 min to re-isolate mitochondria and remove the supernatant.
6. Wash the pellet with SEM buffer.
7. Centrifuge at 13,000 × *g* at 4 °C for 10 min to re-isolate mitochondria, discard supernatant.
8. For analysis on blue native electrophoresis, mitochondrial pellets are resuspended with solubilization buffer containing 1% (w/v) digitonin in a final protein concentration of 1 mg/mL.
9. Incubate on ice for 15 min to lyse the mitochondrial membranes.
10. Centrifuge at 4 °C for 10 min, at 13,000 × *g* to remove non-solubilized materials.
11. Transfer supernatant to a new microcentrifuge tube and add 5 μL 10× BN-loading dye to the samples for blue native electrophoresis.

3.5 Blue Native Electrophoresis

The blue native electrophoresis has been established as a powerful tool to analyse protein complexes, including mitochondrial respiratory chain complexes and protein translocases [21, 42, 66–69] (*see also* Chapter 9 of this book). For blue native electrophoresis, mitochondrial membranes are solubilized with a nonionic detergent like digitonin to maintain the integrity of large mitochondrial protein assemblies. The solubilized sample is mixed with a BN-loading dye containing ε-amino-n-caproic acid and the negatively charged Coomassie dye. Coomassie binds to proteins thereby conferring the negative charge on the proteins to ensure proper running behavior during native electrophoresis.

Gel solutions for blue native electrophoresis can be prepared in advance and stored for up to a week at 4 °C (without the addition of APS and TEMED) (for hints for the preparation of the solutions *see* **Notes 6–9**). The acrylamide concentration used for the blue native gels depends on the size range of complexes to be analyzed. For example, blue native gels with an acrylamide concentration of 6–13% are typically used to study the assembly of the TOM complex or porin complexes.

Solutions for blue native gels:

For one large gel	4%	6%	13%	16.5%	Stacking (3.5%)
3× gel buffer (mL)	3	3	3	3	2.5
Acrylamide solution (mL)	0.73	1.07	2.35	3.05	0.6
Glycerol (mL)	–	–	1.8	1.8	–
10% APS (μL)	38	38	30	30	60
TEMED (μL)	3.8	3.8	3	3	6
H_2O (mL)	5.228	4.89	1.817	1.117	4.367

A gradient maker with a pump is used to cast a polyacrylamide gradient gel. Once the resolving gel is polymerized, a stacking gel is cast on top (*see* **Note 10**). The gel should be cooled to 4 °C before use.

Blue native electrophoresis:

1. The blue native gel chamber is placed on ice or precooled at 4 °C containing anode buffer. Cathode buffer with 0.02% (w/v) Coomassie G-250 should be used.
2. Samples are loaded on the gel using a Hamilton syringe. For lanes with no samples, lysis buffer or lysed old mitochondria may be loaded to ensure straight running of the samples.
3. Electrophoresis is run at 100 V until the samples enter the separation gel.
4. For the remaining run of the gel, the voltage is increased to 600 V.
5. After the run, destain the gel until the bands of the complexes become visible.
6. Dry the gel on Whatman paper by using a gel dryer with a vacuum pump.
7. Analyze the radioactive signals via autoradiography.

3.6 Proteinase K Treatment

After import into isolated mitochondria, proteinase K treatment can be used to assess import efficiency. The proteinase K degrades all non-imported precursor proteins. Under the used conditions, the protease cannot penetrate the membrane to digest imported proteins. Therefore, resistance against proteinase K indicates protein import into mitochondria. For proteins integrated into the outer membrane such as β-barrel proteins, domains exposed to the cytosol will be cleaved, resulting in characteristic fragments of the integrated proteins, which can be used to assess import efficiency.

1. After the import reaction, the mitochondrial pellet is resuspended in 100 μL SEM buffer and divided into two samples of 45 μL.
2. One half is treated with 1 μL proteinase K (of 2.5 mg/mL stock) for 15 min, while the other is stored on ice until the next step.
3. To inactivate proteinase K, 1 μL 200 mM PMSF is added to the reaction mix and samples are incubated on ice for 10 min.
4. Centrifuge at 13,000 × *g* at 4 °C for 10 min to reisolate mitochondria.
5. Discard the supernatant, and wash the pellet with 100 μL SEM buffer. Centrifuge at 13,000 × *g* at 4 °C for 5 min.
6. Discard supernatant, and prepare samples for SDS-PAGE by a resuspending pellet in Laemmli buffer [70].
7. Heat samples for 5 min at 95 °C to denature proteins, and separate proteins by denaturing SDS-PAGE.
8. The gel is stained with Coomassie and dried on Whatman paper by using a gel dryer with a vacuum pump. The radiolabelled proteins can then be detected by autoradiography.

3.7 Carbonate Extraction

Carbonate extraction can be used to check membrane integration of the imported protein. After the import, mitochondria are incubated with 100 mM Na_2CO_3 pH 11.5. Under the basic pH, peripherally membrane-associated proteins are removed from the membrane, while integral membrane proteins remain unaffected [71, 72]. A limitation of these assays is that aggregation of hydrophobic precursor proteins may occur and mask the signal of inserted proteins. A preincubation of the precursor proteins in import buffer reduces the amount of aggregation-prone precursor proteins (*see* **Note 11**).

1. After import, pellet mitochondria at 13,000 × *g* at 4 °C for 15 min and wash the pellet with SEM buffer.
2. The supernatant is removed and the pellet is resuspended into 200 μL 0.1 M Na_2CO_3 pH 11.5, followed by 30 min incubation on ice (*see* **Note 12**).

3. Transfer the samples into microcentrifuge tubes that can be used in an ultracentrifuge.
4. Centrifuge at 125,000 × *g* for 30 min at 4 °C. Collect supernatant and pellet separately.
5. Protein precipitation with 10% (w/v) trichloroacetic acid (TCA) of the supernatant can be performed to allow analysis of the proteins on SDS-PAGE.
6. Resuspend pellet in Laemmli sample buffer [70], and incubate under vigorous shaking conditions for 10 min at 95 °C.
7. The total pellet and supernatant fractions are analyzed by SDS-PAGE and autoradiography.

3.8 Import of Chemical Amounts of β-Barrel Proteins

The import of radiolabelled proteins allows the detection of small amounts of proteins, which is a suitable approach to follow import, folding (*see* **Note 13**) and assembly of mitochondrial β-barrel proteins. To enrich import intermediates, import of recombinant amounts is a powerful assay. This approach combined with blue native electrophoresis or affinity purification enables the investigation of import and assembly intermediates via immunoblotting [50–52, 54]. For this purpose, large amounts of the proteins are generated in a cell-free translation system based on wheat germ extract. This system is based on a coupled transcription and translation reaction. Affinity tags like HA-, His-, or Flag-tag can be easily fused to the proteins of interest to allow their detection and affinity purification. The import into isolated yeast mitochondria follows the protocol described for the import of radiolabelled proteins as described above. This approach was used for different purposes. First, it can be used to perform in vitro complementation in mitochondria from single deletion strains of outer membrane proteins like yeast porin [54]. Second, it was successfully used to study the assembly of Tom40 with the small Tom proteins at the SAM complex [50]. Third, the import of His-tagged Tom40 allows the purification of import intermediates such as the SAM complex [24, 52]. Thus, the import of chemical amounts of β-barrel proteins is a powerful tool to analyze intermediate stages and to monitor different assembly steps of mitochondrial β-barrel proteins.

1. Prepare the import reaction mix as follows:

Component	Amount (μL)
Import buffer	81
NADH 200 mM	1
ATP 200 mM	1
Creatine phosphate 500 mM	1

(continued)

Component	Amount (μL)
Creatine kinase 10 mg/mL	1
50 μg mitochondria (10 mg/mL protein concentration)	5
Total volume	90

2. Prewarm import samples at 25 °C for 2 min.
3. To start the import reaction, add 2–10 μL of wheat germ lysate with the precursor proteins to the import reaction mix.
4. Perform the import at 25 °C. The import reaction will be stopped on ice.
5. Centrifuge at 13,000 × *g* at 4 °C for 10 min to re-isolate mitochondria and remove the supernatant.
6. Wash the pellet with SEM buffer.
7. Centrifuge at 13,000 × *g* at 4 °C for 10 min to re-isolate mitochondria, discard supernatant.
8. Subject mitochondria to further analysis.

4 Notes

1. Caution: PMSF is highly toxic. Solutions containing PMSF must be handled with appropriate safety measures.
2. Sucrose is important to maintain mitochondrial membrane integrity, while methionine in the import buffer prevents unspecific labeling of endogenous proteins with radiolabeled methionine from the lysate. Cold methionine should be added to quench the labeling reaction and to minimize unspecific attachment of [^{35}S]methionine to proteins [55].
3. Due to its highly toxic nature gloves must be used while preparing acrylamide. Work should be performed under a chemical safety hood.
4. The outcome of the in vitro translation could be poor. In this case, different salt concentrations (KOAc or $MgCl_2$) can be tested or the amount of RNA can be titrated.
5. To test efficient spheroplast formation, 50 μL of the yeast cell suspension can be diluted in 2 mL water. In case the cell walls are not properly degraded, the solution remains turbid and the treatment needs to be extended.
6. ε-amino-n-caproic acid is difficult to dissolve and can be stirred overnight.

7. A solution of 40× anode buffer may also be prepared.
8. Use the water-soluble Coomassie blue G-250 for blue native electrophoresis and not Coomassie blue R.
9. Acrylamide is generally prepared by overnight stirring in a safety hood at room temperature to ensure better homogenization and protection from light. The solution is filtered to remove unsolved particles.
10. The freshly prepared gradient gel should be topped with an isopropanol layer and allowed to polymerize to ensure gel running.
11. While importing a radioactively labeled protein, hydrophobic precursor proteins frequently aggregate in the reaction buffer. Such aggregated proteins can mask the signal of membrane-inserted proteins that are detected in the pellet fraction of a carbonate extraction. To minimize such unspecific aggregates, a pre-incubation of the lysate in the import reaction without mitochondria (**Steps 1–4** of the import protocol in Chapter 3, Subheading 3.5) can be performed.
12. For carbonate extraction, 100 mM Na_2CO_3 with pH 10.8 may also be used instead of the standard pH 11.5. Some membrane proteins like Pam18 are extracted from membranes using standard conditions with pH 11.5 [73].
13. The protein import assay detects whether or not a protein is imported into mitochondria. The protein import assay was also successfully modified to study the folding of β-barrel proteins in mitochondria [43]. To this end, specific variants of β-barrel proteins like Tom40 were imported into mitochondria. In these variants, all endogenous cysteines were genetically removed and a pair of cysteines was placed at specific sites of the β-barrel. One cysteine residue was introduced in the first β-strand, while a second cysteine was placed in the last β-strand. The rationale behind this strategy is that once the β-barrel is formed the first and last β-strand are directly adjacent to each other bringing the two cysteine residues in close proximity. In this folded stage, a disulfide bond can be formed between the two cysteines via oxidation, resulting in a migration shift on non-reducing SDS-PAGE [43]. This elegant assay was successfully used to uncover the role of the SAM complex in the folding of the β-barrel in mitochondria [43]. Furthermore, this approach was utilized to reveal the lateral gate of Sam50 by the formation of intermolecular disulfide bridges between Sam50 and the imported β-barrel protein [23]. Altogether, the import of variants with cysteine residues at specific locations combined with chemical oxidation allows to monitor the folding of β-barrel proteins within mitochondria.

Acknowledgments

We thank Nicole Zufall for expert technical assistance. This work was supported by grants of the Deutsche Forschungsgemeinschaft (DFG), BE4679/2-2 project ID 269424439 (to T.B.) and SFB1218 project ID 269925409 (to T.B.).

References

1. Archibald JM (2015) Endosymbiosis and eukaryotic cell evolution. Curr Biol 25:911–921
2. Zimorski V, Ku C, Martin WF, Gould SB (2014) Endosymbiotic theory for organelle origins. Curr Opin Microbiol 22:38–48
3. Spinelli JB, Haigis MC (2018) The multifaceted contributions of mitochondria to cellular metabolism. Nat Cell Biol 20:745–754
4. Pfanner N, Warscheid B, Wiedemann N (2019) Mitochondrial proteins: from biogenesis to functional networks. Nat Rev Mol Cell Biol 20:267–284
5. Morgenstern M, Stiller SB, Lübbert P, Peikert CD, Dannenmaier S, Drepper F et al (2017) Definition of a high-confidence mitochondrial proteome at quantitative scale. Cell Rep 19: 2836–2852
6. Morgenstern M, Peikert CD, Lübbert P, Suppanz I, Klemm C, Alka A et al (2021) Quantitative high-confidence human mitochondrial proteome and its dynamics in cellular context. Cell Metab 33:2464–2483
7. Rath S, Sharma R, Gupta R, Ast T, Chan C, Durham TJ et al (2021) MitoCarta3.0: an updated mitochondrial proteome now with sub-organelle localization and pathway annotations. Nucleic Acids Res 49:1541–1547
8. Hansen KG, Herrmann JM (2019) Transport of proteins into mitochondria. Protein J 38: 330–342
9. Araiso Y, Imai K, Endo T (2022) Role of the TOM complex in protein import into mitochondria: structural views. Annu Rev Biochem 91:679–703
10. Kutik S, Stojanovski D, Becker L, Becker T, Meinecke M, Krüger V et al (2008) Dissecting membrane insertion of mitochondrial β-barrel proteins. Cell 132:1011–1024
11. Zeth K (2010) Structure and evolution of mitochondrial outer membrane proteins of β-barrel topology. Biochim Biophys Acta 1797:1292–1299
12. Webb CT, Heinz E, Lithgow T (2012) Evolution of the β-barrel assembly machinery. Trends Microbiol 20:612–620
13. Hill K, Model K, Ryan MT, Dietmeier K, Martin F, Wagner R et al (1998) Tom40 forms the hydrophilic channel of the mitochondrial import pore for precursor proteins. Nature 395:516–521
14. Becker L, Bannwarth M, Meisinger C, Hill K, Model K, Krimmer T et al (2005) Precursor protein translocase of the outer mitochondrial membrane: reconstituted Tom40 forms a characteristic TOM pore. J Mol Biol 353:1011–1020
15. Suzuki H, Kadowaki T, Maeda M, Sasaki H, Nabekura J, Sakaguchi M et al (2004) Membrane-embedded C-terminal segment of rat mitochondrial TOM40 constitutes protein-conducting pore with enriched β-structure. J Biol Chem 279:50619–50629
16. Shiota T, Imai K, Qiu J, Hewitt VL, Tan K, Shen HH et al (2015) Molecular architecture of the active mitochondrial protein gate. Science 349:1544–1548
17. Bausewein T, Mills DJ, Langer JD, Nitschke B, Nussberger S, Kühlbrandt W (2017) Cryo-EM structure of the TOM core complex from *Neurospora crassa*. Cell 170:693–700
18. Araiso Y, Tsutsumi A, Qiu J, Imai K, Shiota T, Song J et al (2019) Structure of the mitochondrial import gate reveals distinct precursor protein paths. Nature 575:395–401
19. Tucker K, Park E (2019) Cryo-EM structure of the mitochondrial protein-import channel TOM complex at near-atomic resolution. Nat Struct Mol Biol 26:1158–1166
20. Gentle I, Gabriel K, Beech P, Waller R, Lithgow T (2004) The Omp85 family of proteins is essential for outer membrane biogenesis in mitochondria and bacteria. J Cell Biol 164: 19–24
21. Paschen SA, Waizenegger T, Stan T, Preuss M, Cyrklaff M, Hell K et al (2003) Evolutionary conservation of biogenesis of β-barrel membrane proteins. Nature 426:862–866
22. Kozjak V, Wiedemann N, Milenkovic D, Lohaus C, Meyer HE, Guiard B et al (2003) An essential role of Sam50 in the protein sorting and assembly machinery of the

mitochondrial outer membrane. J Biol Chem 278:48520–48523

23. Höhr AIC, Lindau C, Wirth C, Qiu J, Stroud DA, Kutik S et al (2018) Membrane protein insertion through a mitochondrial β-barrel gate. Science 359:eaah6834
24. Takeda H, Tsutsumi A, Nishizawa T, Lindau C, Busto JV, Wenz LS et al (2021) Mitochondrial sorting and assembly machinery operates by β-barrel switching. Nature 590:163–169
25. Takeda H, Busto JV, Lindau C, Tsutsumi A, Tomii K, Imai K et al (2023) A multipoint guidance mechanism for β-barrel folding on the SAM complex. Nat Struct Mol Biol 30: 176–187
26. Colombini M (2012) Mitochondrial outer membrane channels. Chem Rev 112:6373–6387
27. Campo ML, Peixoto PM, Martínez-Caballero S (2017) Revisiting trends on mitochondrial mega-channels for the import of proteins and nucleic acids. J Bioenerg Biomembr 49:75–99
28. Becker T, Wagner R (2018) Mitochondrial outer membrane channels: emerging diversity in transport processes. BioEssays 40:e1800013
29. Meisinger C, Rissler M, Chacinska A, Szklarz LK, Milenkovic D, Kozjak V et al (2004) The mitochondrial morphology protein Mdm10 functions in assembly of the precursor protein translocase of the outer membrane. Dev Cell 7: 61–71
30. Meisinger C, Pfannschmidt S, Rissler M, Milenkovic D, Becker T, Stjanovski D et al (2007) The morphology proteins Mdm12/Mmm1 function in the major β-barrel assembly pathway of mitochondria. EMBO J 26:2229–2239
31. Kornmann B, Currie E, Collins SR, Schuldiner M, Nunnari J, Weissman JS et al (2009) An ER-mitochondria tethering complex revealed by a synthetic biology screen. Science 325:477–481
32. Yamano K, Tanaka-Yamano S, Endo T (2010) Tom7 regulates Mdm10-mediated assembly of the mitochondrial import channel protein Tom40. J Biol Chem 285:41222–41231
33. Klein A, Israel L, Lackey SW, Nargang FE, Imhof A, Baumeister W et al (2012) Characterization of the insertase for β-barrel proteins of the outer mitochondrial membrane. J Cell Biol 199:599–611
34. Ellenrieder L, Opaliński Ł, Becker L, Krüger V, Mirus O, Straub SP et al (2016) Separating mitochondrial protein assembly and endoplasmic reticulum tethering by selective coupling of Mdm10. Nat Commun 7:13021
35. Hiller S, Garces RG, Malia TJ, Orekhov VY, Colombini M, Wagner G (2008) Solution structure of the integral human membrane protein VDAC-1 in detergent micelles. Science 321:1206–1210
36. Bayrhuber M, Meins T, Habeck M, Becker S, Giller K, Villinger S et al (2008) Structure of the human voltage-dependent anion channel. Proc Natl Acad Sci U S A 105:15370–15375
37. Ujwal R, Cascio D, Colletier JP, Faham S, Zhang J, Toro L et al (2008) The crystal structure of mouse VDAC1 at 2.3 A resolution reveals mechanistic insights into metabolite gating. Proc Natl Acad Sci 105:17742–17747
38. Flinner N, Ellenrieder L, Stiller SB, Becker T, Schleiff E, Mirus O (2013) Mdm10 is an ancient eukaryotic porin co-occurring with the ERMES complex. Biochim Biophys Acta 1833:3314–3325
39. Diederichs KA, Ni X, Rollauer SE, Botos I, Tan X, King MS et al (2020) Structural insight into mitochondrial β-barrel outer membrane protein biogenesis. Nat Commun 11:3290
40. Jores T, Lawatscheck J, Beke V, Franz-Wachtel-M, Yunoki K, Fitzgerald JC et al (2018) Cytosolic Hsp70 and Hsp40 chaperones enable the biogenesis of mitochondrial β-barrel proteins. J Cell Biol 217:3091–3108
41. Jores T, Klinger A, Groß LE, Kawano S, Flinner N, Duchardt-Ferner E et al (2016) Characterization of the targeting signal in mitochondrial β-barrel proteins. Nat Commun 7:12036
42. Wiedemann N, Kozjak V, Chacinska A, Schönfisch B, Rospert S, Ryan MT et al (2003) Machinery for protein sorting and assembly in the mitochondrial outer membrane. Nature 424:565–571
43. Qiu J, Wenz LS, Zerbes RM, Oeljeklaus S, Bohnert M, Stroud DA et al (2013) Coupling of mitochondrial import and export translocases by receptor-mediated supercomplex formation. Cell 154:596–608
44. Wenz LS, Ellenrieder L, Qiu J, Bohnert M, Zufall N, van der Laan M et al (2015) Sam37 is crucial for formation of the mitochondrial TOM-SAM supercomplex, thereby promoting β-barrel biogenesis. J Cell Biol 210:1047–1054
45. Wiedemann N, Truscott KN, Pfannschmidt S, Guiard B, Meisinger C, Pfanner N (2004) Biogenesis of the protein import channel Tom40 of the mitochondrial outer membrane: intermembrane space components are involved in an early stage of the assembly pathway. J Biol Chem 279:18188–18194
46. Hoppins SC, Nargang FE (2004) The Tim8-Tim13 complex of *Neurospora crassa* functions

in the assembly of proteins into both mitochondrial membranes. J Biol Chem 279: 12396–12405
47. Weinhäupl K, Lindau C, Hessel A, Wang Y, Schütze C, Jores T et al (2018) Structural basis of membrane protein chaperoning through the mitochondrial intermembrane space. Cell 175:1365–1379
48. Noinaj N, Gumbart JC, Buchanan SK (2017) The β-barrel assembly machinery in motion. Nat Rev Microbiol 15:197–204
49. Wang Q, Guan Z, Qi L, Zhuang J, Wang C, Hong S, Yan L et al (2021) Structural insight into the SAM-mediated assembly of the mitochondrial TOM core complex. Science 373: 1377–1381
50. Becker T, Guiard B, Thornton N, Zufall N, Stroud DA, Wiedemann N et al (2010) Assembly of the mitochondrial protein import channel: role of Tom5 in two-stage interaction of Tom40 with the SAM complex. Mol Biol Cell 21:3106–3113
51. Thornton N, Stroud DA, Milenkovic D, Guard B, Pfanner N, Becker T (2010) Two modular forms of the mitochondrial sorting and assembly machinery are involved in biogenesis of α-helical outer membrane proteins. J Mol Biol 396:540–549
52. Becker T, Wenz LS, Thornton N, Stroud D, Meisinger C, Wiedemann N et al (2011) Biogenesis of mitochondria: dual role of Tom7 in modulating assembly of the precursor protein translocase of the outer membrane. J Mol Biol 405:113–124
53. Rapaport D, Neupert W (1999) Biogenesis of Tom40, core component of the TOM complex of mitochondria. J Cell Biol 146:321–331
54. Stroud DA, Becker T, Qiu J, Stojanovski D, Pfannschmidt S, Wirth C et al (2011) Biogenesis of mitochondrial β-barrel proteins: the POTRA domain is involved in precursor release from the SAM complex. Mol Biol Cell 22: 2823–2833
55. Priesnitz C, Pfanner N, Becker T (2020) Studying protein import into mitochondria. Methods Cell Biol 155:45–79
56. Chen LC, Casadevall A (1999) Labeling of proteins with [^{35}S]methionine and/or [^{35}S] cysteine in the absence of cells. Anal Biochem 269:179–188
57. Wang F, Brown EC, Mak G, Zhuang J, Denic V (2010) A chaperone cascade sorts proteins for posttranslational membrane insertion into the endoplasmic reticulum. Mol Cell 40:159–171
58. Fünfschilling U, Rospert S (1999) Nascent polypeptide-associated complex stimulates protein import into yeast mitochondria. Mol Biol Cell 10:3289–3299
59. Schuldiner M, Metz J, Schmid V, Denic V, Rakwalska M, Schmitt HD et al (2008) The GET complex mediates insertion of tail-anchored proteins into the ER membrane. Cell 134:634–645
60. Pelham HR, Jackson RJ (1976) An efficient mRNA-dependent translation system from reticulocyte lysates. Eur J Biochem 67:247–256
61. Meisinger C, Pfanner N, Truscott KN (2006) Isolation of yeast mitochondria. Methods Mol Biol 313:33–39
62. Mokranjac D, Neupert W (2007) Protein import into isolated mitochondria. Methods Mol Biol 372:277–286
63. Krimmer T, Rapaport D, Ryan MT, Meisinger C, Kassenbrock CK, Blachly-Dyson E et al (2001) Biogenesis of porin of the outer mitochondrial membrane involves an import pathway via receptors and the general import pore of the TOM complex. J Cell Biol 152: 289–300
64. Habib SJ, Waizenegger T, Lech M, Neupert W, Rapaport D (2005) Assembly of the TOB complex of mitochondria. J Biol Chem 280:6434–6440
65. Humphries AD, Streinmann IC, Stojanovski D, Johnston AJ, Yano M, Hoogenraad NJ et al (2005) Dissection of the mitochondrial import and assembly pathway of human Tom40. J Biol Chem 280:11535–11543
66. Schägger H, von Jagow G (1991) Blue native electrophoresis for isolation of membrane protein complexes in enzymatically active form. Anal Biochem 199:223–231
67. Schulte U, den Brave F, Haupt A, Gupta A, Song J, Müller CS et al (2023) Mitochondrial complexome reveals quality-control pathways of protein import. Nature 614:153–159
68. Schägger H, Pfeiffer K (2000) Supercomplexes in the respiratory chains of yeast and mammalian mitochondria. EMBO J 19:1777–1783
69. Dekker PJ, Martin F, Maarse AC, Bömer U, Müller H, Guiard B et al (1997) The Tim core complex defines the number of mitochondrial translocation contact sites and can hold arrested precursor proteins in the absence of matrix Hsp70-Tim44. EMBO J 16:5408–5419
70. Laemmli UK (1970) Cleavage of structural proteins during the assembly of the head of bacteriophage T4. Nature 227:680–685
71. Fujiki Y, Hubbard AL, Fowler S, Lazarow PB (1982) Isolation of intracellular membranes by

means of sodium carbonate treatment: application to endoplasmic reticulum. J Cell Biol 93: 97–102

72. Doan KN, Grevel A, Mårtensson CU, Ellenrieder L, Thornton N, Wenz LS et al (2020) The mitochondrial import complex MIM functions as main translocase for α-helical outer membrane proteins. Cell Rep 31:107567

73. Truscott KN, Pfannschmidt S, Guiard B, Meisinger C, Pfanner N (2003) A J-protein is an essential subunit of the presequence translocase-associated protein import motor of mitochondria. J Cell Biol 163:707–713

Chapter 14

Tracking the Activity and Position of Mitochondrial β-Barrel Proteins

Shuo Wang and Stephan Nussberger

Abstract

Total interference reflection fluorescence (TIRF) microscopy of lipid bilayers is an effective technique for studying the lateral movement and ion channel activity of single integral membrane proteins. Here we describe how to integrate the mitochondrial outer membrane preprotein translocase TOM-CC and its β-barrel protein-conducting channel Tom40 into supported lipid bilayers to identify possible relationships between movement and channel activity. We propose that our approach can be readily applied to membrane protein channels where transient tethering to either membrane-proximal or intramembrane structures is accompanied by a change in channel permeation.

Key words TOM, β-barrel membrane protein, Channels, Mitochondria, Protein translocation, Single molecule, TIRF microscopy

1 Introduction

The functional properties of ion channels and β-barrel porins in biological membranes are traditionally studied by patch-clamp methods and by measurements of the ionic conductivity of purified proteins integrated into planar lipid membranes [1–3]. The analyses of their current/voltage characteristics have provided valuable insights into the channel kinetics, substrate selectivity, sensitivity to molecular ligands, accessory proteins, pH, temperature, and mechanical strain [4, 5]. Such analyses have proved successful not only in characterizing the large class of bacterial and eukaryotic β-barrel porins in terms of diffusion of ions and small metabolites through their water-filled pores [6, 7], but also in terms of defining the specificity of new potential antibiotics [8, 9]. Inspired by the success of nanopore sequencing of DNA and RNA [10–12], β-barrel pores have recently found their way into single-molecule protein sequencing [13–15] and the detection of posttranslational protein modifications [16]. Using the mitochondrial protein

Raffaele Ieva (ed.), *Transmembrane β-Barrel Proteins: Methods and Protocols*, Methods in Molecular Biology, vol. 2778,
https://doi.org/10.1007/978-1-0716-3734-0_14,

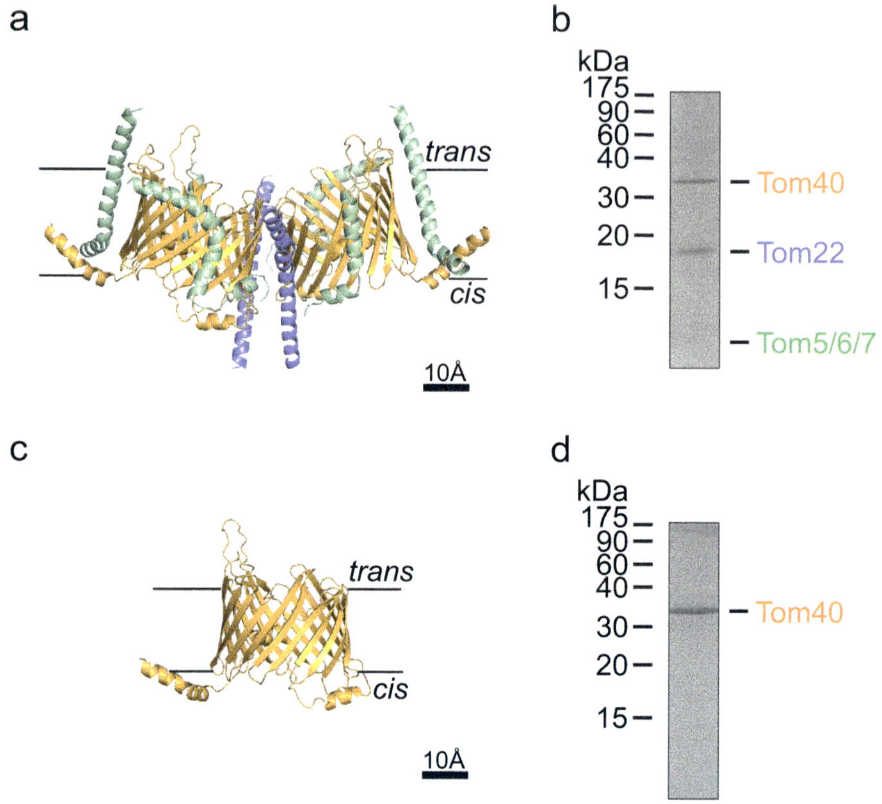

Fig. 1 Mitochondrial β-barrel channels TOM-CC and Tom40. (**a**) Cryo-EM structure and (**b**) SDS-PAGE of *N. crassa* TOM-CC (EMDB, EMD-3761; [18–20]). (**c**) Cryo-EM structure and (**d**) SDS-PAGE of *N. crassa* Tom40 (EMDB, EMD-3761; [17–19])

translocase (the TOM core complex or TOM-CC) and its pore-forming subunit Tom40 (Fig. 1) as examples [17–20], we describe here how β-barrel pore-forming proteins can be studied optically [21]. Thus, the method is complementary to those of classical electrophysiology [17, 22–27].

We describe how droplet-interface bilayers (DIBs) supported by an ultrathin agarose hydrogel [21, 28–32] are formed by lipid self-assembly through the contact of aqueous droplets with hydrogel-supported substrates in a lipid/oil phase (Fig. 2). Next, we show, using TIRF microscopy, how tracking and monitoring of channel activity is performed by measuring the fluorescence of Ca^{2+}-sensitive dyes, which respond to the Ca^{2+} ion flux through individual ion channels (Fig. 3). Finally, we describe how open-source software [33, 34] can be used to evaluate video sequences of individual channels to visualize their trajectories and channel activities over time (Fig. 4).

The main advantage of combining TIRF microscopy and DIBs over other techniques is that it allows simultaneous observation of lateral movement and opening and closing activity of individual

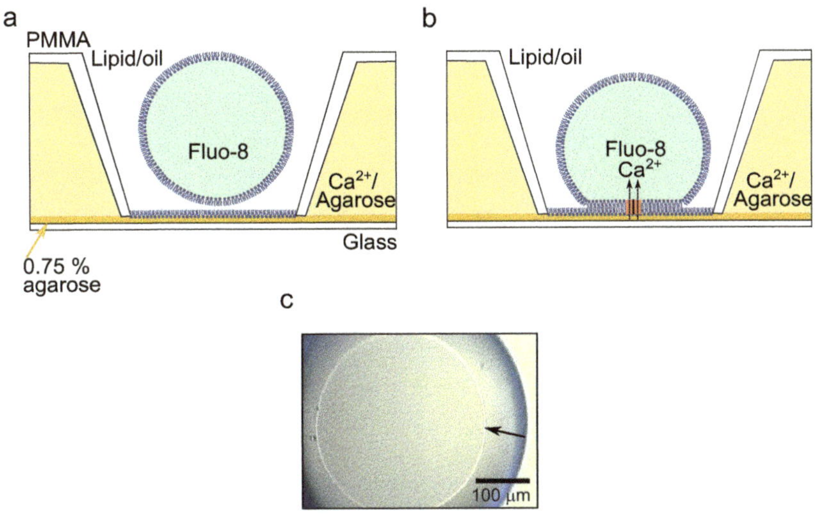

Fig. 2 Schematic representation of a droplet-interface bilayer supported by an agarose hydrogel. (**a**) An aqueous droplet with a lipid monolayer at the oil/water interface sinks onto a lipid monolayer on an ultrathin agarose film. (**b**) TIRF imaging of Ca^{2+} flux through membrane-integrated ion channels using a Ca^{2+}-sensitive fluorescent dye (Fluo-8) in trans. (**c**) DIB membrane imaged with Hoffmann modulation contrast microscopy. Arrow, contact area between hydrogel and droplet

channels without the need for fluorescent labeling. In addition, the application of an external membrane voltage that may interfere with lateral movement and channel function is not required. The approach described here complements optical methods, such as single particle tracking [35–37] and fluorescence correlation spectroscopy [38–42], which have provided valuable insights into membrane organization and membrane-dependent reactions [43–45]. In contrast to the classical voltage-induced gating observed in electrophysiological measurements, the opening of a channel is caused only by random thermal motion and interactions with other structures in or in the immediate vicinity below and/or above the membrane. Depending on how far individual protein domains protrude from the membrane, the mobility of the proteins can range from mobile to completely immobile. The combination of imaging channel activity and tracking of the protein in the membrane plane [21] allows us to decipher the potential interplay between the type of lateral mobility (moving vs immobile) and function (open vs. closed) of the protein. DIBs with integrated membrane proteins could thus serve as a model system for investigating mechanosensitive processes in which transient inclusion of membrane proteins in nanodomains and supercomplexes, as shown, e.g., in vivo for the mitochondrial protein translocases TOM and TIM23 during mitochondrial protein import [46–48] is of physiological importance.

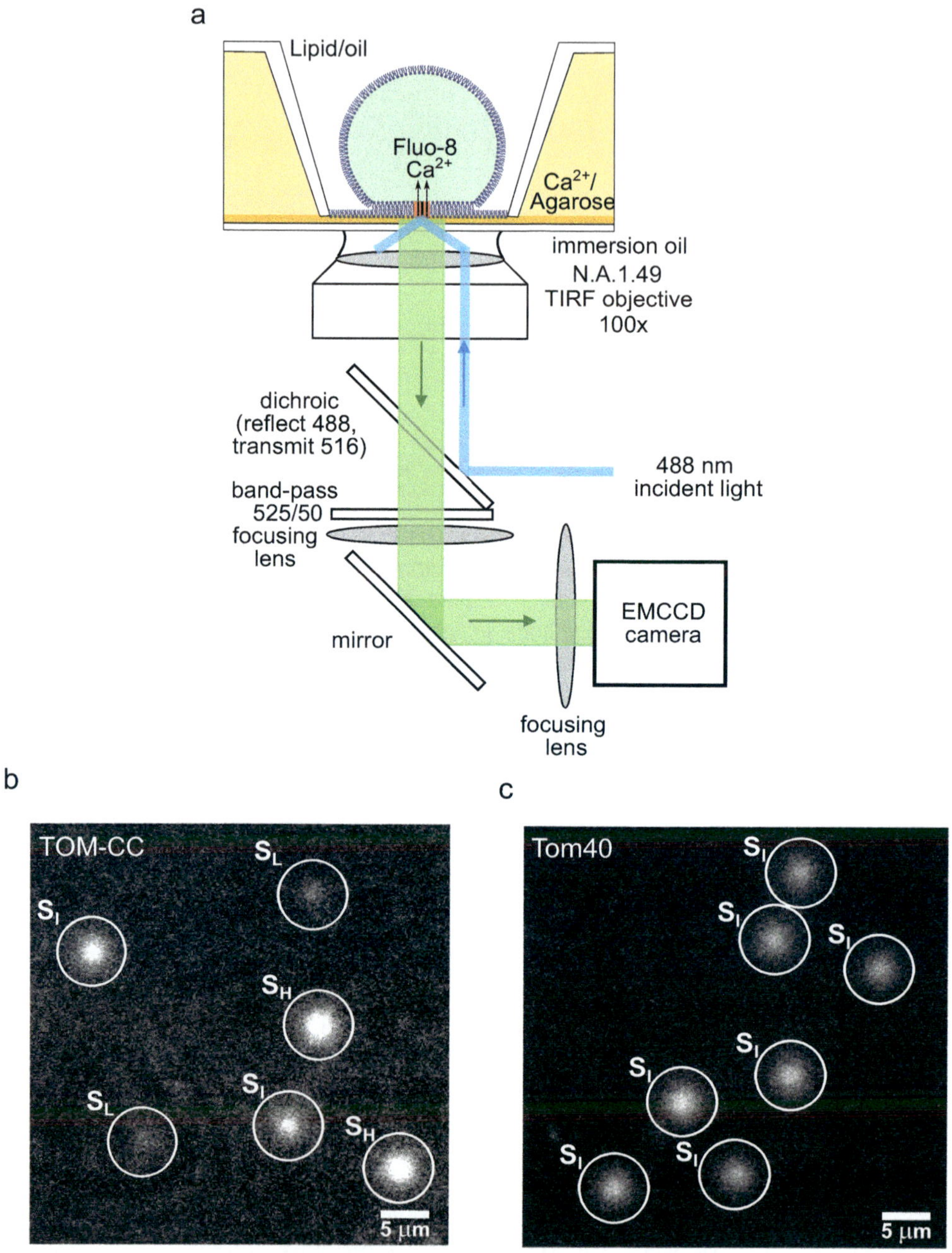

Fig. 3 TIRF microscopy experiment for visualization of mitochondrial TOM-CC or Tom40 channels in a DIB membrane. (**a**) Schematic diagram of the TIRF microscope used for excitation of Fluo-8 in close proximity to the membrane. (**b**) DIB membrane with TOM-CC molecules showing fluorescent spots of high (S_H), intermediate (S_I), and low intensity (S_L). (**c**) DIB membrane with Tom40 channels showing fluorescent spots of medium (S_I)

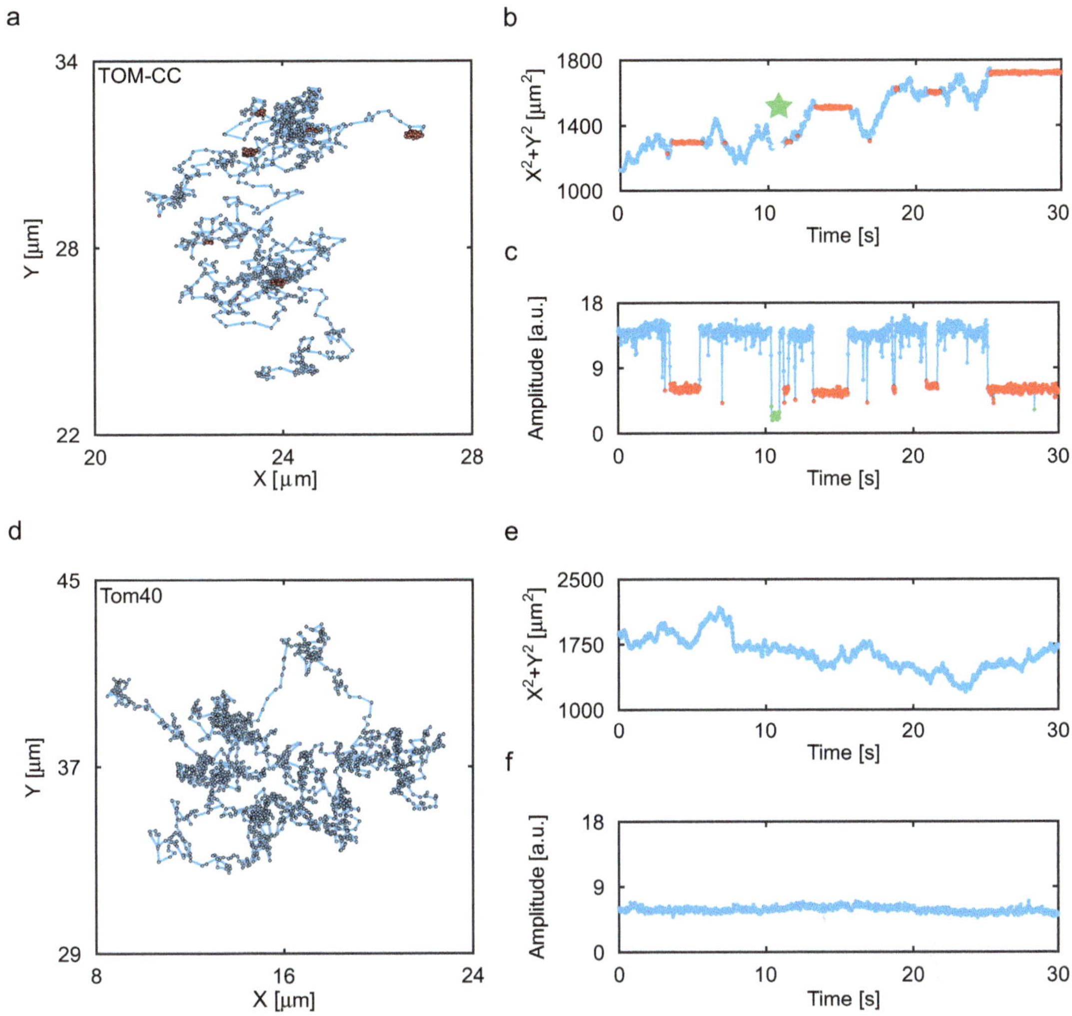

Fig. 4 Trajectories and open/closed activities of TOM-CC and Tom40. A lateral stop of freely moving TOM-CC molecules is accompanied by a partial or complete closure of the TOM-CC channel. This is not the case with Tom40. (**a**) Trajectory of a TOM-CC molecule in a DIB membrane during a period of 30 s. The coordinates $x(t)$, $y(t)$ of a fluorescent spot were determined for each frame of a movie obtained with the Laplacian of Gaussian (LoG) filter implemented in TrackMate [34, 49]. (**b**) Sum of squares displacement $x(t)^2 + y(t)^2$ of the channel shown in (**a**) over time. (**c**) The intensity of a fluorescent spot corresponding to the Ca^{2+} flux through a channel and the trajectory shown in (**a**) were determined for each frame of a film using TrackMate. (**d**) Trajectory of a Tom40 molecule in a DIB membrane. (**e, f**) The corresponding square displacements and the intensities of the fluorescence spots over time were determined as in (**b**) and (**c**). The trajectories, squared displacements, and intensities of the spots corresponding to the moving and non-moving TOM-CC or Tom40 molecules are shown in blue and red, respectively. The fluorescence intensity of TOM-CC in the fully closed state is marked in green. However, its position could not be determined with certainty because the signal-to-noise ratio is too low (green star)

2 Materials

2.1 Reagents

1. Acetone.
2. Chelating resin to bind metal ions (Ca^{2+}).
3. Fluorescent Ca^{2+}-sensitive dye: Fluo-8 dye (35 μM, λ_{Ex} ~ 494 nm, λ_{Em} ~ 517 nm) in DMSO. Store at −20 °C in the dark.
4. Hexadecane.
5. Lipid stock solution: diphytanoyl-sn-glycero-3-phosphocholine (DPhPC) dissolved in chloroform.
6. Nitrogen gas.
7. Oxygen gas.
8. Silicone oil.
9. Spin-coating hydrogel: dissolve 0.75% (w/v) low-melting agarose (T_m < 65 °C) in double-distilled water pretreated with a chelating resin that binds multivalent metal ions (Ca^{2+}). The agarose can be stored at room temperature for several weeks.
10. DIB droplet buffer containing TOM-CC: 400 mM EDTA, 8.78 mM HEPES (pH 7.2), 1.32 KCl, ~30 nM TOM-CC. Prepare fresh-from-stock solutions before use.
11. DIB droplet buffer containing Tom40: 400 mM EDTA, 8.78 mM HEPES (pH 7.2), 1.32 KCl, ~30 nM Tom40. Prepare fresh from stock solutions before use.
12. Purified proteins or protein complexes of interest. This example protocol uses TOM-CC purified from *Neurospora crassa* mitochondria in 20 mM HEPES pH 7.2, 0.1% (w/v) n-dodecyl-β-D-maltoside (DDM), 1 M KCl, 2% (v/v) dimethyl sulfoxide (DMSO) and Tom40 purified from *Neurospora crassa* mitochondria in 20 mM Tris, pH 8.5, 1% (w/v) n-octyl β-D-glucopyranoside (OG), 2% (v/v) DMSO (*see* **Note 1**).
13. Calcium-containing hydrogel: Dissolve 2.5% (w/v) low melting agarose (T_m < 65 °C) in double-distilled water with 0.66 M $CaCl_2$, 8.8 mM HEPES (pH 7.2). The agarose can be stored at room temperature for several weeks.

2.2 Equipment

1. Eppendorf single channel pipettes (0.1–2.5 μL and 1–10 μL).
2. Glass capillaries (inner diameter 0.53 mm; outer diameter 1.14 mm; length 8.9 cm).
3. Glass coverslip holder.
4. Glass coverslips (40 mm × 24 mm × 0.13 mm).
5. Glass vials.
6. Hamilton glass syringes (200 μL, 500 μL).

7. Heating block.
8. Heating plate.
9. ImageJ/Fiji software program with TrackMate v7.10.0 [33, 34, 49].
10. Inverted microscope (Nikon Eclipse TS100) with a NA 0.25 10 × Hoffmann modulation contrast objective.
11. Inverted fluorescence microscope (Nikon Ti-E) with apochromat NA 1.49 100 × oil objective, a 488 nm laser (P_{max} = 100 mW, Visitron), quad-band 446/523/600/677 HC filter set (AHF) and an electron multiplying charged coupled device (EMCCD) camera (iXon Ultra 897, Andor).
12. MATLAB software.
13. Micropipette puller.
14. Piezo-driven nanoinjector.
15. Plasma cleaner.
16. Polymethyl methacrylate (PMMA) chamber fabricated according to Lepthin et al. (*see* **Note 2**).
17. Polymethyl methacrylate (PMMA) well plate with small recesses of 40 mm × 30 mm × 3.5 mm according to [30] (*see* **Note 3**).
18. Spin-coater.
19. Ultrasonic cleaner.
20. Vacuum pump.

3 Methods

3.1 Fabrication of Agarose-Supported Droplet-Interface Bilayers

3.1.1 Lipid Preparation

1. Take an aliquot of lipid stock solution containing 1,2-diphytanoyl-*sn*-glycero-3-phosphocholine (e.g., 25 mg/mL DPhPC) dissolved in chloroform from the freezer at −20 °C and warm to room temperature to avoid condensation of water on the inner surfaces of the lipid vials (*see* **Note 4**).
2. Rinse and clean a Hamilton glass microliter syringe thoroughly with methanol and chloroform.
3. Transfer 9.5 mg DPhPC from stock solution into a glass vial by using the Hamilton glass microliter syringe.
4. Dry lipid under a stream of nitrogen and subsequently under vacuum (~2.0 mbar) overnight to remove residual organic solvents (*see* **Note 5**).
5. Dissolve the dried lipid film in hexadecane/silicone oil solution by adding equal volumes of hexadecane and silicone oil to a final lipid concentration of 9.5 mg/mL (*see* **Note 6**). The lipid solution can be stored at room temperature for up to about 3 weeks.

3.1.2 Formation of Lipid Monolayers at the Oil-Water Interface Around Aqueous Droplets

1. Mount a glass capillary onto a vertical or horizontal micropipette puller and make a glass capillary needle with a tip opening diameter of ~20 μm. The parameters for the heating filament to melt the glass and for the pulling force should be determined experimentally in advance, according to the specifications of the manufacturer of the puller and the type of capillary (*see* **Note 7**).
2. Fill the microcapillary glass needle with ~5 μL of aqueous injection solution containing 8.8 mM HEPES (pH 7.2), 7 μM Fluo-8, 400 μM EDTA, 1.32 M KCl, and ~30 nM TOM-CC (Fig. 1a) or Tom40 (Fig. 1b) using a fine and flexible microcapillary pipette tip (*see* **Note 8**).
3. Attach the microcapillary glass needle with the aqueous injection solution to a piezo-driven nanoinjector.
4. Pipette ~20 μL of lipid hexadecane/silicone oil solution containing 9.5 mg/mL DPhPC to each of several microfabricated wells in a PMMA plate and inject 100–200 nL of aqueous droplets containing 8.8 mM HEPES (pH 7.2), 7 μM Fluo-8, 400 μM EDTA, 1.32 M KCl, and 30 nM TOM-CC or Tom40 into the lipid hexadecane/silicone oil solution using the nanoinjector. The droplets should not touch each other. Otherwise, they will fuse into larger droplets if lipid monolayers have not yet formed at the oil-droplet interface (*see* **Notes 9** and **10**).
5. Allow formation of a lipid monolayer at the oil-droplet interface for about 2 h by keeping the droplet incubation chambers covered on a hot plate heated to 35 °C.

3.1.3 PMMA Chamber Assembly and Formation of a Lipid Monolayer at the Oil-Hydrogel Interface

1. Arrange several glass coverslips vertically in a stainless-steel coverslip holder in a glass jar and sonicate under acetone in an ultrasonic cleaner for ~5 min. For safety reasons, use only enough acetone to immerse the cover glasses in the organic solvent. Cover the beaker loosely with a glass plate (*see* **Note 11**).
2. Remove the rack from the acetone, rinse the glass coverslips with double deionized water, and dry under a stream of nitrogen gas (*see* **Note 12**).
3. To chemically remove other organic contaminants and make the coverslips hydrophilic, expose them to oxygen radicals in a plasma cleaner for ~5 min (p_{oxygen} ~ 0.5 mbar) (*see* **Note 13**).
4. Melt 0.75% (w/v) low-melting agarose by heating to 85 °C.
5. Coat a plasma-cleaned coverslip with a thin film of agarose by slowly adding 140 μL of 0.75% (w/v) low-melting agarose for 30 s using a spin coater at 3000 rpm (*see* **Note 14**).

6. Attach the spin-coated coverslip with the thin layer of the agarose hydrogel to the bottom of the PMMA chamber, and fix the edges of the coverslip to the PMMA with transparent adhesive tape.
7. Melt 2.5% (w/v) low melting agarose dissolved in 0.66 M $CaCl_2$, 8.8 mM HEPES (pH 7.2) by heating to 85 °C (*see* **Note 15**).
8. Place the PMMA device with the spin-coated coverslip on a hot plate heated to 35 °C and gently pipette ~200 μL of 2.5% agarose solution containing 0.66 M $CaCl_2$, 8.8 mM HEPES, pH 7.2, preheated to 85 °C, into the inlet of the PMMA chamber. Make sure that the 2.5% agarose solution spreads around the spin-coated agarose hydrogel, but not over it, which can be avoided by carefully pressing the chamber on the heating plate.
9. Overlay the wells containing the ultrathin 0.75% agarose hydrogel of the PMMA chamber with a total of ~60 μL hexadecane/silicone oil solution containing 9.5 mg/mL DPhPC to initiate the formation of a lipid monolayer by self-assembly at the oil-agarose interface. For this purpose, keep the chamber covered on a hot plate at 35 °C for about 2 h.

3.1.4 Formation of DIB Membranes

1. Transfer single aqueous droplets from the wells of the droplet incubation chamber into the wells of the PMMA chamber under a stereomicroscope using a pipette with a 10 μL tip and allow the droplets to sink onto the lipid monolayers (Fig. 2a). Lipid bilayers usually form within 5 min between the lipid monolayers lining the agarose hydrogel and the oil-droplet interface (Fig. 2b).
2. Assess lipid bilayer formation using a standard inverted microscope equipped with a 10 × Hoffman modulation contrast objective. DIB membrane formation is indicated by a clear white ring (Fig. 2c) at the interface between the droplet and the hydrogel (*see* **Note 16**).

3.2 Recording DIB Images

1. Focus on the edge of a 488 nm illuminated DIB membrane in TIRF mode using an inverted fluorescence microscope (Fig. 3a) equipped with a 100×/N.A. 1.49 apochromat oil TIRF objective and a back-illuminated electron-multiplying CCD camera (>95% QE, 512 × 512 pixels, pixel size ~0.16 μm).
2. Locate single fluorescent spots (Fig. 3b, c) corresponding to Ca^{2+}-flux through TOM-CC or Tom40 channels within the field of view.

3. Set the laser intensity on the objective to 8–10 mW, the TIRF angle, and the EMCCD camera gain so that open ion channels appear as high-contrast fluorescent spots on a dark background (*see* **Note 17**).
4. Make sure that the spots, corresponding to Ca^{2+}-flux through single ion channels have a round shape, with high intensity in the center and gradually decreasing toward the periphery (Fig. 3b, c). Membranes without channels should show only background fluorescence.
5. Check that the fluorescent spots remain in focus as they move. If they do not, this may indicate that the signal is not coming from an open channel in the membrane through which Ca^{2+} ions flow, but rather fluorescent debris dipping in and out of the exciting 488 nm evanescent field near the membrane (*see* **Note 18**).
6. Acquire a stack of membrane N images that allows tracking of the position and monitoring of the open-closed states of individual channels. To reliably determine the type of lateral mobility (e.g., free vs channeled, restricted or hop diffusion; [45]) and the state of channel activity (e.g. open or closed), high positional accuracy (<10 nm) and temporal resolution with sufficiently long (e.g., > 30 s) and well-sampled trajectories (e.g., frame rate > 48 s^{-1}) are advantageous (*see* **Note 19**).

3.3 Data Analysis

To gain insight into the spatiotemporal dynamics of proteins such as TOM-CC and Tom40 in DIB membranes (Fig. 3b, c), the raw image stacks acquired in Subheading 3.2 need to be converted into a series of trajectories containing the coordinates and fluorescence intensities of each particle as a function of time t (Fig. 4). This can be achieved with custom-made programs [21] based on commercial software (e.g., MATLAB), or with the open-source platform ImageJ/Fiji [33] (*see* **Note 20**). Here, we present the use of TrackMate [34, 49], which is implemented into ImageJ/Fiji because it quickly provides an initial insight into possible correlations between lateral channel motion and channel activity. A demo stack (Video S1) can be downloaded from the University of Stuttgart Data Repository (DaRUS): S.W. and S.N. 2023, Data for: Tracking the activity and position of mitochondrial β-barrel proteins (https://doi.org/10.18419/darus-3573).

3.3.1 Background Correction

1. Read TIRF video stack into ImageJ/Fiji.
2. Subtract background using the rolling ball algorithm to remove large spatial variations of intensities in the *Process › Subtract background* menu. The radius should be larger than the largest object (e.g., 50 pixels).
3. Adjust brightness and contrast settings through *Image › Adjust › Brightness/Contrast* if the signal to noise ratio is too low (not required when analyzing the demo video).

3.3.2 Tracking Individual Channels

1. Open and start *TrackMate* plugin by selecting *Plugins › tracking › TrackMate* in the menu bar.
2. Verify the spatial and temporal calibration of your data.
3. Define region of interest (ROI) in which to track fluorescent spots by selecting *Edit › Selection › Specify* in the menu bar (e.g., width, 170; height, 170; *x*-coordinate 63; *y*-coordinate 90; T, 0 to 1423).
4. Use the *LoG* (Laplacian of Gaussian) detector to segment the image. It is optimal for medium spot sizes between ~5 and ~20 pixels in diameter. For detection of fluorescent spots corresponding to TOM-CC and Tom40 (e.g., 20 pixels) (*see* **Note 21**).
5. Use the *Preview* function to check the parameters of the current data and adjust settings if necessary. Otherwise, continue the program.
6. Set initial spot filter settings (e.g., "quality") that only well-defined spots are taken from the pool of detected spots. For this purpose, manually select a threshold on the filter settings (e.g. 'quality') histogram to discard all spots with a value below this value (e.g. 3.10).
7. Check by scrolling though the image stack to ensure that the best spots are retained.
8. Select and run *LAP tracker* [50] which links spots from frame to frame to build track segments and assembles the segments into complete trajectories. Configure "frame to frame linking max distance" (e.g., 2.0 μm), "track segment gap-closing max distance" (e.g., 10.0 μm), and "track segment gap-closing max frame gap" (e.g., 10).
9. Filter tracks by selecting a filter function until you are satisfied with the remaining tracks (e.g., track displacement, 5.7).
10. Go to *Display options* and save the track table.
11. To show a possible interplay between lateral motion and ion permeability, the coordinates $x(t)$ are plotted against coordinates $y(t)$ as well as the sum of square displacements $x(t)^2 + y(t)^2$ and the amplitudes $I(t)$ corresponding to the quality value are plotted as a function of time (Fig. 4). In case of free lateral channel movement, the lateral diffusion coefficient D can be determined by linear regression of the mean square displacement of the spots according to $D = \langle|r(x, y, \tau) - r(x_0, y_0, \tau)|^2\rangle/4\tau$, where r represents the position of a spot and τ the time delay. Typical diffusion coefficients for TOM-CC and Tom40 freely moving in DIB membranes range between ~0.5 and ~1.5 $\mu m^2 s^{-1}$. Molecules whose diffusion constant is less than 0.01 $\mu m^2 s^{-1}$ can be defined as immobilized [21].

4 Notes

1. Experimenters are likely to use their own proteins or protein complexes of interest, which require dedicated and optimized purification procedures. In the case of this example protocol, mitochondrial outer membrane protein translocase TOM-CC was isolated according to refs. 18, 22, whereas Tom40 was isolated according to ref. 17. The best reconstitution of proteins into the DIB membranes can be achieved at stock protein concentrations greater than 0.4 mg/mL.
2. A sketch with all dimensions for microfabrication of a PMMA chamber with 12 wells, allowing preparation of 12 hydrogel-supported DIBs, is described in [30]. The wells have a diameter of 0.5 mm and a depth of 1.8 mm.
3. A sketch with all dimensions for making a PMMA container with small recesses of 40 mm × 30 mm × 3.5 mm is described in [30].
4. Diphytanoyl lipids have branched fatty acid chains that give the membranes unique properties, e.g., no clear phase transition at low temperatures and high stability [51]. They are, therefore, commonly used as model membranes in electrophysiology and when preparing DIB membranes. But any other lipid that maintains its liquid environment at low temperatures may also be suitable for this application.
5. Chloroform is toxic. Therefore, all steps in handling lipids in chloroform should be carried out under a fume hood in accordance with the relevant safety regulations.
6. It is advisable to vary the lipid concentration if DIB membranes do not form. Should lipids other than DPhPC be used, the appropriate lipid concentration should be optimized experimentally.
7. If no puller for glass capillaries is available, these can also be drawn over a gas flame, although the reproducibility of this is not necessarily given.
8. It is recommended to optimize the concentrations of proteins added to the DIB droplet buffer. Very high concentrations of membrane proteins may destabilize the DIB membranes when integrated from a detergent solution, as too high detergents dissolve the membranes. Very high protein densities in DIB membranes are more complicated to track when the corresponding trajectories overlap in time. If proteins do not enter the membrane from the droplet side, they can first be reconstituted into small unilamellar vesicles (SUVs), which are then fused to the DIB membranes from the droplet side [52].

9. Suitable containers for the lipid hexadecane/silicone oil solution are small cutouts in a thin PMMA plate and are described including dimensions (40 mm × 30 mm × 3.5 mm) in [30].
10. If no glass capillary needles and no nanoinjector are available, the droplets can alternatively be prepared manually with a mechanical Eppendorf single channel pipette for volumes between 0.1 and 2.5 μL [30]. In this case, however, the droplet volumes are not as accurate.
11. Acetone can form a flammable and explosive vapor-air mixture if ventilation is inadequate. Observe relevant safety precautions and perform cleaning under a fume hood.
12. Coverslips cleaned in this way can be stored in a Petri dish for several weeks.
13. For hydrogel coating of the cover glass and single molecule microscopy, it is important that the cover glasses are hydrophilic and very clean. Since the hydrophilic effect of the plasma cleaning wears off over time, the plasma cleaning should therefore be carried out immediately before spin coating and assembly of the PMMA chamber.
14. Low-melting agarose functionalized with ligands (e.g., Ni-NTA) [21] can be used to locally adjust the lateral mobility by binding proteins to the polymer matrix. In this case, however, care should be taken that the agarose for spin coating should not be chemically cross-linked. Finally, the concentration of the agarose must be optimized experimentally, whereby concentrations of up to 50% (w/v) are possible. The thickness of the agarose film should be less than ~200 nm and can be determined by atomic force microscopy (AFM) [21]. Possible parameters that influence the thickness of the coating are the concentration of the agarose, its temperature, the applied volume, and the angular velocity with which the spin coater is operated.
15. The Ca^{2+} concentration is surprisingly high. This value, however, is correct and is fully in line with that used in other "optical" patch-clamp work using $CaCl_2$ concentrations of 0.72 M [53] and 0.75 M [54]. The high concentrations are due to the fact that a high concentration gradient must be built up across the DIB membrane in order to obtain a sufficiently large flux and a sufficient signal-to-noise ratio.
16. Alternatively, the DIBs can be visualized at 10× magnification by phase contrast or differential interference contrast microscopy.
17. As there is often no quantitative information about the laser intensity at the lens, the laser intensity settings can be calibrated beforehand in separate measurements with an optical laser power meter. When doing so, be aware of the possible dangers of high laser light intensity and observe all safety regulations.

18. Successful integration of proteins into a DIB membrane can also be demonstrated by visualizing the proteins after chemical labeling with fluorescent dyes.
19. The positional accuracy Δx can be estimated after conversion of the EMCCD camera readout into photon numbers according to [55–57]:

$$\left\langle (\Delta x)^2 \right\rangle = \frac{s^2 + a^2/12}{N} + \frac{8\pi s^4 b^2}{a^2 N^2}$$

where s is the standard deviation of the 2D Gaussian function fitting a fluorescent spot, a is the pixel size (here: 0.16 μm), N is the total number of photons in the spot, and b is the background noise. Typical positional accuracies of intermediate and high-intensity spots, as shown in Fig. 3b, c, are between 5 and 10 nm [21].
20. For a more detailed evaluation, the use of self-written programs (e.g., in MATLAB; [21]) is recommended, especially if the positional accuracy for the calculation of the lateral diffusion coefficients is required.
21. For spot sizes below 5 pixels corresponding, e.g., to fluorescently labeled proteins, the DoG detector can be used. It is based on a Gaussian difference approach to approximate a LoG filter by the difference of two Gaussian filters [34].

Acknowledgments

We thank Robin Ghosh for critically reviewing the manuscript. This work was supported in part by Laura Na Liu (2nd Physics Institute, University of Stuttgart) and by a grant from the Baden Württemberg Foundation (BiofMO-6 to SN).

References

1. Sigworth FJ (1994) Voltage gating of ion channels. Q Rev Biophys 27:1–40
2. Delcour AH (2003) Solute uptake through general porins. Front Biosci 8:d1055–d1071
3. Benz R (2021) Historical perspective of pore-forming activity studies of voltage-dependent anion channel (Eukaryotic or Mitochondrial Porin) since its discovery in the 70th of the last century. Front Physiol 12:734226
4. Dunlop J, Bowlby M, Peri R et al (2008) High-throughput electrophysiology: an emerging paradigm for ion-channel screening and physiology. Nat Rev Drug Discov 7:358–368
5. Liu C, Li T, Chen J (2019) Role of high-throughput electrophysiology in drug discovery. Curr Protoc Pharmacol 87:e69
6. Benz R (2006) Bacterial and eukaryotic porins: structure, function, mechanism. John Wiley & Sons, Hoboken
7. Zeth K, Thein M (2010) Porins in prokaryotes and eukaryotes: common themes and variations. Biochem J 431:13–22
8. Nestorovich EM, Danelon C, Winterhalter M et al (2002) Designed to penetrate: time-resolved interaction of single antibiotic molecules with bacterial pores. Proc Natl Acad Sci 99:9789–9794

9. Pagès J-M, James CE, Winterhalter M (2008) The porin and the permeating antibiotic: a selective diffusion barrier in Gram-negative bacteria. Nat Rev Microbiol 6:893–903
10. Kasianowicz JJ, Brandin E, Branton D et al (1996) Characterization of individual polynucleotide molecules using a membrane channel. Proc Natl Acad Sci 93:13770–13773
11. Branton D, Deamer DW, Marziali A et al (2008) The potential and challenges of nanopore sequencing. Nat Biotechnol 26:1146–1153
12. Deamer D, Akeson M, Branton D (2016) Three decades of nanopore sequencing. Nat Biotechnol 34:518–524
13. Restrepo-Pérez L, Joo C, Dekker C (2018) Paving the way to single-molecule protein sequencing. Nat Nanotechnol 13:786–796
14. Alfaro JA, Bohländer P, Dai M et al (2021) The emerging landscape of single-molecule protein sequencing technologies. Nat Methods 18: 604–617
15. Hu Z-L, Huo M-Z, Ying Y-L et al (2021) Biological nanopore approach for single-molecule protein sequencing. Angew Chem Int Ed Engl 60:14738–14749
16. Ensslen T, Sarthak K, Aksimentiev A et al (2022) Resolving isomeric posttranslational modifications using a biological nanopore as a sensor of molecular shape. J Am Chem Soc 144:16060–16068
17. Ahting U, Thieffry M, Engelhardt H et al (2001) Tom40, the pore-forming component of the protein-conducting TOM channel in the outer membrane of mitochondria. J Cell Biol 153:1151–1160
18. Bausewein T, Mills DJ, Langer JD et al (2017) Cryo-EM structure of the TOM core complex from Neurospora crassa. Cell 170:693–700
19. Bausewein T, Naveed H, Liang J et al (2020) The structure of the TOM core complex in the mitochondrial outer membrane. Biol Chem 401:687–697
20. Ornelas P, Bausewein T, Martin J et al (2023) Two conformations of the Tom20 preprotein receptor in the TOM holo complex. Proc Natl Acad Sci 120:e2301447120
21. Wang S, Findeisen L, Leptihn S et al (2022) Spatiotemporal stop-and-go dynamics of the mitochondrial TOM core complex correlates with channel activity. Commun Biol 5:1–11
22. Ahting U, Thun C, Hegerl R et al (1999) The TOM core complex: the general protein import pore of the outer membrane of mitochondria. J Cell Biol 147:959–968
23. Poynor M, Eckert R, Nussberger S (2008) Dynamics of the preprotein translocation channel of the outer membrane of mitochondria. Biophys J 95:1511–1522
24. Romero-Ruiz M, Mahendran KR, Eckert R et al (2010) Interactions of mitochondrial presequence peptides with the mitochondrial outer membrane preprotein translocase TOM. Biophys J 99:774–781
25. Gessmann D, Flinner N, Pfannstiel J et al (2011) Structural elements of the mitochondrial preprotein-conducting channel Tom40 dissolved by bioinformatics and mass spectrometry. Biochim Biophys Acta 1807:1647–1657
26. Mahendran KR, Romero-Ruiz M, Schlösinger A et al (2012) Protein translocation through Tom40: kinetics of peptide release. Biophys J 102:39–47
27. Mahendran KR, Lamichhane U, Romero-Ruiz M et al (2013) Polypeptide translocation through the mitochondrial TOM channel: temperature-dependent rates at the single-molecule level. J Phys Chem Lett 4:78–82
28. Thompson JR, Heron AJ, Santoso Y et al (2007) Enhanced stability and fluidity in droplet on hydrogel bilayers for measuring membrane protein diffusion. Nano Lett 7:3875–3878
29. Bayley H, Cronin B, Heron A et al (2008) Droplet interface bilayers. Mol BioSyst 4: 1191–1208
30. Leptihn S, Castell OK, Cronin B et al (2013) Constructing droplet interface bilayers from the contact of aqueous droplets in oil. Nat Protoc 8:1048–1057
31. Huang S, Romero-Ruiz M, Castell OK et al (2015) High-throughput optical sensing of nucleic acids in a nanopore array. Nat Nanotechnol 10:986–991
32. Manafirad A (2021) Single ion-channel analysis in droplet interface bilayer. In: Fahie MAV (ed) Nanopore technology: methods and protocols. Springer US, New York, pp 187–195
33. Schindelin J, Arganda-Carreras I, Frise E et al (2012) Fiji – an Open Source platform for biological image analysis. Nat Methods 9: 676–68
34. Tinevez J-Y, Perry N, Schindelin J et al (2017) TrackMate: an open and extensible platform for single-particle tracking. Methods 115:80–90
35. Qian H, Sheetz MP, Elson EL (1991) Single particle tracking. Analysis of diffusion and flow in two-dimensional systems. Biophys J 60: 910–921
36. Kusumi A, Nakada C, Ritchie K et al (2005) Paradigm shift of the plasma membrane concept from the two-dimensional continuum

fluid to the partitioned fluid: high-speed single-molecule tracking of membrane molecules. Annu Rev Biophys Biomol Struct 34:351–378
37. Kusumi A, Tsunoyama TA, Hirosawa KM et al (2014) Tracking single molecules at work in living cells. Nat Chem Biol 10:524–532
38. Schafer DA (2002) Coupling actin dynamics and membrane dynamics during endocytosis. Curr Opin Cell Biol 14:76–81
39. Rajendran L, Simons K (2005) Lipid rafts and membrane dynamics. J Cell Sci 118:1099–1102
40. Kusumi A, Shirai YM, Koyama-Honda I et al (2010) Hierarchical organization of the plasma membrane: investigations by single-molecule tracking vs. fluorescence correlation spectroscopy. FEBS Lett 584:1814–1823
41. Betaneli V, Schwille P (2013) Fluorescence correlation spectroscopy to examine protein–lipid interactions in membranes. In: Kleinschmidt JH (ed) Lipid-protein interactions: methods and protocols. Humana Press, Totowa, pp 253–278
42. Schink KO, Tan K-W, Stenmark H (2016) Phosphoinositides in control of membrane dynamics. Annu Rev Cell Dev Biol 32:143–171
43. Engelman DM (2005) Membranes are more mosaic than fluid. Nature 438:578–580
44. Lingwood D, Ries J, Schwille P et al (2008) Plasma membranes are poised for activation of raft phase coalescence at physiological temperature. Proc Natl Acad Sci 105:10005–10010
45. Jacobson K, Liu P, Lagerholm BC (2019) The lateral organization and mobility of plasma membrane components. Cell 177:806–819
46. Mokranjac D, Sichting M, Popov-Čeleketić D et al (2009) Role of Tim50 in the transfer of precursor proteins from the outer to the inner membrane of mitochondria. Mol Biol Cell 20: 1400–1407
47. Waegemann K, Popov-Čeleketić D, Neupert W et al (2015) Cooperation of TOM and TIM23 complexes during translocation of proteins into mitochondria. J Mol Biol 427:1075–1084
48. Gomkale R, Linden A, Neumann P et al (2021) Mapping protein interactions in the active TOM-TIM23 supercomplex. Nat Commun 12:5715
49. Ershov D, Phan M-S, Pylvänäinen JW et al (2022) TrackMate 7: integrating state-of-the-art segmentation algorithms into tracking pipelines. Nat Methods 19:829–832
50. Jaqaman K, Loerke D, Mettlen M et al (2008) Robust single-particle tracking in live-cell time-lapse sequences. Nat Methods 5:695–702
51. Lindsey H, Petersen NO, Chan SI (1979) Physicochemical characterization of 1,2-diphytanoyl-sn-glycero-3-phosphocholine in model membrane systems. Biochim Biophys Acta Biomembr 555:147–167
52. Rosholm KR, Baker MAB, Ridone P et al (2017) Activation of the mechanosensitive ion channel MscL by mechanical stimulation of supported Droplet-Hydrogel bilayers. Sci Rep 7:45180
53. Wang Y, Yan S, Zhang P et al (2018) Osmosis-driven motion-type modulation of biological nanopores for parallel optical nucleic acid sensing. ACS Appl Mater Interfaces 10:7788–7797
54. Heron AJ, Thompson JR, Cronin B et al (2009) Simultaneous measurement of ionic current and fluorescence from single protein pores. J Am Chem Soc 131:1652–1653
55. Thompson RE, Larson DR, Webb WW (2002) Precise nanometer localization analysis for individual fluorescent probes. Biophys J 82:2775–2783
56. Selvin PR, Lougheed T, Hoffman MT et al (2008) In vitro & in vivo FIONA and other acronyms for watching molecular motors walk. In: Selvin PR, Ha T (eds) Single-molecule techniques: a laboratory manual. Cold Spring Harbor Laboratory Press, Cold Spring Harbor, pp 37–71
57. Hirsch M, Wareham RJ, Martin-Fernandez ML et al (2013) A stochastic model for electron multiplication charge-coupled devices – from theory to practice. PLoS One 8:e53671

Chapter 15

Conformational Heterogeneity of β-Barrel Membrane Proteins Observed In Situ Using Orthogonal Spin Labels and Pulsed ESR Spectroscopy

Sophie Ketter, Aathira Gopinath, and Benesh Joseph

Abstract

Outer membrane proteins (OMPs) of Gram-negative bacteria are involved in many essential functions of the cell. They are tightly packed in the outer membrane, which is an asymmetric lipid bilayer. Electron spin resonance (ESR) spectroscopic techniques combined with site-directed spin labeling (SDSL) enable observation of structure and conformational dynamics of these proteins directly in their native environments. Here we depict a protocol for site-directed spin labeling of β-barrel membrane proteins in isolated outer membranes and intact *E. coli* using nitroxide, triarylmethyl (trityl), and Gd^{3+}-based spin tags. Furthermore, subsequent continuous wave (CW) and orthogonal pulsed electron-electron double resonance (PELDOR) measurements are described along with experimental setup at Q-band (34 GHz), the data analysis, and interpretation.

Key words Outer membrane proteins, Site-directed spin labeling, Electron spin resonance spectroscopy, Orthogonal spin labels, Continuous wave ESR, PELDOR or DEER

1 Introduction

Membrane β-barrel proteins are primarily found in the highly asymmetric outer membrane of Gram-negative bacteria, which directly interfaces the cell with the extracellular environment. Outer membrane proteins (OMPs) fulfill a variety of fundamental functions crucial for cell viability including nutrient uptake, signal transduction, OMP folding and insertion, and biogenesis of the outer membrane itself. Hence, membrane-embedded proteins are considered the most sought-after target for a vast number of present and future drugs. Over the past years, a critical dependence of membrane protein structure, function and dynamics on the native cellular environment and lipid composition has been emphasized

Authors Sophie Ketter and Aathira Gopinath have equally contributed to this chapter.

Raffaele Ieva (ed.), *Transmembrane β-Barrel Proteins: Methods and Protocols*, Methods in Molecular Biology, vol. 2778, https://doi.org/10.1007/978-1-0716-3734-0_15,

[1–6]. Thus, the understanding of a protein's conformational dynamics in its native surroundings poses a highly indispensable yet challenging task.

Electron spin resonance (ESR) spectroscopy combined with site-directed spin labeling (SDSL) has received great attention as a versatile tool to investigate biomolecules of unlimited size and directly in the native cellular and lipid environments [7–15]. Most commonly, spin labeling of the target protein is achieved via cysteine substitution and disulfide linkage with a thiosulfonate functionalized nitroxide spin probe (methanethiosulfonate label, MTSL) [16]. Subsequent continuous wave (CW) ESR spectroscopy techniques allow to follow protein structural dynamics at physiological temperatures [7]. Also, it can provide information on spin labeling efficiency or off-target labeling that can occur in in situ approaches [17, 18]. Importantly, by introducing two spin labels, long-range distances between two target sites can be accessed using dipolar ESR techniques such as pulsed electron-electron double resonance (PELDOR or double electron-electron resonance, DEER) spectroscopy [19].

Instability of the common nitroxide spin labels and the disulfide linking them to the protein are major challenges for in cell/in situ applications. Consequently, reduction-resistant spin probes including sterically shielded nitroxides, triarylmethyl (trityl) radicals, and those based on Cu^{2+}, Gd^{3+}, and Mn^{2+} were introduced [11, 20–27]. Further, substitution of thiosulfonate with maleimide functionalization was shown to improve the redox stability of the bioconjugation [28–30] (Fig. 1). Moreover, the availability of spectroscopically orthogonal spin probes opens new possibilities to observe protein conformation with high sensitivity and selectivity, especially for heterooligomeric protein complexes even within the native surroundings [18, 31–36].

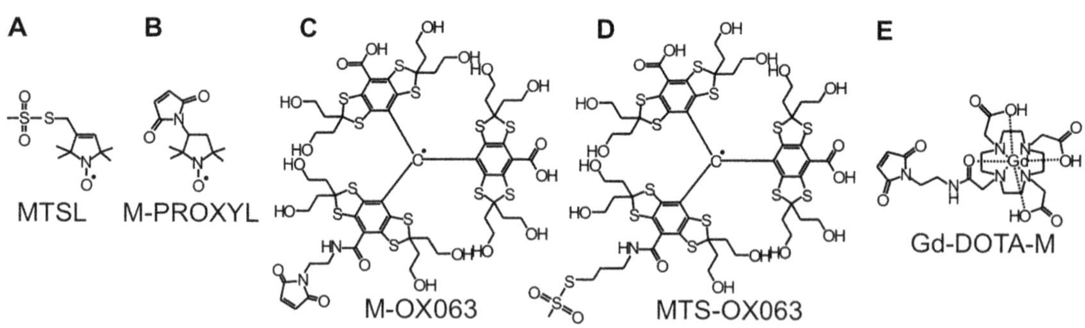

Fig. 1 Chemical structure of the spin labels discussed in this protocol. (**a**) The structure of the methanethiosulfonate functionalized nitroxide spin label MTSL bearing a pyrroline ring and (**b**) that of the maleimide functionalized analog M-PROXYL based on a pyrrolidine ring. (**c**) Structure of maleimide functionalized OX063 trityl spin label (M-OX063) and (**d**) of its methanethiosulfonate functionalized analog (MTS-OX063). Both tags differ only in their linker attaching it to the protein. (**e**) The maleimide functionalized Gd-DOTA-M spin label

We used in situ spin labeling and ESR experiments to observe conformational dynamics of the outer membrane β-barrel protein BamA, the core component of the β-barrel assembly machinery (BAM) [37], and of the cobalamin transporter BtuB [17], in isolated native outer membranes (OM) and directly in *E. coli*. The OM is an asymmetric bilayer consisting of phospholipids and lipopolysaccharides and carries numerous β-barrel proteins (known as outer membrane proteins or OMPs) performing crucial biological functions. OMPs are folded and inserted into the OM by the BAM complex [38]. Altogether OMPs perform a wide range of functions, including transport of nutrients, signaling, motility, membrane biogenesis, protein import and secretion among others [39, 40]. We further demonstrated the advantage of PELDOR spectroscopy combined with orthogonal spin labels to achieve enhanced sensitivity, in particular for in situ studies of the hetero-oligomeric BtuB-TonB-cobalamin complex in the native lipid bilayers [18, 31, 33].

Here we report a strategy for protein expression and in situ site-directed spin labeling of membrane β-barrel proteins exemplified on BamABCDE (~200 kDa) and BtuB (~70 kDa) using thiosulfonate and maleimide functionalized nitroxides, trityl, and Gd^{3+}-based spin labels. We cover the case of a double-cysteine variant to investigate protein conformation (Figs. 2a and 3) and interaction of a single-cysteine variant with a spin-labeled ligand (Figs. 2 and 4). We depict a protocol for CW and PELDOR measurements involving commonly used nitroxide spin pairs and further orthogonal spin label combinations. We briefly discuss in situ PELDOR data analysis in light of existing approaches and how this strategy can be used to study conformational dynamics of membrane proteins in the native environments.

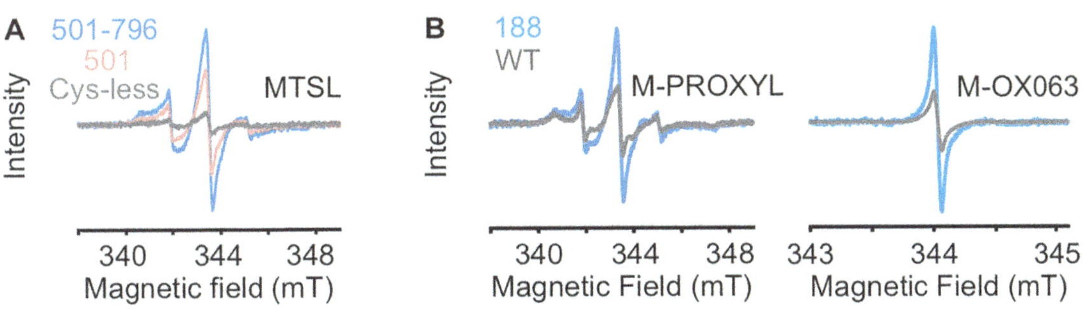

Fig. 2 Room temperature continuous wave (CW) ESR spectra of spin-labeled BamABCDE in *E. coli* and BtuB in isolated outer membranes recorded at X-band (~9 GHz) frequency. (**a**) Spectrum of the MTSL labeled double cysteine variant $BamA^{501C\text{-}796C}BCDE$ (blue) compared to the spectrum of the corresponding single cysteine variant $BamA^{501C}BCDE$ (light pink) and the cysteine-less (Cys-less) protein (grey) in *E. coli*. The spectra are normalized to the same OD_{600} value to depict the amount of background labeling. (**b**) Left, spectra of M-PROXYL labeled $BtuB^{188C}$ (blue) and the WT protein (grey) in isolated outer membranes. On the right, the corresponding spectra for BtuB WT and the 188C variant after spin labeling with M-OX063. The spectra are normalized to the same amount of outer membranes to depict the amount of signal coming from background labeling (WT)

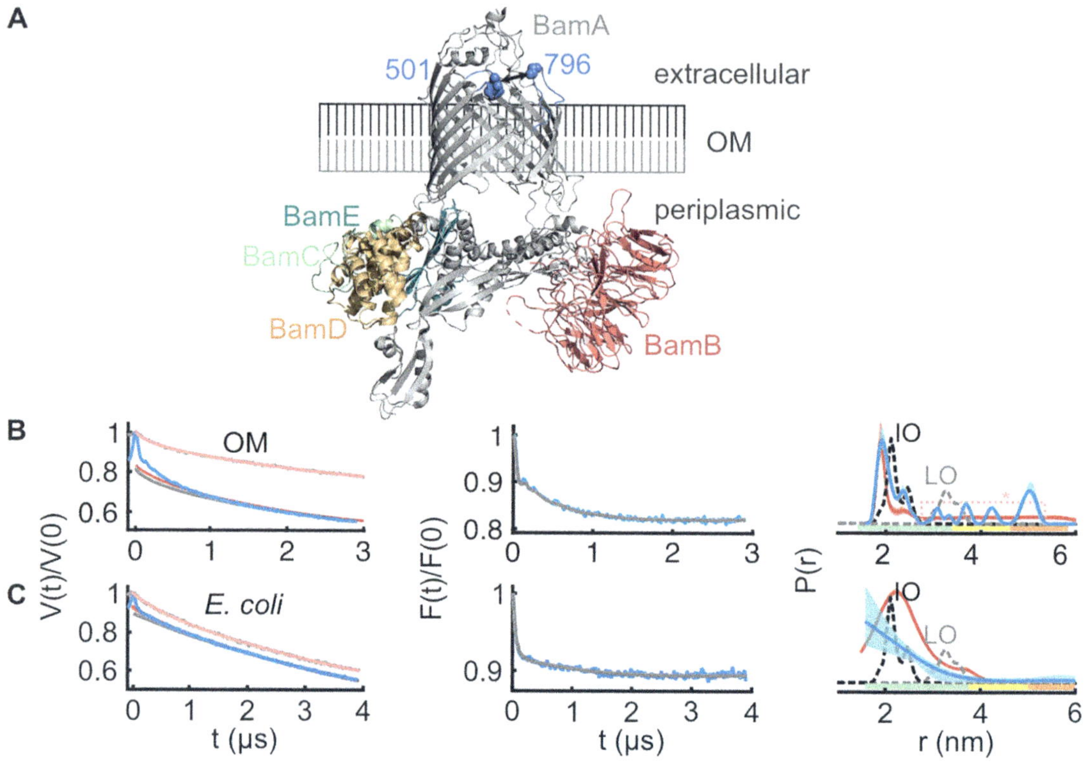

Fig. 3 Nitroxide-nitroxide (NO-NO) 4-pulse PELDOR measurements of MTSL labeled BamA$^{501C\text{-}796C}$ in native outer membranes and in *E. coli* cells. (**a**) Structure of BamABCDE (PDB 5D0O) comprising of BamA (grey), BamB (salmon), BamC (green), BamD (yellow) and BamE (teal). The spin labelled positions L501C on the extracellular loop3 and G796C on the extracellular loop8 of BamA are highlighted as blue spheres. (**b**) 4-pulse PELDOR measurement of spin-labeled BamA$^{501C\text{-}796C}$ in native outer membrane [16] and (**c**) BamA$^{501C\text{-}796C}$BCDE in intact *E. coli* cells. Left panels show the primary data (in blue) with the intermolecular (background) contribution overlaid (in grey from two-step Tikhonov regularization and in red from DEERNet analysis). Middle panels represent the background corrected form factors overlaid with the corresponding fit from Tikhonov regularization. Right panels show the distance distributions. Distances marked with a dotted pink line suggest the presence of longer distances [16]. The primary data (light grey) shown in left panel depicts the single cysteine (L501C) variant and fits to an exponential decay function (light pink, $d = 1.7$ in (**b**) and $d = 2.3$ in (**c**)) devoid of any distances. The error bounds (light blue) for the distribution obtained from Tikhonov regularization represent variations of the probability distribution corresponding to the uncertainty in the background function. The reliability of shape, width, and mean distance is related to the maximum dipolar evolution time and represented with the color code as implemented in the DeerAnalysis software. The distance distribution obtained from simulation on existing structures using the MMM software is overlaid in dotted lines: inward-open (IO, in black dotted lines, PDB 5D0O) and lateral-open (LO, grey dotted lines, PDB 5LJO) conformations, respectively of the BAM complex. A broader distribution for BamABCDE (**c**) suggests that the lipoproteins affect the dynamic behavior of the central BamA barrel

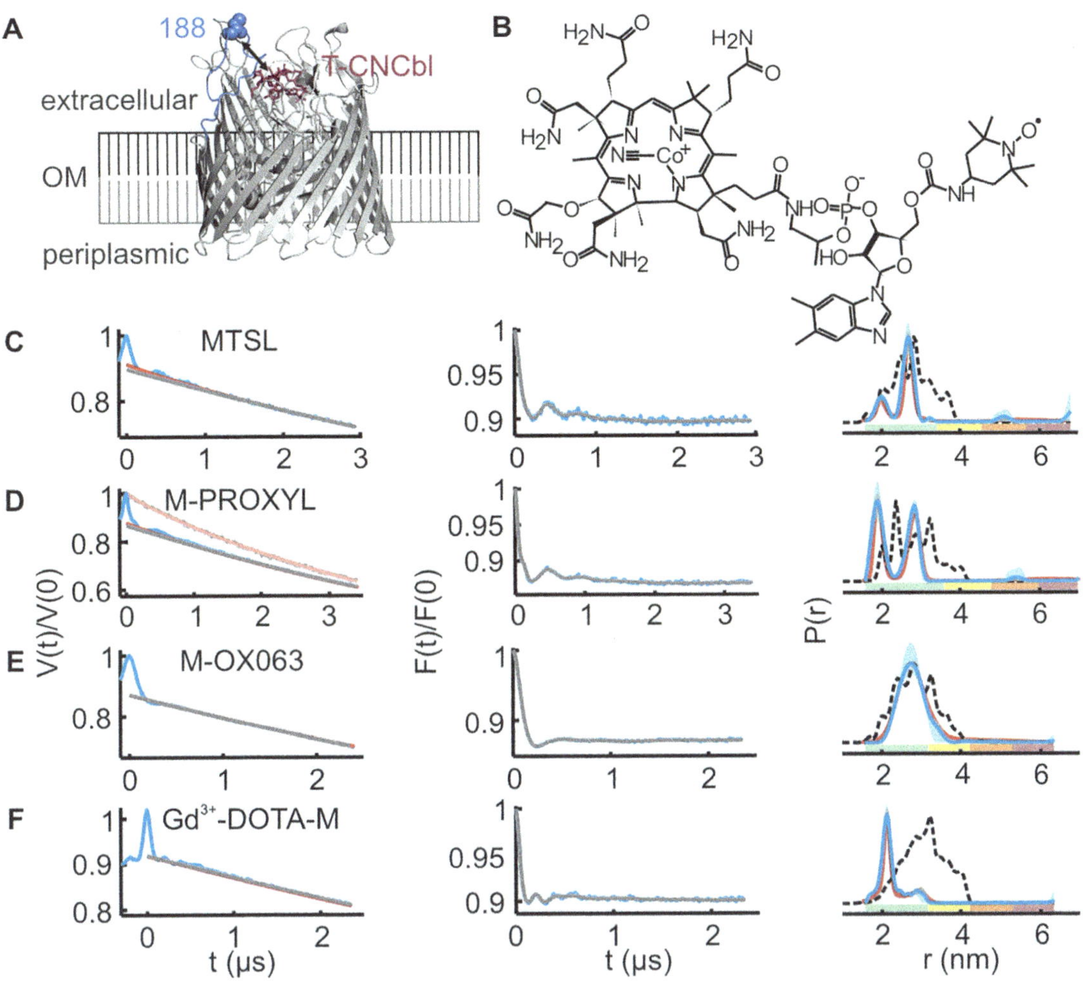

Fig. 4 Orthogonal 4-pulse PELDOR measurements between spin-labeled $BtuB^{188C}$ and its 2,2,6,6-tetramethyl-1-piperidinyloxy (TEMPO) modified cobalamin (T-CNCbl) ligand in isolated outer membranes (OM). (**a**) Structure of BtuB (PDB 1NQH) showing the spin-labeled position 188C located on the second extracellular loop (in blue) and the bound cobalamin ligand (in red). (**b**) Chemical structure of the TEMPO-modified cobalamin. (**c**–**f**) 4-pulse PELDOR data measured between spin-labeled $BtuB^{188C}$ with the indicated spin tag and T-CNCbl. Panel description as in Fig. 3. Background functions obtained from Tikhonov regularization (TR, in grey) or DEERNet analysis (in red) are overlaid in the first panel. The primary data (light grey) shown in the left panel in (**d**) corresponds to the measurement in the absence of T-CNCbl and fits entirely into an exponential decay function (light pink, $d = 2.7$) devoid of any distances. The simulations on the crystal structure of BtuB (PDB 1NQH) are overlaid in dark grey dotted lines with the distributions obtained with TR (in blue) or DEERNet analysis (in red) [44]

2 Materials

2.1 Site-Directed Mutagenesis and Transformation

1. Plasmid DNA carrying the protein gene of interest (here we use pJH114 carrying *bamABCDE* [41], and pAG1carrying *btuB* [42]).
2. Site-directed mutagenesis kit.
3. *E. coli* competent cells (here we use BL21(DE3) and RK5016 (MC4100 derivative, *metE70 argH recA btuB*)) [43].
4. Lysogeny broth (LB) media: 10 g/L tryptone, 5 g/L yeast extract and 10 g/L NaCl. Autoclave at 121 °C for 20 min.
5. LB agar plates containing the desired antibiotic: LB media supplemented with 15 g/L agar. Autoclave at 121 °C for 20 min. When the temperature of the product reaches ~50 ° C, add the antibiotic. Commonly, the working concentrations are 100 μg/mL for ampicillin, 50 μg/mL for streptomycin, 50 μg/mL for kanamycin, and 20–34 μg/mL for chloramphenicol. Pour the mixture into sterile Petri dishes and store at 4 °C.

2.2 Protein Expression

1. 2× YT growth media: 16 g/L tryptone, 10 g/L yeast extract, 5 g/L NaCl. Autoclave at 121 °C for 20 min.
2. 10× stock of minimal growth media mineral salts: 105 g/L K_2HPO_4, 45 g/L KH_2PO_4, 10 g/L $(NH_4)_2SO_4$ and 5 g/L $C_6H_5Na_3O_7 \cdot 2H_2O$. Autoclave at 121 °C for 20 min.
3. Stock solution of 10% (w/v) L-arginine, 5% (w/v) L-methionine, 40% (w/v) glucose, 3 mM thiamine, 2.5 M $MgSO_4$, and 1 M $CaCl_2$. Filter sterilize each solution.
4. Minimal media: 1× minimal media mineral salts, 0.01% (w/v) L-arginine, 0.01% (w/v) L-methionine, 0.2% (w/v) glucose, 150 μM thiamine, 3 mM $MgSO_4$, 300 μM $CaCl_2$. Store at room temperature (RT).
5. Antibiotic solution: Prepare a stock solution of the desired antibiotic. Filter-sterilize and store at −20 °C.
6. Isopropyl-*β*-D-thiogalacto-pyranoside (IPTG) stock solution: 1 M IPTG. Filter-sterilize and store at −20 °C.

2.3 Isolation of Outer Membrane

1. Spin labeling buffer: 50 mM MOPS, 60 mM NaCl. Adjust the pH to 7.5 or 6.8 using NaOH. Store at 4 °C.
2. Phenylmethanesulfonyl fluoride (PMSF) stock solution: 100 mM PMSF in isopropanol. Store at −20 °C.
3. Deoxyribonuclease I from bovine pancreas (DNase I).
4. Lysozyme stock solution: 50 mg/mL lysozyme. Store at −20 °C.
5. N-Lauroylsarcosine sodium salt solution: Prepare 20% stock solution in water and store at RT.

2.4 Spin Labeling in Outer Membrane and E. coli

1. Spin labeling buffer: As described in Subheading 2.3.
2. 1-oxyl-2,2,5,5-tetramethyl-3-pyrroline-3-methyl methanethiosulfonate (MTSL, Toronto Research Chemicals) stock solution: 100 mM MTSL in DMSO. Store at −20 °C.
3. 3-Maleimido-2,2,5,5-tetramethyl-1-pyrrolidinyloxy (M-PROXYL, Sigma-Aldrich) stock solution: 100 mM M-PROXYL in DMSO. Store at −20 °C.
4. MTS-OX063 stock solution: 25 mM MTS-OX063 in DMSO. Store at −20 °C in the dark [44].
5. M-OX063 stock solution: 25 mM M-OX063 in DMSO. Store at −20 °C in the dark [18].
6. Gd-DOTA-M solution: Dissolve maleimido-mono-amide-DOTA (available from Macrocyclics) and Gadolinium(III) chloride in a ratio of 1:1.1 by mol in water. Stir the mixture at RT for ~8 h while slowly adjusting the pH to 5.5–6 using NaOH and further incubate overnight at RT. Freeze-dry the mixture and dissolve the resulting powder in anhydrous DMF to reach a concentration of 25 mM [18]. Store under argon atmosphere and at −20 °C.
7. Micropipettes Intramark and micropipette controller (Blaubrand).
8. Deuterated glycerol (glycerol-d_8).
9. Quartz ESR tubes with a 1.6 mm outer diameter (Suprasil, Wilmad LabGlass).
10. Liquid nitrogen.
11. Argon.

2.5 Equipment

1. Thermomixer.
2. Plate incubator.
3. Incubator shaker.
4. Optical density meter.
5. Centrifuges.
6. Cell lysis device.
7. X-band ESR spectrometer (here we use a Bruker EMXnano Benchtop Spectrometer).
8. Q-band ESR spectrometer (here we use a Bruker Elexsys E580 Q-Band Pulsed ESR spectrometer equipped with an arbitrary waveform generator (SpinJet-AWG)).

2.6 Software Packages

1. MMM (Multiscale Modelling of Macromolecules) software package [45].
2. Bruker Xenon software.
3. DeerAnalysis software [46].
4. DeerLab software [47].

3 Methods

3.1 Plasmid Construction and Site-Directed Mutagenesis

1. Clone the protein gene of interest into an appropriate vector. In the case of BamABCDE complex described here, pTrc99A-based plasmid (pJH114) wherein the genes are under the control of a *trc* promoter is used for IPTG inducible protein expression. In the case of BtuB, the plasmid pAG1 carrying the gene under the control of its native promoter is used for constitutive protein expression.
2. Introduce cysteine codon substitutions for spin labeling at desired sites using a site-directed mutagenesis kit according to the protocol of the manufacturer (*see* **Note 1**). Spin label rotamer populations and PELDOR simulations can be obtained on the protein structure using the MATLAB-based software package MMM [45] and provide a useful tool for choosing suitable cysteine substitution positions.

3.2 Transformation

1. Add 50–100 ng of the plasmid to a 50–100 μL aliquot of competent cells of a suitable *E. coli* strain in a sterile Eppendorf tube. In the example described here, BL21(DE3) and RK5016 cells were used for BamABCDE and BtuB, respectively.
2. Incubate the cells on ice for 30 min.
3. Transfer the cells to 42 °C for 30–45 s as required for the competent cell in use.
4. Place the cells on ice for 5 min.
5. Add 200 μL of sterile LB media to the cells.
6. Incubate at 37 °C for 1 h under shaking.
7. Spread the cells onto LB agar plates supplemented with an appropriate concentration of the antibiotic based on the plasmid used. In our case, ampicillin was used for both pJH114 and pAG1.
8. Incubate the plate at 37 °C overnight.

3.3 Protein Expression

The culture volume described here is sufficient for multiple spin labeling experiments in *E. coli* (*see* Subheading 3.6). For isolation of and spin labeling in native outer membranes (*see* Subheadings 3.4 and 3.5), the culture volume should be scaled up appropriately.

3.3.1 Inducible Expression of BamABCDE

1. Prepare a preculture by transferring one or multiple colonies of transformed cells to 20 mL of LB growth media supplemented with ampicillin in a sterile Erlenmeyer flask.
2. Incubate the culture at 37 °C with shaking until the OD_{600} (optical density measured at a wavelength of 600 nm) reaches 0.6–0.8.
3. Add a suitable volume of the preculture to 100 mL of 2× YT growth media supplemented with the antibiotic to an $OD_{600} < 0.01$ in a sterile Erlenmeyer flask. Incubate the culture at 37 °C with shaking until the OD_{600} reaches ~0.7.
4. Induce protein expression by addition of IPTG to a final concentration of 0.4 mM.
5. Incubate the culture under shaking at 37 °C for 1.5 h and determine the OD_{600} value (*see* **Note 2**).
6. For spin labeling in *E. coli* (*see* Subheading 3.6), store the culture at 4 °C. For isolation of outer membranes (*see* Subheading 3.4), pellet the cells at 8000× *g* for 10 min at 4 °C and store at −80 °C until further use.

3.3.2 Constitutive Expression of BtuB

1. Prepare a preculture by transferring one or multiple colonies of transformed cells to 20 mL of minimal growth media supplemented with ampicillin in a sterile Erlenmeyer flask.
2. Incubate the culture at 33 °C with shaking until the OD_{600} reaches 0.3–0.5.
3. Add a suitable volume of the preculture to 100 mL of minimal growth media supplemented with the antibiotic to an $OD_{600} < 0.01$ in a sterile Erlenmeyer flask. Incubate the culture at 33 °C overnight for ~13 h with shaking until the OD_{600} reaches 0.3–0.5 (*see* **Note 2**).
4. Proceed as described in **Step 6** of Subheading 3.3.1.

3.4 Isolation of Outer Membrane

1. Thaw the cells on ice and suspend in 100 mL of ice-cold spin labeling buffer (pH 7.5) per 5 g of cell and supplement the cell suspension with 10 μg/mL DNase I, 1 mM PMSF, and 100 μg/mL lysozyme.
2. Lyse the cell suspension by sonication on ice.
3. Centrifuge the cell lysate at 15,000× *g* for 20 min at 4 °C.
4. Treat the supernatant containing the total membrane with 0.5% (w/v) N-lauroylsarcosine sodium salt for 10 min at RT to solubilize the inner membrane.
5. Collect the outer membrane fraction by ultracentrifugation at 200,000× *g* for 1.5 h at 4 °C.

3.5 Spin Labeling in Outer Membrane

1. Suspend the outer membrane pellet homogeneously in a proportion of 0.1 g per 10 mL of spin labeling buffer. Use the buffer at a pH of 7.5 for spin labeling with methanethiosulfonate functionalized and the buffer at a pH of 6.8 for maleimide functionalized spin labels (*see* **Note 3**).
2. Add the spin label and incubate at RT with mixing for 1 h. Refer to Table 1 for further details regarding the labeling conditions for different spin labels (Fig. 1).

Table 1
Spin labeling conditions in isolated outer membranes and in *E. coli* cells with methanethiosulfonate and maleimide functionalized nitroxides, trityl, and Gd^{3+} based spin tags

Spin label	Labeling condition in isolated outer membranes	Labeling condition in *E. coli*	Notes
MTSL	Final concentration: 10–100 μM Time: 1 h Temperature: RT Buffer pH: 7–8	Final concentration: 10–100 μM Time: 10 min – 1 h Temperature: RT Buffer pH: 7–8 OD_{600}: 0.5–5	*See* **Note 4**
M-PROXYL	Final concentration: 10 μM Time: 1 h Temperature: RT Buffer pH: 6.8	Final concentration: 15 μM Time: 1 h Temperature: RT Buffer pH: 6.8 OD_{600}: 0.5	*See* **Note 5**
MTS-OX063	Final concentration: 10–25 μM Time: 1 h Temperature: RT Buffer pH: 7–8 Multiple cycles of freezing and thawing during spin labeling and washing are needed to introduce the spin label into outer membrane vesicles (*see* **Note 6**).	Final concentration: 10–100 μM Time: 10 min – 1 h Temperature: RT Buffer pH: 7–8 OD_{600}: 0.5–5	*See* **Notes 4** and **7**
M-OX063	Final concentration: 10–25 μM Time: 1 h Temperature: RT Buffer pH: 6.8 Multiple cycles of freezing and thawing (*see* **Note 6**).	Final concentration: 15 μM Time: 1 h Temperature: RT Buffer pH: 6.8 OD_{600}: 0.5	*See* **Notes 5** and **7**
Gd-DOTA-M	Final concentration: 10 μM Time: 1 h Temperature: RT Buffer pH: 6.8 Multiple cycles of freezing and thawing (see above).	Under our experimental conditions, no successful spin labeling was observed	*See* **Note 8**

Where multiple conditions were tested successfully, a range is given

3. To remove excess spin label, dilute the outer membrane suspension five-fold and centrifuge at 200,000× *g* for 1.5 h at 4 °C.
4. Collect the membrane fraction and suspend in buffer with the same dilution factor as described above.
5. Repeat the ultracentrifugation and suspension step twice.
6. Finally, suspend the outer membrane pellet homogeneously in a proportion of approximately 0.1 g per 100 μL of buffer.
7. For the CW ESR measurement, transfer the outer membrane to a micropipette and proceed with Subheading 3.7 (Fig. 2).
8. For PELDOR experiments, add 15–20% deuterated glycerol to the outer membrane, transfer to a Suprasil quartz ESR tube, and subsequently flash freeze in liquid nitrogen. If required, a spin-labeled ligand can be added to the membrane as well. Store the sample at −80 °C until further use. Proceed with Subheading 3.8 for the PELDOR measurement.

3.6 Spin Labeling in E. coli

1. Pellet the culture from Subheading 3.3 at 6000× *g* for 10 min at 4 °C.
2. Suspend the cells gently in a falcon tube in the appropriate spin labeling buffer to reach an OD_{600} value as discussed in Table 1.
3. Add the spin label to the cell suspension and incubate at RT with mixing (*see* **Note 9**). The spin labeling conditions are as specified in Table 1.
4. To remove excess spin label, dilute the cells in spin labeling buffer and centrifuge at 6000× *g* for 10 min at 4 °C.
5. Suspend the cells in a 1.5 mL Eppendorf tube and centrifuge at 6000× *g* for 5 min at 4 °C. Repeat this washing step.
6. Suspend the cells in buffer in a smaller volume to reach an OD_{600} value of around 500.
7. For the CW ESR measurement, transfer the cell suspension to a micropipette and proceed with Subheading 3.7 (Fig. 2a).
8. For sample preparation for PELDOR experiments, add 15–20% deuterated glycerol to the sample from **Step 6**, transfer it to a Suprasil quartz ESR tube, and flash freeze in liquid nitrogen. Another spin-labeled molecule can be added to the cell suspension before freezing. Store the sample at −80 °C until further use. Proceed with Subheading 3.8 for the PELDOR measurement.

3.7 CW ESR

The CW ESR measurements described here are carried out at RT using a Bruker EMXnano Benchtop Spectrometer running the Xenon software and operating at X-band frequency (9.4 GHz). The reader is referred to the EMXnano User and Application Manuals for a detailed description on spectrometer operation.

1. Insert the sample and tune the microwave resonator.
2. Acquire the ESR spectrum of the spin-labelled protein. A sample containing the wildtype or cysteine-less protein can be measured to estimate the amount of non-specific (background) labeling (*see* **Note 10**). The parameters for nitroxide spectra are as follows: 100 kHz modulation frequency, 0.15 mT modulation amplitude, 1.995 mW microwave power, 5.12 ms time constant, 22.5 ms conversion time, and 18 mT sweep width. For trityl labeled samples, record spectra with 100 kHz modulation, 0.02 mT modulation amplitude, 1 mW microwave power, 1.28 ms time constant, 4.8 ms conversion time, and 5 mT sweep width.
3. Quantify the spin concentration using the reference-free concentration determination of paramagnetic species as implemented in the Xenon software. Alternatively, the spin concentration can be determined from an external calibration curve recorded with the spin label in use.

3.8 PELDOR

The 4-pulse PELDOR measurements are performed on a Bruker Elexsys E580 Q-Band Pulsed spectrometer equipped with an arbitrary waveform generator (SpinJet-AWG), a continuous-flow helium cryostat, a temperature control system (Oxford Instruments), a 50 W solid state amplifier, and a Bruker EN5107D2 dielectric resonator. For general and safe spectrometer operation, the reader is referred to the corresponding manual of their device. The protocol described hereinafter focuses solely on settings and parameters required to acquire field-swept spectra, to determine phase-memory time (T_M) values, and to perform 4-pulse PELDOR experiments with the dead-time free sequence $\pi/2_{obs} - \tau_1 - \pi_{obs} - t_1 - \pi_{pump} - (\tau_1 + \tau_2 - t_1) - \pi_{obs} - \tau_2$. The protocol serves as a guide to perform PELDOR experiments involving nitroxide (NO), trityl, and Gd^{3+}-based spin tags and their orthogonal combinations thereof. Modifications to the protocol may be needed depending on the configuration of the spectrometer in use.

1. Cool the cryostat of the device to 10–100 K as required using liquid helium.
2. Insert the frozen sample into the resonator. For NO – NO and $trityl_{pump} - NO_{obs}$ PELDOR experiments, measurements are performed at 50 K. For the $NO_{pump} - trityl_{obs}$ experiment, set the temperature to 100 K. For measurements involving Gd^{3+}, that is $NO_{pump} - Gd^{3+}{}_{obs}$, $trityl_{pump} - Gd^{3+}{}_{obs}$, and $Gd^{3+} - Gd^{3+}$, cool the cryostat further to 10 K. Wait until the temperature is stable (*see* **Note 11**).
3. For NO – NO and $NO_{pump} - trityl_{obs}$ experiments, optionally change the frequency of the resonance dip by +40 MHz or by −40 MHz for $trityl_{pump} - NO_{obs}$ from an over-coupled resonator mode for sensitivity consideration.

4. Setup a two pulse Hahn echo sequence $\pi/2 - \tau - \pi - \tau$ with Gaussian pulses and an appropriate inter-pulse delay (τ, say 400 ns). Set the $\pi/2$ and π pulses between 32–48 ns and for observing NO (50 K) the shot repetition time (SRT) is set to 2 ms. For the trityl – NO pair, the SRT is set to 10 ms and 2 ms at 50 K and 100 K, respectively. In the case of the NO – Gd^{3+} pair (10 K), the SRT is set to 10 ms for the detection of NO and reduced to 1–2 ms for observing Gd^{3+}. Similar settings also work for the trityl – Gd^{3+} pair. For the Gd^{3+} – Gd^{3+} case (10 K), the SRT is set to 1–2 ms. Optimize the Hahn echo intensity by varying the magnetic field and the amplitude of the pulses and phase the signal. For Gd^{3+} – Gd^{3+} PELDOR, it is recommended to set the main attenuator at 10–12 dB to suppress undesired artifacts [48].
5. Record an echo-detected field-swept spectrum with an appropriate sweep range. For NO_{pump} – ${Gd^{3+}}_{obs}$ and $trityl_{pump}$ – ${Gd^{3+}}_{obs}$ record two spectra optimized for each spin as discussed in **step 4**. Determine the magnetic field corresponding to the maximum of the pump spin and set the magnetic field to this value.
6. Optimize the pump (ELDOR) pulse. Insert a Gaussian pulse with an appropriate delay before the two pulse Hahn echo sequence as optimized in **step 4**. As a reference, the following pump pulse lengths may be used: 38 ns for NO – NO, NO_{pump} – $trityl_{obs}$, and $trityl_{pump}$ – ${Gd^{3+}}_{obs}$, 48 ns for $trityl_{pump}$ – NO_{obs}, 24 ns for the NO_{pump} – ${Gd^{3+}}_{obs}$ experiment, and 32 ns for the Gd^{3+} – Gd^{3+} case. A stronger pump pulse may be used when possible provided that a sufficient frequency offset can be achieved for the observer pulses to avoid overlap of the excitation bandwidths. Gradually increase the amplitude of the pump pulse until a maximum inversion of the echo is observed.
7. Test for any phase offset of the AWG-generated pulses. Set up four pulses with a sufficient inter-pulse delay (e.g., 200 ns rectangular pulses with a 400 ns delay) and defined phases of 0°, 90°, 180°, and 270°. Observe the pulses on the SpecJet by switching to the transmission mode (TM). Adjust the phase of the first pulse and then correct the phase of the other pulses as required.
8. Change the microwave frequency to the observer frequency by –80 MHz for NO – NO, –90 MHz for NO_{pump} – $trityl_{obs}$, +90 MHz for $trityl_{pump}$ – NO_{obs}, –190 MHz for $trityl_{pump}$ – ${Gd^{3+}}_{obs}$, –280 MHz in case of NO_{pump} – ${Gd^{3+}}_{obs}$ experiment, and –100 MHz for the Gd^{3+} – Gd^{3+} case.
9. Repeat as described in **Step 7** to determine the phase offset at observer frequency.

10. Set the observer pulses using the $\pi/2 - \tau - \pi - \tau$ sequence with Gaussian pulses. For NO – NO experiments, set the $\pi/2$ and π pulses to 48 ns with a SRT of 2 ms. For trityl$_{pump}$ – NO$_{obs}$ (50 K) use 48 ns pulses with 2 ms SRT. For the NO$_{pump}$ – trityl$_{obs}$ pair (100 K), use 48 – 64 ns pulses with 2–3 ms SRT. In the case of NO$_{pump}$ – $Gd^{3+}{}_{obs}$, trityl$_{pump}$ – $Gd^{3+}{}_{obs}$, and Gd^{3+} – Gd^{3+} experiments, set the pulse lengths to 32–48 ns and the SRT to 1–2 ms. Optimize the echo intensity by changing the amplitude of the pulses while adjusting the phase.
11. Determine the phase memory time (T_M) of the observer spin. The T_M is a useful parameter to estimate the maximum observable dipolar evolution time window ($t_{max} \sim 2 \times T_M$). Increase the inter-pulse delay τ in steps of 4 ns and shift the acquisition trigger by 8 ns with a sufficient number of points to observe the signal decay to zero with a two-step phase cycling. Fit the data (after a baseline correction if needed) with an appropriate exponential decay function to extract the value for T_M.
12. Set up the 4-pulse PELDOR experiment with the observer pulse sequence $\pi/2 - \tau_1 - \pi - \tau_2 - \pi - \tau_2$. Load the appropriate PulseSpel program and variable definition files (optionally available from Bruker). Enter the Gaussian pump and observer pulse lengths with the optimized amplitudes from **Steps 6** and **10**, the SRT from **Step 10**, the frequency offset from **Step 8**, and the phase values from **Steps** 7 and **9**. Set the first inter-pulse delay τ_1 value to 150–200 ns while observing nitroxide or trityl and to 400 ns for observing Gd^{3+}. To average out the deuterium modulations, increase τ_1 by 16 ns for eight steps. Adjust the τ_2 value according to the phase memory time from **step 11** and the expected distance (if available from a simulation). The value should be chosen long enough to observe at least 1.25–2 full periods of oscillations. Record the PELDOR trace with a 16-step phase cycling (x[x][xp]x) [49]. Accumulate enough scans until a good signal-to-noise ratio (SNR) is obtained (~6–24 h depending on the total spin concentration and spin labeling efficiency of the sample).

3.9 PELDOR Data Analysis and Interpretation

1. The currently recommended approach for PELDOR data analysis is a single-step method using deep neural networks (DEERNet) [50], one-step Tikhonov analysis (DeerLab) [47], or a combined approach as implemented in the ComparativeDeerAnalyzer (CDA) software package [51]. The conventional way of transforming primary PELDOR data ($V(t)/V(0)$) into the wanted distance distribution between the two spin labeled target sites is by using the two-step Tikhonov regularization approach as implemented in the MATLAB-based software package DeerAnalysis [46]. When the data quality is sufficient, this is a particularly useful approach for data analysis

with an experimentally determined value for the dimensionality of spin distribution (*d*, *see* Figs. 3 and 4). The reader is referred to the corresponding manuals for a more detailed description of the analysis process (*see* **Note 12**). The modulation depth parameter (Δ) provides direct information on effective spin labeling efficiency of the target molecule in the native surroundings (*see* **Note 13**). The resulting distance distribution can be compared with the corresponding simulation. If structures of the protein in various conformational states are available, simulations on those structures can be used to assign the experimental distance distribution (*see* **Note 14**).

4 Notes

1. Outer membrane proteins rarely have reactive cysteines. When present, native cysteines should be substituted (provided it does not impair protein function) to eliminate off-target labeling. Extracellular loops might offer better accessibility and more efficient labeling. Under our experimental conditions, direct spin labeling in *E. coli* cells is limited to extracellular sites, while isolated outer membranes provide bidirectional accessibility for spin labels or ligands.
2. For spin labeling in *E. coli*, protein expression level is a highly critical factor. While a certain amount of protein expression is needed for sensitivity reasons, an expression level that is too high could potentially be harmful to the cell and lead to spatial clustering or aggregation of the protein. The expression level of the target protein could be qualitatively analyzed using sodium dodecyl sulfate polyacrylamide gel electrophoresis (SDS-PAGE) and staining of protein bands.
3. To maintain a high reaction rate and chemoselectivity of maleimide for thiols, the pH value should be kept lower and around 6.8. At pH values above 7.0, the reaction rate of the maleimide-thiol reaction decreases and the maleimide moiety begins reacting competitively with free primary amines (e.g. of the amino acid lysine) [52].
4. The concentration and incubation time could be increased to improve labeling, especially at buried sites. However, this could be accompanied by an increase in non-specific (background) labeling. Generally, we observe more non-specific (background) labeling in isolated outer membranes compared to *E. coli* cells, which is likely due to an enhanced interaction with the membrane surface.
5. Maleimide functionalized spin labels yield a higher labeling efficiency with 1 h incubation in contrast to their MTS functionalized analogs, which give comparable labeling efficiency with shorter incubation times.

6. Outer membrane vesicles we isolated here might mostly exist in right-side-out orientation. Cycles of freezing and thawing (5×) can help to introduce spin labels into the lumen of the vesicles to improve spin labeling efficiency, especially for periplasmic positions. This is especially useful when the target spin label exceeds the exclusion limit of the porins (~600 Da), e.g., in the case of the trityl and Gd^{3+}-based spin labels described here. Freezing and thawing should also be used during the following washing steps to remove free spin label.
7. With the OX063 labels, we observe a tendency for aggregation. Working with a lower spin label concentration (10–25 μM) is ideal. Add the spin label in small aliquots, mix in between, and perform labeling in the dark.
8. With the spin labeling conditions tested, we could not observe successful labeling of the protein with Gd3+ in *E. coli* cells. This could be due to a release of the Gd^{3+} from the DOTA chelator [18]. A higher spin label concentration or a shorter incubation time could be tested, although a higher concentration could lead to an increase in non-specific labeling.
9. The common nitroxide spin labels (MTSL, M-PROXYL) tend to reduce to the ESR-silent hydroxylamine in the cellular (reducing) environments. For this reason and to maintain cell viability, the spin labeling procedure in *E. coli* should be carried out as gentle and fast as possible.
10. Due to the spin label reduction as well as to maintain cell viability (as discussed in **Note 9**), the CW spectra in *E. coli* should not be accumulated over too many scans. We found that averaging over 40 scans with a sweep time of 27 s for NO is sufficient to record spectra with a reasonable signal-to-noise ratio (SNR) under the experimental conditions specified. The CW spectra in isolated outer membranes however can be accumulated over more scans (usually up to 150 scans with a sweep time of 27 s for NO). Spin labeling in *E. coli* or isolated outer membranes overexpressing the WT protein is useful to estimate the amount of signal arising from nonspecific labeling. When working with a double cysteine variant, the corresponding single cysteine variants can be used to roughly estimate the amount of labeling arising from the individual sites (*see* Fig. 2a).
11. For trityl – NO PELDOR experiment, placing the pump (ELDOR) pulse on trityl yield a much larger modulation depth. However, if the nitroxide spin is in excess over the trityl labeled component (or there is a significant fraction of randomly distributed spins arising from background labeling), it can be useful to place the pump pulse on the nitroxide. When pumping nitroxide, it is useful to increase the temperature to 100 K as the transverse relaxation time (T_1) of trityl becomes more favorable (shorter) and allows faster data acquisition.

12. The experimental PELDOR time trace ($V(t)/V(0)$) contains contributions from intermolecular dipolar interaction ($B(t)/B(0)$) between spins randomly distributed on neighboring molecules and from the wanted intramolecular dipolar interaction ($F(t)/F(0)$) between spins located on the same biomolecule of interest. The intermolecular interaction is described as an exponential decay function of the form $B(t) = \exp(-C\lambda_B(kt)^{d/3})$ defined by the concentration C of the observer spin, the inversion efficiency λ_B of the pump (ELDOR) pulse, and a factor d for the spatial distribution of the spins. For spin-labeled proteins purified in detergent micelles, usually a homogenous spatial distribution of the spins is observed ($d = 3$) and the intermolecular contribution takes the form of a mono-exponential decay [53]. However, for spin-labeled membrane proteins in *E. coli* or isolated outer membranes, often a spatial distribution of fractional dimension ($1.5 \leq d < 3$) is observed (12, 37). Since the parameter d needed to perform precise PELDOR data analysis is often unknown for the more complex in situ samples, it is helpful to measure a PELDOR trace of spin-labeled *E. coli* cells or isolated outer membranes expressing the WT protein. The experimental data is devoid of any intramolecular contributions and can thus be fitted entirely with an exponential decay function yielding the desired value for d. Similarly, when the target protein is a double cysteine variant, the corresponding spin-labeled single cysteine variants can be measured to determine the values for d. When the data quality is sufficient, the validation tool implemented in the DeerAnalysis software allows to vary dimensionality for spin distribution in a range provided by the user and therefore fully covers its effect on the uncertainty for the distance distribution.

13. The modulation depth (Δ) of the PELDOR data is a critical parameter defined as the product of the inversion efficiency of the pump pulse λ_B and the labeling efficiency (L.E.) of the protein. Consequently, the labeling efficiency can be estimated if the λ_B is known. Under the experimental conditions described in Subheading 3.8 and when pumping the nitroxide spin, the inversion efficiency of the pump pulse can be estimated as ~30% yielding a Δ of ~30% for a 100% labeling case. When pumping the trityl spectrum, an inversion efficiency of ~100% could be achieved due to its very narrow spectral feature. Deviation of Δ from the theoretical maximum towards lower values can result from a comparatively low labeling efficiency of one or both the target sites, low labeling efficiency of an interacting ligand/protein partner and or a lower affinity for their binding (when the trityl spectrum is fully excited).

14. It should be noted that the DeerAnalysis software provides a color code with the resulting distance distribution depicting its reliability, which heavily depends on the chosen τ_2 period of the experiment. In the green region, the shape, mean distance, and width are reliable; in the yellow region, reliability of the shape is lost; in the orange region only the mean distance is reliable, while distances lying in the red region cannot be reliably quantified. The resulting distance distribution of the PELDOR experiment can vary from a narrow single distance peak to a broad multimodal distribution. Most commonly, conformational dynamics of proteins are investigated using nitroxide spin labels as they are small in size, have the least interference with the protein structure and their conformational properties are well characterized. Accordingly, the following discussion relates primarily to proteins spin-labeled with nitroxides. Broad and multimodal distributions may indicate a high intrinsic flexibility of the target site(s), e.g., if both or one of the chosen positions on the protein are located on a flexible loop region. Furthermore, broad distributions can also denote conformational heterogeneity of the target site(s) as a result of large conformational changes that the protein undergoes during function. As indicated in Subheading 3.9, a comparison of the experimental data with the simulation of the distance distribution on the protein structure can be immensely useful in such a case. If structures of the protein in various conformational states are available, the experimental distance distribution may be assigned to the simulations on those structures (*see* Fig. 3). However, the simulations carried out using the MMM software are performed on the static protein structure and usually in the absence of a lipid bilayer. Therefore, under in situ conditions as described here, deviations from the simulation do not solely reflect a dynamic behavior of the protein but importantly can result from the protein being embedded in its native cellular and membrane bilayer environment. The above discussion may be transferred to interpret distance distributions from orthogonal PELDOR experiments involving trityl or Gd^{3+} labels. However, care should be taken as these spin labels have so far been used less often to study protein conformation and yet await a more detailed characterization of their conformational properties. It should also be ruled out that the larger size of these spin labels compared to the common nitroxides does not interfere with the protein structure of the labeling site and its function. Moreover, due to their much more complex chemical structures, it is conceivable that interactions with the surrounding membrane environment could take place and affect the distance distribution (18). Ideally and if possible, a control measurement with nitroxide labeled system should be performed beforehand.

Acknowledgments

This work was financially supported from the Deutsche Forschungsgemeinschaft via the Emmy Noether program (JO 1428/1−1), SFB 1507 – "Membrane-associated Protein Assemblies, Machineries, and Supercomplexes" and a large equipment fund (438280639) to B.J. Trityl labels were provided by Victor M. Tormyshev and Elena G. Bagryanskaya.

References

1. Gessmann D et al (2014) Outer membrane β-barrel protein folding is physically controlled by periplasmic lipid head groups and BamA. Proc Natl Acad Sci U S A 111(16):5878–5883
2. Gupta K et al (2017) The role of interfacial lipids in stabilizing membrane protein oligomers. Nature 541(7637):421–424
3. Laganowsky A et al (2014) Membrane proteins bind lipids selectively to modulate their structure and function. Nature 510(7503):172–175
4. Dowhan W, Bogdanov M (2011) Lipid-protein interactions as determinants of membrane protein structure and function. Biochem Soc Trans 39(3):767–774
5. Balusek C, Gumbart JC (2016) Role of the native outer-membrane environment on the transporter BtuB. Biophys J 111(7):1409–1417
6. Tomasek D et al (2020) Structure of a nascent membrane protein as it folds on the BAM complex. Nature 583(7816):473–478
7. Hubbell WL, Altenbach C (1994) Investigation of structure and dynamics in membrane proteins using site-directed spin labeling. Curr Opin Struct Biol 4(4):566–573
8. Galazzo L et al (2022) The ABC transporter MsbA adopts the wide inward-open conformation in *E. coli* cells. Sci Adv 8(41):eabn6845
9. Goldfarb D (2022) Exploring protein conformations in vitro and in cell with EPR distance measurements. Curr Opin Struct Biol 75:102398
10. Kugele A et al (2021) In situ EPR spectroscopy of a bacterial membrane transporter using an expanded genetic code. Chem Commun 57(96):12980–12983
11. Karthikeyan G et al (2018) A bioresistant nitroxide spin label for in-cell EPR spectroscopy: in vitro and in oocytes protein structural dynamics studies. Angew Chem Int Ed 57(5):1366–1370
12. Joseph B et al (2019) In situ observation of conformational dynamics and protein ligand-substrate interactions in outer-membrane proteins with DEER/PELDOR spectroscopy. Nat Protoc 14(8):2344–2369
13. Azarkh M et al (2013) Site-directed spin-labeling of nucleotides and the use of in-cell EPR to determine long-range distances in a biologically relevant environment. Nat Protoc 8(1):131–147
14. Krstic I et al (2011) Long-range distance measurements on nucleic acids in cells by pulsed EPR spectroscopy. Angew Chem Int Ed 50(22):5070–5074
15. Pierro A et al (2022) Probing the structural dynamics of a bacterial chaperone in its native environment by nitroxide-based EPR spectroscopy. Chem Eur J 28(66):e202202249
16. Altenbach C, Flitsch SL, Khorana HG, Hubbell WL (1989) Structural studies on transmembrane proteins. 2. Spin labeling of bacteriorhodopsin mutants at unique cysteines. Biochemistry 28(19):7806–7812
17. Joseph B et al (2015) Distance measurement on an endogenous membrane transporter in *E. coli* cells and native membranes using EPR spectroscopy. Angew Chem Int Ed 54(21):6196–6199
18. Ketter S et al (2022) In situ distance measurements in a membrane transporter using maleimide functionalized orthogonal spin labels and 5-pulse electron-electron double resonance spectroscopy. J Magn Reson Open:100041
19. Martin RE et al (1998) Determination of end-to-end distances in a series of TEMPO diradicals of up to 2.8 nm length with a new four-pulse double electron electron resonance experiment. Angew Chem Int Ed 7(20):2833–2837
20. Bleicken S et al (2019) Gem-diethyl pyrroline nitroxide spin labels: synthesis, EPR characterization, rotamer libraries and biocompatibility. ChemistryOpen 8(8):1057–1065
21. Braun TS et al (2020) Isoindoline-based nitroxides as bioresistant spin labels for protein labeling through cysteines and alkyne-bearing

noncanonical amino acids. Chembiochem 21(7):958–962

22. Reginsson GW, Kunjir NC, Sigurdsson ST, Schiemann O (2012) Trityl radicals: spin labels for nanometer-distance measurements. Chem Eur J 18(43):13580–13584
23. Prokopiou G et al (2018) Small Gd(III) tags for Gd(III)-Gd(III) distance measurements in proteins by EPR spectroscopy. Inorg Chem 57(9):5048–5059
24. Shevelev GY et al (2015) Triarylmethyl labels: toward improving the accuracy of EPR nanoscale distance measurements in DNAs. J Phys Chem B 119(43):13641–13648
25. Cunningham TF et al (2015) The double-histidine Cu2+-binding motif: a highly rigid, site-specific spin probe for electron spin resonance distance measurements. Angew Chem Int Ed 54(21):6330–6334
26. Wu ZY et al (2017) Selective distance measurements using triple spin labeling with Gd3+, Mn2+, and a nitroxide. J Phys Chem Lett 8(21):5277–5282
27. Rudolph M, Tampe R, Joseph B (2023) Time-resolved Mn(2+) -NO and NO-NO distance measurements reveal that catalytic asymmetry regulates alternating access in an ABC transporter. Angew Chem Int Ed 62:e202307091
28. Igarashi R et al (2010) Distance determination in proteins inside Xenopus laevis oocytes by double electron-electron resonance experiments. J Am Chem Soc 132(24):8228–8229
29. Hasanbasri Z et al (2021) Cleavage-resistant protein labeling with hydrophilic trityl enables distance measurements in-cell. J Phys Chem B 125(20):5265–5274
30. Jassoy JJ et al (2019) Site selective and efficient spin labeling of proteins with a maleimide-functionalized trityl radical for pulsed dipolar EPR spectroscopy. Molecules 24(15):2735
31. Ketter S et al (2021) In situ labeling and distance measurements of membrane proteins in *E. coli* using Finland and OX063 trityl labels. Chem Eur J 27(7):2299–2304
32. Giannoulis A et al (2019) DEER distance measurements on trityl/trityl and Gd(iii)/trityl labelled proteins. Phys Chem Chem Phys 21(20):10217–10227
33. Ketter S, Joseph B (2023) Gd(3+)-trityl-nitroxide triple labeling and distance measurements in the heterooligomeric cobalamin transport complex in the native lipid bilayers. J Am Chem Soc 145(2):960
34. Galazzo L, Teucher M, Bordignon E (2022) Orthogonal spin labeling and pulsed dipolar spectroscopy for protein studies. Methods Enzymol 666:79–119
35. Narr E, Godt A, Jeschke G (2002) Selective measurements of a nitroxide-nitroxide separation of 5 nm and a nitroxide-copper separation of 2.5 nm in a terpyridine-based copper (II) complex by pulse EPR spectroscopy. Angew Chem Int Ed 41(20):3907–3910
36. Wort JL et al (2019) Sub-micromolar pulse dipolar EPR spectroscopy reveals increasing Cu(II) -labelling of double-histidine motifs with lower temperature. Angew Chem Int Ed 58(34):11681–11685
37. Gopinath A, Joseph B (2022) Conformational flexibility of the protein insertase BamA in the native asymmetric bilayer elucidated by ESR spectroscopy. Angew Chem Int Ed 61(2): e202113448
38. Wu T et al (2005) Identification of a multicomponent complex required for outer membrane biogenesis in *Escherichia coli*. Cell 121(2): 235–245
39. Fairman JW, Noinaj N, Buchanan SK (2011) The structural biology of β-barrel membrane proteins: a summary of recent reports. Curr Opin Struct Biol 21(4):523–531
40. Walther DM, Rapaport D, Tommassen J (2009) Biogenesis of β-barrel membrane proteins in bacteria and eukaryotes: evolutionary conservation and divergence. Cell Mol Life Sci 66(17):2789–2804
41. Roman-Hernandez G, Peterson JH, Bernstein HD (2014) Reconstitution of bacterial autotransporter assembly using purified components. elife 3:e04234
42. Fuller-Schaefer CA, Kadner RJ (2005) Multiple extracellular loops contribute to substrate binding and transport by the *Escherichia coli* cobalamin transporter BtuB. J Bacteriol 187(5):1732–1739
43. Heller K, Mann BJ, Kadner RJ (1985) Cloning and expression of the gene for the vitamin B12 receptor protein in the outer membrane of *Escherichia coli*. J Bacteriol 161(3):896–903
44. Tormyshev VM et al (2020) Methanethiosulfonate derivative of OX063 trityl: a promising and efficient reagent for side-directed spin labeling of proteins. Chem Eur J 26(12): 2705–2712
45. Jeschke G (2021) MMM: Integrative ensemble modeling and ensemble analysis. Protein Sci 30(1):125–135
46. Jeschke G et al (2006) DeerAnalysis2006 – a comprehensive software package for analyzing pulsed ELDOR data. Appl Magn Reson 30: 473–498
47. Fabregas Ibanez L, Jeschke G, Stoll S (2020) DeerLab: a comprehensive software package for analyzing dipolar electron paramagnetic

resonance spectroscopy data. Magn Reson 1(2):209–224

48. Galazzo L et al (2020) Spin-labeled nanobodies as protein conformational reporters for electron paramagnetic resonance in cellular membranes. Proc Natl Acad Sci U S A 117(5):2441–2448
49. Tait CE, Stoll S (2016) Coherent pump pulses in double electron electron resonance spectroscopy. Phys Chem Chem Phys 18(27): 18470–18485
50. Worswick SG, Spencer JA, Jeschke G, Kuprov I (2018) Deep neural network processing of DEER data. Sci Adv 4(8):eaat5218
51. Schiemann O et al (2021) Benchmark test and guidelines for DEER/PELDOR experiments on nitroxide-labeled biomolecules. J Am Chem Soc 143(43):17875–17890
52. Hermanson GT (2013) Bioconjugate techniques, 3rd edn, pp 1–1146
53. Jeschke G (2012) DEER distance measurements on proteins. Annu Rev Phys Chem 63: 419–446

Chapter 16

Characterization of β-Barrel Outer Membrane Proteins and Their Interactions with Chaperones by Chemical-Crosslinking Mass Spectrometry

Antonio N. Calabrese

Abstract

Chemical crosslinking-mass spectrometry (XL-MS) is an established tool that can be used to study the architecture and dynamics of proteins and protein assemblies. Here the application of XL-MS to study outer membrane proteins (OMPs) and their interactions with periplasmic chaperones is described, to inform on the molecular mechanisms underpinning OMP assembly. XL-MS data are especially powerful when used to complement high-resolution structural data, data from structural prediction or to drive molecular modeling of proteins and protein assemblies. The approach described here could be applied to the study of any protein assembly (including other membrane proteins).

Key words Chemical crosslinking, Mass spectrometry, Structural proteomics, β-barrel assembly machinery complex, Chaperone, Outer membrane proteins

1 Introduction

Chemical crosslinking-mass spectrometry (XL-MS) can be used to inform on the architecture of proteins and their assemblies and is especially powerful when studying transient and/or weak protein complexes [1, 2]. In a crosslinking experiment, the protein, or protein complex, is treated with an appropriate crosslinking reagent. To account for differences in the amino acid residue composition of protein surfaces (or those of protein-protein interfaces), a range of crosslinkers with different chemistries that target different amino acid sidechains have been developed. However, all crosslinkers comprise the basic architecture of two reactive groups, that covalently react with amino acid side-chains, that are separated by a spacer arm [3]. Typical crosslinkers comprise NHS-ester groups that react predominantly with Lys residues, with other reactive groups deployed in crosslinker designs including diazirines [4, 5] and benzophenones [6], which have the advantage of being

Raffaele Ieva (ed.), *Transmembrane β-Barrel Proteins: Methods and Protocols*, Methods in Molecular Biology, vol. 2778,
https://doi.org/10.1007/978-1-0716-3734-0_16,

photoactivatable, but often these functional groups result in lower yields of crosslinked material [7]. The structure of crosslinking reagents means that proximal residues, within a distance defined by the length of the spacer arm, can be crosslinked. Crosslinkers with a range of spacer arm lengths have been developed so that a range of distance restraints can be obtained [3].

After crosslinking, a conventional sample preparation workflow for bottom-up proteomics can be followed, using either in-gel or in-solution proteolysis. In-gel digestion is a useful strategy to isolate a selected crosslinked species for downstream analysis, for example in situations where contaminating proteins/assemblies may wish to be removed. However, in our experience recovery of crosslinked peptides is often poor for in-gel digestion. In these situations, in-solution digestion may be preferred. Following proteolysis, the location of residues involved in crosslinks can be determined using liquid chromatography-mass spectrometry (LC-MS). Crosslinked residues must be within the distance spanned by the spacer arm, and therefore the data generated can be compared with high-resolution structures, structural models, or used as distance restraints for modelling [2, 8–11]. An alternative strategy for XL-MS is to site-specifically incorporate a crosslinkable group into a protein sequence, either during protein synthesis in vivo [10] or by incorporating specific reactive groups into proteins site-selectively via the design of constructs that incorporate reactive residues at defined sites. For example, reagents have been previously described that label site-selectively at engineered Cys-residue in a Cys-free protein background [7]. These heterobifunctional reagents also comprise a reactive diazirine moiety that reacts non-specifically (but with preference for acidic residues [4]). Using this approach, "targeted" crosslinking experiments can be performed to interrogate the location of a specific residue/region within a protein or protein assembly.

XL-MS is an ensemble-averaged technique, and the distance restraints generated report on all members of the conformational ensemble. For conformationally diverse proteins/assemblies it has been shown that integrative modeling pipelines can be deployed to generate structure(s) that are consistent with the generated data [9, 12], but the relative abundance of each species within an ensemble remains difficult to discern from these data.

Using SurA, an outer membrane protein (OMP) chaperone, as an example here we show how XL-MS has been used to study (i) the architecture/dynamics of SurA [13], (ii) the interaction of unfolded OMPs with periplasmic chaperones, to identify the molecular determinants of OMP:chaperone interactions [7, 13], and (iii) the interaction of the OMP chaperone SurA with the β-barrel assembly machinery complex (BAM) [14], which comprises a central subunit, BamA, itself an OMP. Data from XL-MS, together with biochemical experiments and data from other structural MS tools (e.g., hydrogen-deuterium exchange), have revealed unprecedented insights into the molecular mechanisms underpinning OMP assembly [7, 13, 14].

2 Materials

Prepare all solutions using LC-MS grade water and solvents. Solutions should be prepared freshly and stored at room temperature for up to 1 week, unless otherwise indicated.

2.1 Protein Samples

1. Purified proteins for crosslinking. For example, unfolded OMPs (here OmpX) and SurA, expressed in *E. coli* and purified as described previously (*see* **Note 1**).

2.2 Crosslinking

1. SDS-PAGE gels, running buffer, loading dye. Any commercially available or homemade gels/reagents are suitable (*see* **Note 2**).
2. Appropriate crosslinking reagent, e.g., disuccinimidyl dibutyric urea (DSBU) (Fig. 1a) (Thermo Fisher Scientific) has been used in this study (*see* **Note 3**).
3. Desalting columns (e.g., Zeba, Thermo Fisher Scientific), or other desalting apparatus.
4. Crosslinking buffer: 10 mM potassium phosphate, pH 8.0, or other appropriate (sample dependent) (*see* **Note 4**).
5. OMP denaturation buffer: 10 mM potassium phosphate, pH 8.0, 8 M urea.
6. Quenching buffer: 1 M Tris-HCl, pH 8.0 (*see* **Note 5**).
7. Spectrophotometer for protein concentration measurements (e.g., NanoDrop, Thermo Fisher Scientific).

2.3 In-Gel Digestion

1. Biopsy punch or scalpel.
2. 30% (v/v) ethanol.
3. 100% acetonitrile (LC-MS grade).
4. Trypsin (sequencing grade): 0.02 μg/μL trypsin in 25 mM ammonium bicarbonate pH 8.0. (store unused trypsin solution at −80 °C) (*see* **Note 6**).
5. 10 mM dithiothreitol (DTT) (*see* **Note 7**).
6. 55 mM iodoacetamide (*see* **Note 7**).
7. 60% (v/v) acetonitrile, 5% (v/v) formic acid.
8. Laminar flow hood (*see* **Note 8**).
9. Thermoshaker for 1.5 mL tubes (e.g., PHMT, Grant Instruments).

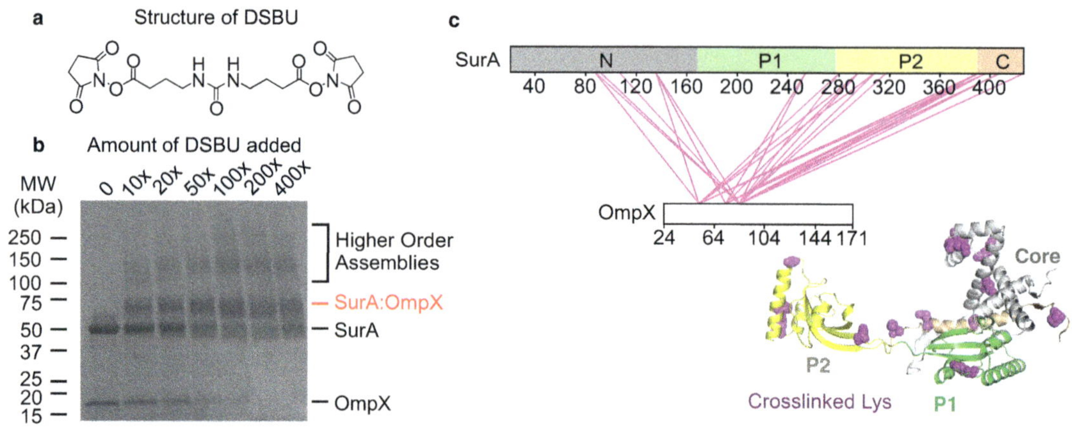

Fig. 1 Chemical crosslinking of the SurA-OmpX complex. (**a**) Structure of DSBU, comprising two NHS-ester reactive groups separated by a spacer arm. The spacer arm contains a urea linkage which is MS cleavable. (**b**) SDS-PAGE of crosslinked SurA-OmpX. Both proteins are at 5 μM concentration in this experiment. An increasing concentration of DSBU was added to samples of SurA-OmpX and the resultant crosslinked assemblies analyzed by SDS-PAGE. Increasing concentrations of crosslinker result in more protein being covalently trapped in complexes. Note the disappearance of the band corresponding to uncrosslinked OmpX at a ca 50× excess of DSBU. Higher-order assemblies are also captured, consistent with multiple copies of SurA binding to an OMP, as we have observed previously. Higher order SurA-OmpX assemblies are also observed, demonstrating why in this case it was necessary to perform in-gel digestion. If in-solution digestion was performed, the detected crosslinks would be derived from both the 1:1 SurA:OmpX complex and the higher-order species. Samples for crosslinking contained 5 μM SurA, 5 μM OmpX, 0.24 M urea, 50 μM–2 mM DSBU, in 10 mM potassium phosphate buffer, pH 8.0, 25 °C. (**c**) Detected inter-molecular SurA-OmpX crosslinks indicated by lines between the two sequences (upper). Crystal structure of SurA (PDB 1M5Y [30]), with purple spheres denoting residues on SurA that were crosslinked to OmpX with DSBU (lower). SurA is colored by domain, with the N-terminal region of the core domain, P1, P2, and the C-terminal region of the core domain colored grey, green, yellow, and orange, respectively

10. Vacuum concentrator (e.g., Concentrator Plus, Eppendorf).
11. Low protein binding 1.5 mL tubes.
12. 2.5% (v/v) acetonitrile, 0.1% (v/v) formic acid.

2.4 Liquid Chromatography-Mass Spectrometry

1. LC-MS system. For example, Vanquish Neo UHPLC system coupled to an EASY-Spray nanoelectrospray ionization source and an Orbitrap Eclipse mass spectrometer (Thermo Fisher Scientific).
2. Total recovery LC vials.
3. HPLC column appropriate for instrument configuration. For example, EASY-Spray C18 HPLC column (75 um × 500 mm, 2 um particle size) with integrated temperature controller and nanoelectrospray emitter. An optional trap column can also be used (e.g. PepMap Neo C18 trap cartridge (Thermo Fisher Scientific, 5 μm particle size, 300 μm × 0.5 cm)).
4. LC-MS grade solvent A: 0.1% (v/v) formic acid in water.
5. LC-MS grade solvent B: 0.1% (v/v) formic acid in acetonitrile.

2.5 Data Analysis and Visualization

1. Software for crosslinking analysis (e.g., XlinkX [15], xiSEARCH [16], MaxLynx [17], pLink [18]). Note that each has different advantages [19]. We routinely use the freely available software MeroX [20]).
2. Software for data file conversion (e.g., MSConvert (https://proteowizard.sourceforge.io/) [21]).
3. Data visualization software (e.g., xiVIEW [22] and xVis [23] to generate maps of crosslinking data or tools that interface with molecular visualization software, e.g., PyXlinkViewer [24] or XMAS [25]).

3 Methods

3.1 Crosslinking

1. Buffer exchange the OMP from storage buffer (typically 25 mM Tris–HCl, 6 M Gdn–HCl, pH 8.0) into 10 mM potassium phosphate, pH 8.0, 8 M urea using desalting spin columns according to the manufacturer's instructions.
2. Measure the protein concentration spectrophotometrically (*see* **Note 9**).
3. Prepare dilutions of the proteins to assemble the OMP-chaperone complexes. Typically, solutions comprise 5 μM of both OMP and chaperone, in 40 μL reactions. Here we exemplify this using OmpX and SurA, which were each present at 5 μM in the reaction mixture (Fig. 1) (*see* **Note 10**).
4. Reconstitute the crosslinker in the appropriate buffer at a concentration of 10 mM. For DSBU (Fig. 1a), and other NHS-ester-based crosslinkers, DMSO is used for solubilization of crosslinkers (*see* **Note 11**).
5. Add the crosslinker to the assembled OMP:chaperone complex. Perform a titration by adding increasing concentrations of crosslinker from 20× molar excess over protein to 400× molar excess.
6. Incubate the crosslinking reactions at room temperature for 30–45 min and quench by adding 1 μL of 1 M Tris-HCl pH 8.0.
7. Analyze the crosslinking reaction mixtures by SDS-PAGE, and gel staining using a mass spectrometry–compatible reagent (*see* **Note 12**).
8. Monitor the appearance of a band at the molecular weight corresponding to the crosslinked OMP:SurA assembly in the SDS-PAGE gel (Fig. 1b). Note that as the concentration of crosslinker is increased, the amount of OmpX remaining uncrosslinked reduces, and the intensity of the band corresponding to the SurA:OmpX assembly increases. Other

bands at higher molecular weight are also observed, likely corresponding to higher-order assemblies (these may reflect higher-order SurA:OMP assemblies as reported previously) (*see* **Note 13**).

9. From the stained gel (Fig. 1b), determine the optimal concentration of crosslinker to be used to capture the native complex (if performing in solution digestion), or identify the band corresponding to the crosslinked species for excision and in-gel digestion (Subheading 3.2).

3.2 In-Gel Digestion

1. In a laminar flood hood, excise the protein bands using either a biopsy punch or scalpel. Excise only the most intense portion of the gel band corresponding to the crosslinked species. Cut the gel pieces so that they are ~1 mm^3 in size. Transfer the gel pieces to a low protein binding 1.5 mL tube.
2. Add 30% (v/v) ethanol (1 mL) to the gel pieces and incubate at 56 °C for 30 min with shaking (200 rpm). Remove the supernatant, and repeat until all the gel pieces have destained (*see* **Note 14**).
3. Perform reduction and alkylation (*see* **Note 7**). First, add 10 mM DTT (50 μL) to the gel pieces and incubate at 56 °C for 1 h with shaking (200 rpm). Cool the reaction mixture and remove the DTT solution. Add 55 mM iodoacetamide (50 μL) and incubate at room temperature in the dark with shaking (200 rpm) for 45 min (cover the thermoshaker with aluminum foil). IAA solution is light sensitive and should be freshly prepared.
4. Remove the supernatant and add 100% acetonitrile (sufficient volume to cover the gel pieces, 50–100 μL). Replace the 100% acetonitrile, and incubate at room temperature for 5 min. At this point, the gel pieces should shrink and become opaque.
5. Remove the supernatant and dry the gel pieces. If using a laminar flow hood, air dry by opening the tube lids open and leaving for 30–60 min. Alternatively, use a vacuum centrifuge (*see* **Note 8**).
6. Cool the gel pieces and trypsin solution on ice.
7. Add sufficient trypsin solution to cover the gel pieces (20–50 μL) and incubate on ice for 30 min to allow them to rehydrate. If the gel pieces remain opaque, then add a further quantity of trypsin and repeat. If there are few Lys or Arg residues in the protein, different proteases may be needed (e.g., Asp-N, Glu-C, chymotrypsin, or others).
8. After rehydration, remove excess trypsin solution, and add 25 mM ammonium bicarbonate to cover the gel pieces (20–50 μL).

9. Incubate the gel pieces at 37 °C for 18 h with shaking (200 rpm).
10. Centrifuge the digest mixtures and remove the supernatant to a clean low protein binding tube labeled "recovered peptides."
11. To the gel pieces, add 30% acetonitrile/5% formic acid (v/v) (75 μL) and incubate at 37 °C, 10 min, 200 rpm. Remove the supernatant and add to the "recovered peptides" tube.
12. To the gel pieces, add 60% acetonitrile/5% formic acid (v/v) (75 μL) and incubate at 37 °C, 10 min, 200 rpm. Remove the supernatant and add to the "recovered peptides" tube.
13. To the gel pieces, add 70% acetonitrile/5% formic acid (v/v), (75 μL) and incubate at 37 °C, 10 min, 200 rpm. Remove the supernatant and add to the 'recovered peptides' tube.
14. Use the vacuum concentrator (according to the manufacturer's instructions) to reduce the volume of the "recovered peptides" solution to ~5–10 μL (*see* **Note 15**).
15. Dilute the concentrated "recovered peptides" to a final volume of ca. 20 μL with 5% acetonitrile/0.1% formic acid in preparation for LC-MS analysis.

3.3 Liquid Chromatography-Mass Spectrometry

1. Any LC-MS system for proteomics applications can be used to acquire crosslinking data. System requirements include a nanoflow HPLC system interfaced with a high-resolution/accuracy mass spectrometer fitted with a nanoelectrospray ionization source.
2. Our current instrument configuration used for analysis of crosslinked peptides comprises a Vanquish Neo UHPLC system coupled to an EASY-Spray nanoelectrospray ionization source and an Orbitrap Eclipse mass spectrometer.
3. Peptides are injected onto the Vanquish Neo LC and trapped using a PepMap Neo C18 trap cartridge (5 μm particle size, 300 μm × 0.5 cm). Separation is achieved using an EASY-spray reverse phase column with integrated nanoelectrospray emitter (2 μm particle size, 75 μm × 500 mm).
4. Typically, 1–5 μL of crosslinked peptides are loaded onto the trap column (depending on the intensity of the band excised from the gel).
5. Next, peptides are eluted and separated using a linear gradient of 2–40% (v/v) solvent B (0.1% (v/v) formic acid in acetonitrile) in solvent A (0.1% (v/v) formic acid in water) over 80 min at 250 nL/min. The gradient length should be tuned based on the complexity of the sample. More complex samples require longer gradient lengths.

6. The data should be acquired in data-dependent analysis (DDA) mode. Each high-resolution full scan (m/z 380–1400, $R = 60{,}000$), should be followed by high-resolution MS2 scans ($R = 30{,}000$), using a cycle time of 3 s. Only select charge states 3–8+ for fragmentation (crosslinked peptides are on average larger and adopt higher charge states than linear peptides, so 2+ charge state precursors should be excluded). Use normalized collision energies of 21%, 26%, and 31% in stepped mode. Exclude precursors with charge states <3+ and >8+. Enable dynamic exclusion (60 s).

3.4 Data Analysis and Visualization

1. Use an appropriate software for crosslinking data analysis. A number of algorithms have been devised (e.g., XlinkX [15], xiSEARCH [16], MaxLynx [17], pLink [18]), and each has different advantages [19]. We routinely use the freely available software MeroX [20] (www.stavrox.com).
2. Convert the raw data from the mass spectrometer into a vendor-neutral file format (e.g., MGF) using ProteoWizard's MSConvert (https://proteowizard.sourceforge.io/) [21].
3. Generate a list of protein sequences in the sample in FASTA file format (https://www.ncbi.nlm.nih.gov/genbank/fastaformat/) (*see* **Note 16**).
4. Search parameters will vary depending on the crosslinker used, but MeroX [20] has a Setup wizard function that guides the user through the configuration and the default parameters are usually satisfactory. For more details *see* Ref. [20].
5. Manually verify all MS2 spectra to ensure satisfactory sequence coverage of both peptides in the crosslinked peptide pair (Fig. 2).
6. Use the export function to export processed data for use in an appropriate data visualization tool (examples of software for this purpose include xiVIEW [22] and xVis [23]), to generate maps of crosslinking data (Fig. 1c) and identify the locations in the sequences of intra-protein crosslinks and OMP-SurA crosslinks (Fig. 1c). Alternatively, these data can be imported into tools that interface with molecular visualization software (e.g., PyXlinkViewer [24] or XMAS [25]) (Fig. 3). Use these software tools to compare the obtained crosslinks with structural data. Crosslinks are consistent with a structure if they are within the distance spanned by the spacer arm (for DSBU this corresponds to a maximum Cα-Cα distance of ca. 30 Å [26]). Overlength crosslinks could be because alternative conformations are being sampled in solution. Data can also be deployed in integrative modeling pipelines [8, 25].

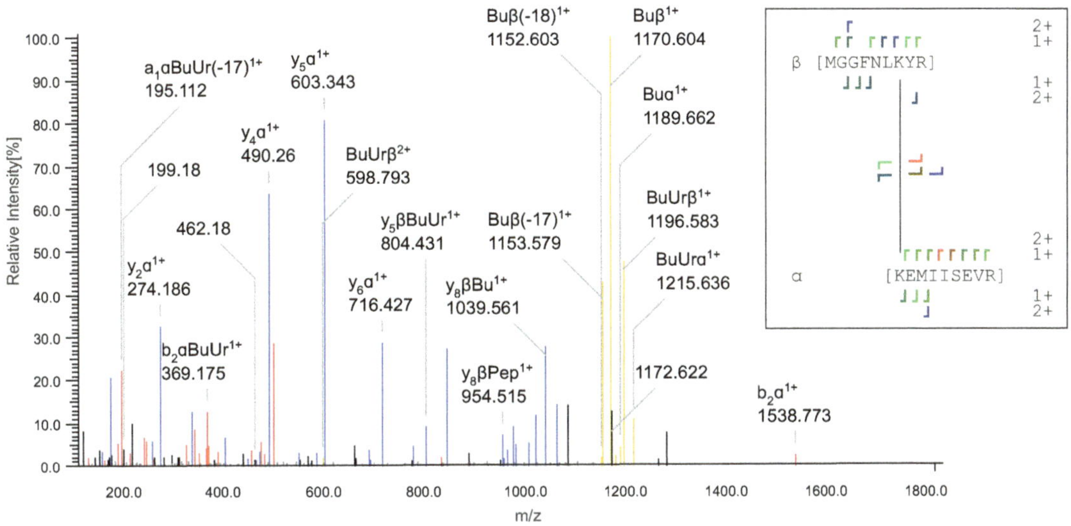

Fig. 2 Representative mass spectrum of a crosslinked dipeptide. Fragment ions corresponding to backbone fragmentation of the two peptides are colored in blue and red. Fragment ions corresponding to cleavage of the DSBU crosslinker are colored in yellow. Inset shows the two peptide sequences and the location of the crosslinked residues (joined by a line). Assigned b and y fragments are shown above/below the peptide sequence. Assigned fragments corresponding to crosslinker fragmentation are indicated along the line connecting the two peptide sequences. Annotated spectrum obtained using MeroX [20]

4 Notes

1. Most XL-MS studies require purified proteins, but an advantage of the XL-MS method is that it can be deployed even on impure (or partially purified) proteins/assemblies or even in lysates or cells. In these more complex environments, it is likely that additional crosslink enrichment steps [27–29] will be needed to obtain sufficient signal from the relatively low abundance crosslinked peptides, relative to the linear peptides in the sample.
2. SDS-PAGE is used to check the yield of the crosslinking reaction and ensure that nonnative higher-order assemblies are not captured. Any mass spectrometry–compatible SDS-PAGE gels/buffers are suitable. If casting homemade gels, ensure all casting apparatus are clean before use (wash gel casting plates well with 70% ethanol) and that all solutions are filtered to minimize the risk of keratin contamination. If possible, once the gel has been run, perform all subsequent processing steps in a laminar flow hood.
3. A range of crosslinker designs, based on different chemistries is available [2, 3]. Choose a crosslinker that is appropriate for the system under investigation and that has properties amenable for detection pipelines or enrichment (if required). We typically

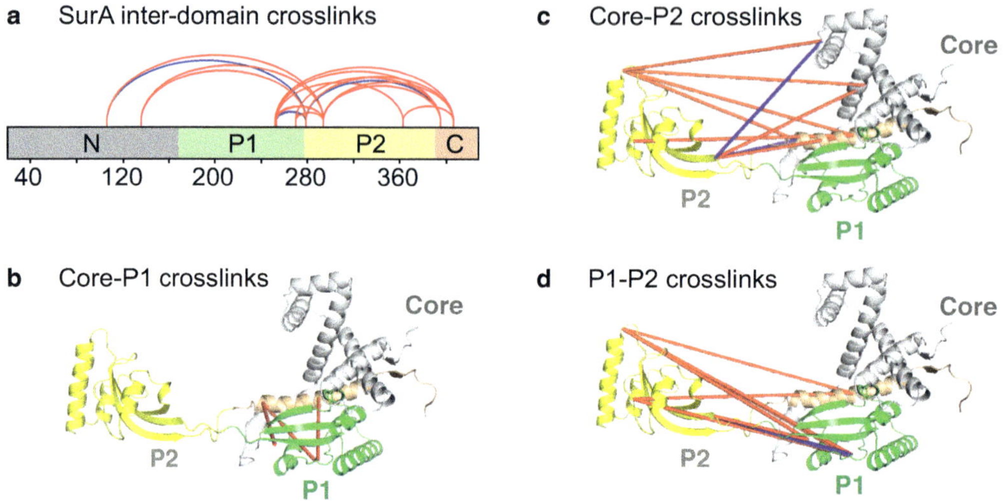

Fig. 3 Chemical crosslinking-mass spectrometry of apo-SurA. (**a**) 19 inter-domain crosslinks were identified in SurA, and only four of these were consistent with the X-ray crystal structure (blue lines), whereas the remaining crosslinks were not consistent with this structure (red lines). (**b–d**) Crystal structure of SurA showing the inter-domain crosslinks between (**b**) core-P1, (**c**) core-P2, and (**d**) P1-P2. Crosslinks consistent with this structure are shown in blue, whereas those that are inconsistent with the crystal structure are shown in red. Congruence of crosslinks with the crystal structure was determined using solvent accessible surface distance (SASD) measurements and a Cα-Cα cutoff distance of 35 Å [31]. A majority of the crosslinks are inconsistent with this crystal structure, suggesting that alternative conformations are sampled in solution. This was confirmed using complementary methods (e.g., single-molecule Förster resonance energy transfer) [13]. (Figure was prepared using the PyXlinkViewer plugin for Pymol [24]). SurA is colored by domain, with the N-terminal region of the core domain, P1, P2, and the C-terminal region of the core domain colored grey, green, yellow, and orange, respectively

employ DSBU as our crosslinker of choice, as this has an MS cleavable linker to facilitate crosslink identification [20]. MS cleavable crosslinkers are not essential for XL-MS studies. DSBU comprises two NHS ester reactive groups that primarily react with Lys residues (and to a lesser extent Ser, Thr, and Tyr). If Lys residues are not abundant in the proteins under investigation other crosslinkers should be explored [2, 3].

4. Buffers that contain primary amines should not be used for crosslinking experiments as they react with the NHS ester groups. Tris-HCl should not be used, and phosphate is often the buffer of choice. Buffer additives, including detergents in the case of membrane proteins or glycerol, for unstable proteins, are well tolerated.

5. This solution is used to quench the crosslinking reaction. This controls the time of the crosslinking pulse and prevents crosslinking during sample manipulation prior to SDS-PAGE. If performing in solution digestion, this quench step also ensures the crosslinker is no longer active during digestion which may give erroneous results.

6. The enzyme used for digestion may need to be optimized. Trypsin cleaves specifically after Lys and Arg residues, and if these are not abundant in the crosslinked proteins then other enzymes may be used. It is important to note that crosslinking at Lys residues prevents subsequent trypsin cleavage at that site, which may further complicate efficient digestion for proteins where Lys/Arg are not abundant. It is also possible to perform digestion with two enzymes, e.g., trypsin and GluC or AspN (which cleave after Glu or before Asp residues, respectively).
7. Reduction (using DTT) and alkylation (using IAA) should only be performed if the crosslinked proteins contain Cys residues. These steps are included here for completeness, but SurA and OmpX do not contain any Cys residues, and so reduction/alkylation was not performed in these experiments.
8. The low quantities of protein used for in-gel digestion mean that the probability of keratin contamination is high. The use of a laminar flow hood is recommended (not essential) when excising gel bands and for all manipulations, apart from the peptide recovery steps (where keratin contamination is no longer an issue as the proteolytic enzymes have been inactivated). When drying the gel pieces after dehydration it is possible to do this in a vacuum centrifuge if a laminar flow hood is not available, but the probability of keratin contamination is high.
9. The extinction coefficient at 280 nm of SurA is 29,450 M^{-1} cm^{-1} and for OmpX it is 34,840 M^{-1} cm^{-1}.
10. When crosslinking unfolded OMP samples a residual amount of urea will remain in the samples. This may quench the reaction somewhat and it is recommended that a titration of increasing crosslinker concentrations be performed (as described) to determine the optimal crosslinker concentration that gives acceptable yields of crosslinked material.
11. Prepare solutions fresh and use immediately. Discard any remaining crosslinker solutions.
12. It is important that mass spectrometry–compatible stain reagents are used. We routinely use colloidal Commassie stain solution, but other staining reagents are suitable. If low protein concentrations are used, it is also possible to use mass spectrometry–compatible silver staining protocols.
13. Crosslinked products must be analyzed by SDS-PAGE to ensure artificial assemblies are not being captured, and that the crosslinked species are of approximately the expected molecular weight. This is especially important when performing in-solution digestion (not described here) as the detected crosslinked peptides will be reflective of all the species present (and may be biased by the presence of non-native artefactual assemblies). It is possible to remove these unwanted

higher-order species from the digestion reaction by performing in gel digestion as described here. Note that the bands of crosslinked species are often more diffuse than uncrosslinked proteins, and may be detected by SDS-PAGE at unexpected molecular weights because crosslinking restricts the unfolding of the protein which is typically achieved during SDS-PAGE. If required, the crosslinking reaction may need optimization, either to promote more crosslinking or reduce over-crosslinking (trapping of higher-order artifactual assemblies). This can be done by reducing the crosslinking time, altering the amount of crosslinker added, altering the pH of the reaction, or altering the temperature (e.g., by performing crosslinking on ice).

14. It is important that the gel pieces are completely destained. If the staining of the gel pieces is heavy, it may be necessary to perform a destain step with 1 M triethylammonium bicarbonate (TEAB), pH 8.5 before adding 30% ethanol. Alternatively, it is possible to dehydrate the gel pieces with 100% acetonitrile before adding the destain solution as this can increase the amount absorbed by the gel pieces and improve the efficiency of the destaining steps.
15. Sample evaporation to dryness may result in loss of peptides and should be avoided if possible. However, this is sometimes not possible. Peptides may be stored frozen, ideally at −80 °C, until analysis.
16. The FASTA file should reflect the protein sequences used in the crosslinking reaction. For example, protein sequences should include any additional residues compared to the native sequences found on Uniprot, e.g., residues that are present as a result of adding purification handles.

Acknowledgments

ANC is supported by the Sir Henry Dale Fellowship, jointly funded by Wellcome and the Royal Society [grant number 220628/Z/20/Z], a research grant from the Royal Society [RGS/R2/222357], and a University Academic Fellowship from the University of Leeds. Instrumentation in the mass spectrometry facility at the University of Leeds has been funded by Wellcome (223810/Z/21/Z, 208385/Z/17/Z), BBSRC (BB/M012573/1) and the University of Leeds. Dr. James Ault and Alice Colyer are thanked for their critical reading of the manuscript.

References

1. Lee K, O'Reilly FJ (2023) Crosslinking mass spectrometry for mapping protein complex topologies in situ. Essays Biochem 67: EBC20220168
2. Piersimoni L, Kastritis PL, Arlt C, Sinz A (2022) Cross-linking mass spectrometry for investigating protein conformations and protein–protein interactions – a method for all seasons. Chem Rev 122(8):7500–7531
3. Belsom A, Rappsilber J (2021) Anatomy of a crosslinker. Curr Opin Chem Biol 60:39–46
4. West AV, Muncipinto G, Wu H-Y, Huang AC, Labenski MT, Jones LH et al (2021) Labeling preferences of diazirines with protein biomolecules. J Am Chem Soc 143(17):6691–6700
5. Iacobucci C, Götze M, Piotrowski C, Arlt C, Rehkamp A, Ihling C et al (2018) Carboxyl-photo-reactive MS-cleavable cross-linkers: unveiling a hidden aspect of diazirine-based reagents. Anal Chem 90(4):2805–2809
6. Belsom A, Mudd G, Giese S, Auer M, Rappsilber J (2017) Complementary benzophenone cross-linking/mass spectrometry photochemistry. Anal Chem 89(10):5319–5324
7. Horne JE, Walko M, Calabrese AN, Levenstein MA, Brockwell DJ, Kapur N et al (2018) Rapid mapping of protein interactions using tag-transfer photocrosslinkers. Angew Chem Int Ed 57(51):16688–16692
8. Britt HM, Cragnolini T, Thalassinos K (2022) Integration of mass spectrometry data for structural biology. Chem Rev 122(8): 7952–7986
9. Ding Y-H, Gong Z, Dong X, Liu K, Liu Z, Liu C et al (2017) Modeling protein excited-state structures from "over-length" chemical cross-links. J Biol Chem 292(4):1187–1196
10. Stahl K, Graziadei A, Dau T, Brock O, Rappsilber J (2023) Protein structure prediction with in-cell photo-crosslinking mass spectrometry and deep learning. Nat Biotechnol 41: 1810. https://doi.org/10.1038/s41587-023-01704-z
11. O'Reilly FJ, Graziadei A, Forbrig C, Bremenkamp R, Charles K, Lenz S et al (2023) Protein complexes in cells by AI-assisted structural proteomics. Mol Syst Biol 19:e11544
12. Gong Z, Ding YH, Dong X, Liu N, Zhang EE, Dong MQ et al (2015) Visualizing the ensemble structures of protein complexes using chemical cross-linking coupled with mass spectrometry. Biophys Rep 1:127–138
13. Calabrese AN, Schiffrin B, Watson M, Karamanos TK, Walko M, Humes JR et al (2020) Inter-domain dynamics in the chaperone SurA and multi-site binding to its outer membrane protein clients. Nat Commun 11(1):2155
14. Schiffrin B, Machin JM, Karamanos TK, Zhuravleva A, Brockwell DJ, Radford SE et al (2022) Dynamic interplay between the periplasmic chaperone SurA and the BAM complex in outer membrane protein folding. Commun Biol 5(1):560
15. Liu F, Rijkers DTS, Post H, Heck AJR (2015) Proteome-wide profiling of protein assemblies by cross-linking mass spectrometry. Nat Methods 12(12):1179–1184
16. Mendes ML, Fischer L, Chen ZA, Barbon M, O'Reilly FJ, Giese SH et al (2019) An integrated workflow for crosslinking mass spectrometry. Mol Syst Biol 15(9):e8994
17. Yılmaz Ş, Busch F, Nagaraj N, Cox J (2022) Accurate and automated high-coverage identification of chemically cross-linked peptides with MaxLynx. Anal Chem 94(3):1608–1617
18. Chen Z-L, Meng J-M, Cao Y, Yin J-L, Fang R-Q, Fan S-B et al (2019) A high-speed search engine pLink 2 with systematic evaluation for proteome-scale identification of cross-linked peptides. Nat Commun 10(1):3404
19. Yılmaz Ş, Shiferaw GA, Rayo J, Economou A, Martens L, Vandermarliere E (2018) Cross-linked peptide identification: a computational forest of algorithms. Mass Spectrom Rev 37(6): 738–749
20. Iacobucci C, Götze M, Ihling CH, Piotrowski C, Arlt C, Schäfer M et al (2018) A cross-linking/mass spectrometry workflow based on MS-cleavable cross-linkers and the MeroX software for studying protein structures and protein–protein interactions. Nat Protoc 13(12):2864–2889
21. Chambers MC, Maclean B, Burke R, Amodei D, Ruderman DL, Neumann S et al (2012) A cross-platform toolkit for mass spectrometry and proteomics. Nat Biotechnol 30(10):918–920
22. Graham M, Combe C, Kolbowski L, Rappsilber J (2019) xiView: a common platform for the downstream analysis of crosslinking mass spectrometry data. BioRxiv. https://doi.org/10.1101/561829
23. Grimm M, Zimniak T, Kahraman A, Herzog F (2015) xVis: a web server for the schematic visualization and interpretation of crosslink-derived spatial restraints. Nucleic Acids Res 43 (W1):W362–W3W9
24. Schiffrin B, Radford SE, Brockwell DJ, Calabrese AN (2020) PyXlinkViewer: a flexible tool

for visualization of protein chemical crosslinking data within the PyMOL molecular graphics system. Protein Sci 29(8):1851–1857

25. Lagerwaard IM, Albanese P, Jankevics A, Scheltema RA (2022) Xlink mapping and analySis (XMAS) – smooth integrative modeling in ChimeraX. BioRxiv. https://doi.org/10.1101/2022.04.21.489026
26. Merkley ED, Rysavy S, Kahraman A, Hafen RP, Daggett V, Adkins JN (2014) Distance restraints from crosslinking mass spectrometry: mining a molecular dynamics simulation database to evaluate lysine-lysine distances. Protein Sci 23(6):747–759
27. Jiang P-L, Wang C, Diehl A, Viner R, Etienne C, Nandhikonda P et al (2022) A membrane-permeable and immobilized metal affinity chromatography (IMAC) enrichable cross-linking reagent to advance in vivo cross-linking mass spectrometry. Angew Chem Int Ed 61(12):e202113937
28. Schmidt R, Sinz A (2017) Improved single-step enrichment methods of cross-linked products for protein structure analysis and protein interaction mapping. Anal Bioanal Chem 409(9):2393–2400
29. Leitner A, Reischl R, Walzthoeni T, Herzog F, Bohn S, Förster F et al (2012) Expanding the chemical cross-linking toolbox by the use of multiple proteases and enrichment by size exclusion chromatography. Mol Cell Proteomics 11(3):M111.014126
30. Bitto E, Mckay DB (2002) Crystallographic structure of SurA, a molecular chaperone that facilitates folding of outer membrane porins. Structure 10(11):1489–1498
31. Bullock JMA, Schwab J, Thalassinos K, Topf M (2016) The importance of non-accessible crosslinks and solvent accessible surface distance in modeling proteins with restraints from crosslinking mass spectrometry. Mol Cell Proteomics 15(7):2491–2500

Chapter 17

Dissecting the Organization of a β-Barrel Assembly Machinery (BAM) Complex by Neutron Reflectometry

Xiangfeng Lai and Hsin-Hui Shen

Abstract

The β-barrel assembly machinery (BAM) complex in Gram-negative bacteria facilitates the assembly of β-barrel proteins into the outer membrane. Understanding the protein-protein interactions within this complex is essential for unravelling its functional mechanisms. Here, we present the use of neutron reflectometry for investigating the organization of β-barrel membrane protein complexes in the membrane environment. The spatial organization, protein positioning, protein-lipid interactions, and conformational changes within the complex can be elucidated by this method.

Key words β-barrel assembly machinery, BAM complex, Gram-negative bacteria, Outer membrane, Neutron reflectometry, Protein interactions, Biomimetic membrane

1 Introduction

Gram-negative bacteria have an outer membrane that safeguards the cell and contains β-barrel proteins for transport and communication [1]. The five-protein β-barrel assembly machinery (BAM) complex (Fig. 1), composed of BamA, BamB, BamC, BamD, and BamE, ensures proper insertion and folding of these proteins [2–4]. BAM complex is crucial for transmembrane transport [5]. Understanding their organization, including spatial arrangement and protein-lipid interactions, is vital for comprehending their stability, dynamics, and molecular recognition [2]. Investigating these factors provides insights into substrate specificity, gating mechanisms, and assembly processes. Current methods for investigating the organization of BAM complex, such as X-ray crystallography and cryo-electron microscopy (Cryo-EM), provide valuable structural insights [6–9]. However, X-ray crystallography requires the formation of crystals, which can be challenging for some proteins. Cryo-EM can also be limited by its image resolution [10, 11]. Moreover, the dynamic nature of membrane protein

Raffaele Ieva (ed.), *Transmembrane β-Barrel Proteins: Methods and Protocols*, Methods in Molecular Biology, vol. 2778, https://doi.org/10.1007/978-1-0716-3734-0_17,

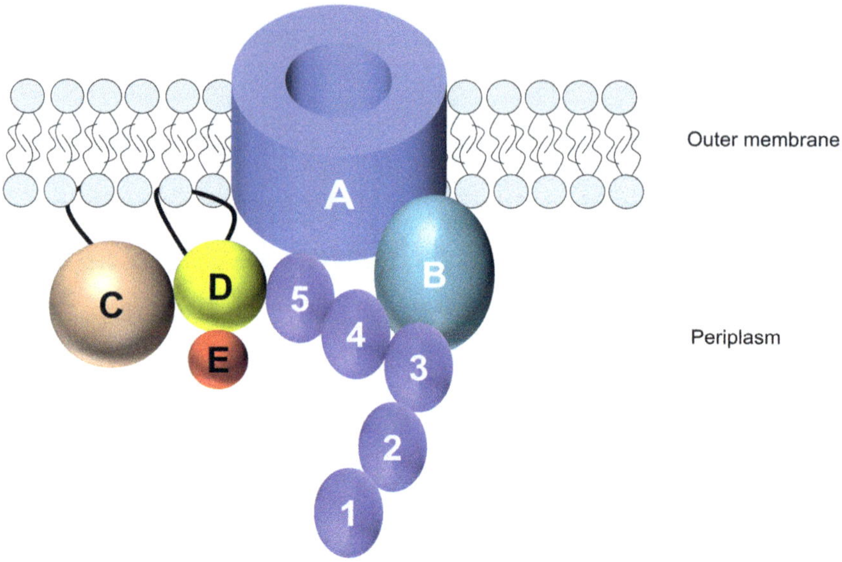

Fig. 1 Schematic of BAM complex model. The BAM complex consists of BamA, an outer membrane β-barrel protein with five N-terminal POTRA domains 1–5, and the lipoproteins BamB–E

complexes and the influence of the lipid bilayer on their organization pose challenges for traditional structural characterization methods.

Here, we demonstrate that neutron reflectometry (NR) can overcome these limitations by allowing us to investigate the organization of BAM complex within model lipid bilayers that mimic the native environment. NR is a highly effective method for studying the organization of BAM complex within lipid membranes. NR provides precise structural information on the location of these protein complexes in relation to the neighbouring membrane. In NR experiments, supported lipid bilayers serve as cell membrane models, enabling measurements to be taken prior to and subsequent to the protein complex interaction. The analysis of NR data reveals whether protein molecules reside predominantly in the membrane or periplasm region. Additionally, NR allows for the characterization of protein complexes that do not interact. The lipid composition of supported lipid bilayers is readily modifiable, enabling the design of NR experiments that investigate selective interactions between BAM complex and particular lipid species.

Thus, we reconstituted each subunit of BAM complex into biomimetic membranes composed of 1-palmitoyl -2-oleoyl-sn-glycero-3-phosphocholine (POPC) on a Ni-NTA modified gold surface. These model membrane systems provided a controlled and reproducible platform for NR experiments. Similar procedures have been used to study the organization of protein complexes within model lipid bilayers [12–14]. By using NR, we were able to show that the embedding of POPC bilayer had no effect on the BamA morphology, and to determine the relative positions of BAM subunits in the membrane environment and investigate the organization, interactions, and conformational changes of the BAM complexes at ångström level. This approach provides a powerful tool for understanding the structure and function of BAM complex.

2 Materials

Prepare all solutions using milli-Q water and analytical grade reagents. Prepare and store all reagents at room temperature (unless indicated otherwise). Diligently follow all waste disposal regulations when disposing waste materials (*see* **Note 1**).

1. 1-Palmitoyl -2-oleoyl-sn-glycero-3-phosphocholine (POPC).
2. n-Dodecyl β-D-maltoside (DDM).
3. 5 mM mercaptopropionic acid in isopropanol. Dissolve 53.07 mg of mercaptopropionic acid in 100 mL of isopropanol (*see* **Note 2**).
4. 50 mM $NiCl_2$ solution. Dissolve 648 mg of $NiCl_2$ in 80 mL of water using a volumetric flask and stirring, then adjust the final volume to 100 mL with water.
5. 75 mM EDC/25 mM NHS mixture. Dissolve 1.16 g of 1-ethyl-3-(3-dimethylaminopropyl) carbodiimide (EDC) and 288 mg of N-Hydroxysuccinimide (NHS) in 100 mL of water.
6. K_2CO_3 buffer: 150 mM N_α, N_α-bis(carboxymethyl)-L-lysine, 0.5 M K_2CO_3, pH 9.8. Dissolve 3.93 g of N_α, N_α-bis(carboxymethyl)-L-lysine and 6.91 g of K_2CO_3 in 80 mL of water, adjust the pH to 9.8 with NaOH, and add more water to reach the desired final volume of 100 mL.
7. Purified BAM complex subunits or proteins of interest (*see* **Note 3**).
8. Tris (Tris(hydroxymethyl)aminomethane) buffer (or H_2O buffer): 20 mM Tris-HCl, 150 mM NaCl, pH 7.5 (*see* **Note 4**).
9. Tris buffer with 0.03% DDM. Supplement Tris buffer with 0.03% (w/v) DDM.
10. D_2O buffer: Tris buffer in 100% D_2O. Prepare as in **step 8** but add 100% D_2O to the flask (*see* **Note 5**).
11. GMW buffer: Tris buffer in 75% D_2O and 25% H_2O(v/v) (*see* **Note 5**). Prepare as in **step 8** but add D_2O: H_2O (75/25, v/v) to the flask.
12. Silicon (Si) block: diameter × thick, 100 × 10 mm.
13. Chromium (Cr)-Gold (Au)-coated silicon (Si) blocks (*see* **Note 6**).

14. 2% Hellmanex solution for NR cell cleaning (*see* **Note 7**).
15. 2.5 mL plastic syringe for manual injection of the samples in the NR cell.
16. HPLC pump for D_2O buffer, GMW buffer, and H_2O buffer injection in the NR cell.

3 Methods

Sample preparation was performed at room temperature, while the NR measurements were performed at 25 °C. Injections of the solutions in the NR solid/liquid cell were performed by manual syringe injection. In particular, samples were injected manually with a syringe, while the buffers were injected using the HPLC pump with a flow rate of 1 mL/min. The experimental data showed here were collected on the PLATYPUS time-of-flight neutron reflectometer [15] at the Australian Nuclear Science and Technology Organisation in Sydney, Australia. Reflectivity was measured at three incident angles of 0.45°, 1.6°, and 4.5° with an illuminated footprint of 60 mm. The overall q-resolution ($\Delta Q/Q$) of ~5% was used.

3.1 Nickel-Nitrilotriacetic Acid (Ni-NTA) Modification

1. Immerse the Cr-Au-coated Si block in the 5 mM mercaptopropionic acid solution overnight (Fig. 2a) (*see* **Note 8**).
2. Rinse the gold substrate with pure isopropanol three times to remove any un-bounded thiol group (redundant mercaptopropionic) and dry the rinsed gold substrate under a stream of nitrogen gas.
3. Place the dried gold substrate into the EDC/NHS mixture and allow the substrate to remain in the mixture for 30 min to form the thio(succinimidylpropionate) (TSP) monolayer (Fig. 2b).
4. Rinse the substrate with milli-Q water to remove any excess EDC and NHS.
5. Immerse the gold substrate in the K_2CO_3 buffer solution for 3 h to form a carboxamide linkage (Fig. 2c).
6. Rinse the substrate with milli-Q water to remove any excess N_α, N_α-bis (carboxymethyl)-L-lysine.
7. Incubate the NTA-modified surface with the 50 mM $NiCl_2$ solution for 1 h to charge the carboxylate groups of NTA with Ni^{2+} through the three carboxylates and the tertiary amine of NTA (Fig. 2d).
8. Rinse the surface with milli-Q water to remove excess $NiCl_2$.
9. Rinse the Ni-NTA surface with the Tris buffer to remove any remaining contaminants and to equilibrate the surface with the desired buffer.

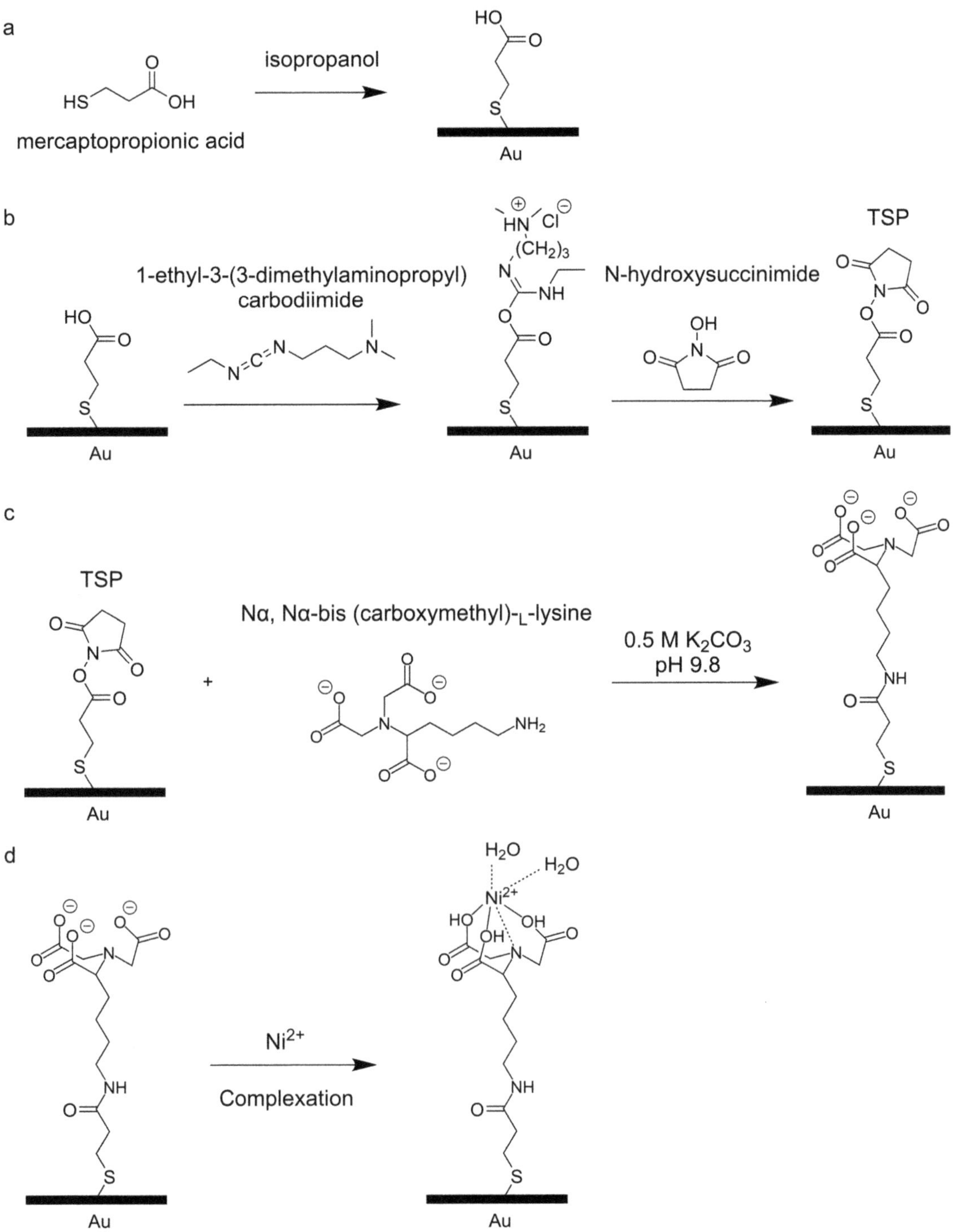

Fig. 2 Method developed for derivatization of gold surfaces. (**a**) Surface modification of the clean gold ("Au") with mercaptopropionic acid. (**b**) TSP monolayer formation on the surface. (**c**) Surface modification of Au with NTA molecules. (**d**) Complexation of NTA with Ni^{2+}

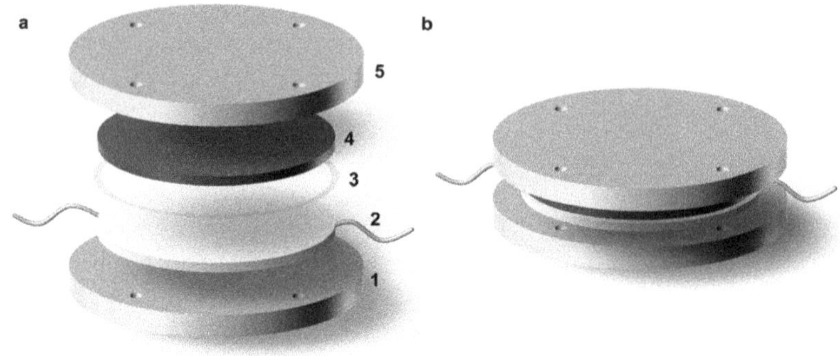

Fig. 3 (**a**) Schematic illustration of the components of a solid/liquid NR cell. (1) and (5) are aluminium plates to immobilize the cell. (2) is a Teflon plate facing the substrate reflective surface. This component is connected to an HLPC pump (not represented in the figure) to inject the isotopic contrasts through the cell. (3) O-ring. (4) Silicon substrate. The cell components (2–4) will come into contact with the sample. (**b**) Illustration of the assembled NR cell

3.2 NR Cell Cleaning and Assembling

1. Clean all the NR cell components that will be in contact with the sample, for example, O-ring and the sample injection tubes, with 2% Hellmanex solution, followed by extensive rinsing with ultrapure water (Fig. 3a).
2. Assemble the NR solid/liquid cell as reported in Fig. 3b and fill the cell with D2O buffer. Collect a first set of NR data for the substrate in contact with D_2O buffer (scattering length density, SLD, $\rho = 6.36 \times 10^{-6}$ Å^{-2}), and subsequently pump 8 mL of H_2O buffer ($\rho = -0.56 \times 10^{-6}$ Å^{-2}) before collecting another dataset (*see* **Note 9**). The main purpose of the NR measurements of the substrate in contact with D_2O and H_2O buffer is to characterize the Ni-NTA array, which is present on the gold surface, and to assess the overall substrate quality (i.e., surface roughness). The Ni-NTA is characterized by a sufficiently different SLD compared to gold (i.e., $\rho = 2.80 \times 10^{-6}$ Å^{-2} and 4.63×10^{-6} Å^{-2} respectively). The difference in ρ between the two components gives rise to a reflectivity signal (*see* **Note 10**). The collected experimental data in the case of Ni-NTA array are reported in Fig. 4a (*see* **Note 11**).

3.3 BamA Assembly on Ni-NTA Surface

1. Prepare BamA with 8 histidine-tag (His_8) at a concentration of 0.1 mg/mL in Tris buffer with 0.03% DDM (*see* **Note 12**).
2. Inject the BamA solution of ten times the cell's void volume into the flow chamber using a syringe.
3. Allow the protein to bind to the Ni-NTA surface for 1 h.
4. Remove any unbound proteins or contaminants by injecting 8 mL of D_2O buffer supplemented with 0.03% (w/v) DDM.

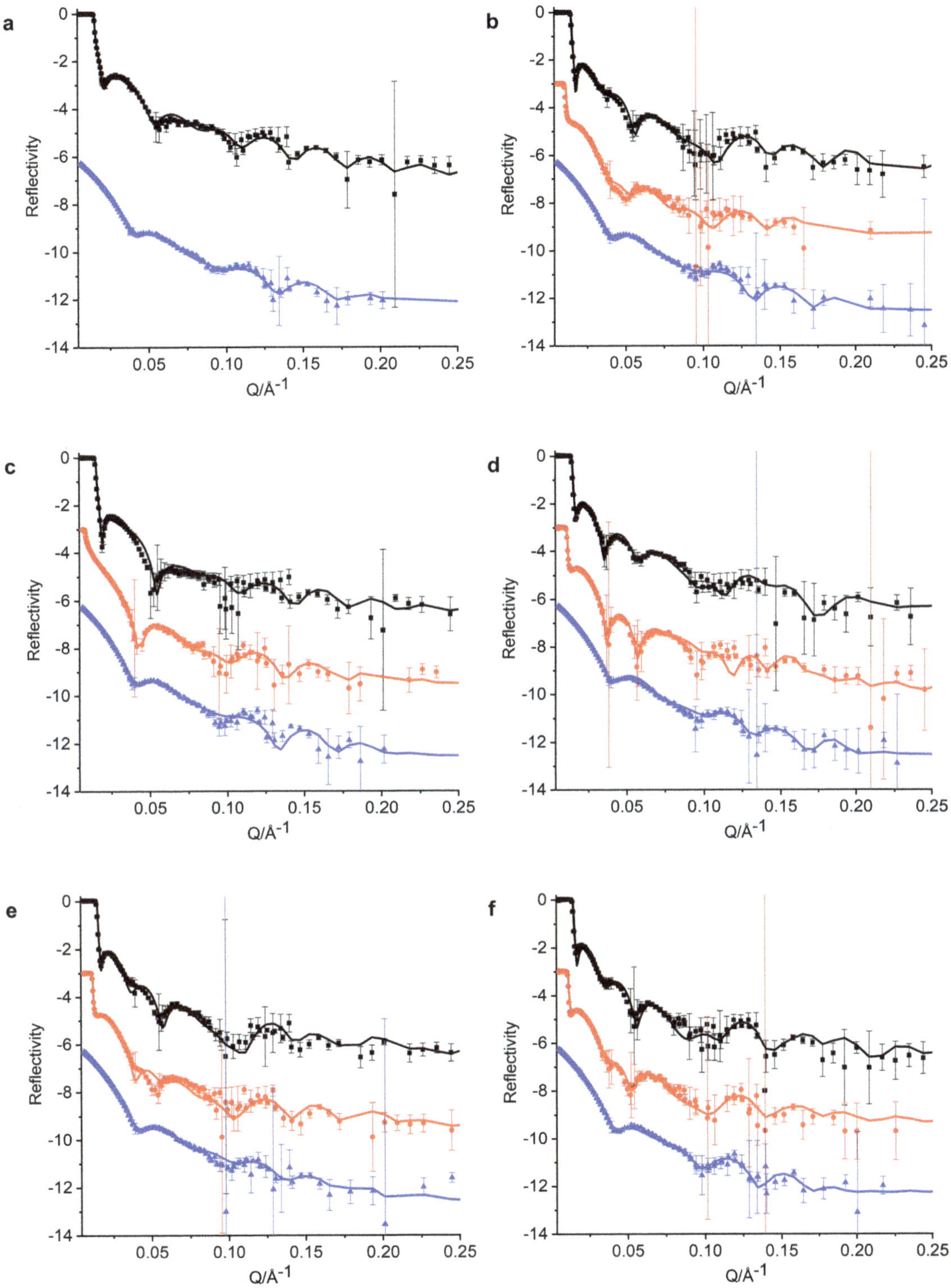

Fig. 4 NR experimental data with the relative fitting curves for (**a**) Ni-NTA array on the gold surface. (**b**) BamA in DDM. (**c**) BamA after removal of DDM. (**d**) BamA in the POPC bilayer. (**e**) BamAB in POPC bilayer and (**f**) BamABCDE in POPC bilayer. The symbols with error bars are the collected data and the solid lines are the fits. Data in D_2O, GMW, and H_2O buffer are offset for clarity. (Reproduced from Refs. [16, 17])

5. The strongest signal is produced when the BamA is in contact with D_2O buffer (*see* **Note 13**). Hence, inject 8 mL of D_2O buffer in DDM to guarantee an efficient isotope exchange.
6. Collect the NR data in three different buffers: D_2O, GMW, and H_2O buffer in DDM. Pumping 8 mL of buffer is sufficient to ensure a proper contrast exchange. The collected experimental data in the case of BamA in DDM are reported in Fig. 4b.

3.4 Removal of DDM

1. Remove residual DDM by pumping ten times the chamber volume of D_2O buffer.
2. Reintroduce the D_2O buffer in the cell before data collection.
3. Collect the NR data in three different buffers: D_2O, GMW, and H_2O buffer. Pumping 8 mL of buffer is sufficient to ensure a proper contrast exchange. The collected experimental data in the case of BamA without DDM are reported in Fig. 4c.

3.5 Introduction of Supported Lipid Bilayer

1. Weigh 10 mg of 1-palmitoyl-2-oleoyl-sn-glycero-3-phosphocholine (POPC) powder into a glass vial. Add 0.5 mL of chloroform to dissolve the lipid completely (*see* **Note 14**).
2. Dry the chloroform under a gentle stream of nitrogen gas to form a lipid film on the bottom of the glass vial. Place the vial under high vacuum overnight to remove any residual solvent.
3. Rehydrate the lipid film with a 1 mL D_2O buffer. The sample will contain POPC vesicles of variable sizes and lamellarity and will appear turbid (*see* **Note 15**).
4. Subject the lipid suspension to 20 rounds of extrusion through polycarbonate filters with a pore size of 100 nm using a mini extruder. This step helps generate unilamellar vesicles of homogeneous size (*see* **Note 16**).
5. Adjust the final concentration of the POPC solution to 1 mg/mL in the D_2O buffer.
6. Use syringe to manually inject the POPC solution into the NR chamber and allow it to adsorb onto the surface for 1 h (*see* **Note 17**).
7. Further adsorb the POPC bilayer by introducing a 10–100 times diluted solution. Allow the system to reach equilibrium for 1 h.
8. Reintroduce the D_2O buffer in the cell before data collection (*see* **Note 5**).
9. Collect the NR data in three different buffers: D_2O, GMW, and H_2O buffer. The collected experimental data in the case of BamA in POPC lipid bilayer are reported in Fig. 4d.

3.6 Formation of BamAB Architecture

1. Prepare BamB at a concentration of 0.1 mg/mL in Tris buffer (*see* **Note 18**).
2. Inject the BamB solution of ten times the cell's void volume into the flow chamber using a syringe.
3. Allow the protein to bind onto BamA for 1 h.
4. Remove any unbound proteins or contaminants by injecting 8 mL of D_2O buffer.
5. Collect the NR data in three different buffers: D_2O, GMW, and H_2O buffer. The collected experimental data in the case of BamAB are reported in Fig. 4e.

3.7 Formation of BamABDE Architecture

1. Prepare BamC, D, and E at a concentration of 0.1 mg/mL in Tris buffer.
2. Inject the BamC, BamD, and finally BamE solution of ten times the cell's void volume into the flow chamber using a syringe.
3. Allow the protein to bind onto BamAB for 1 h.
4. Remove any unbound proteins or contaminants by injecting 8 mL of D_2O buffer.
5. Collect the NR data in three different buffers: D_2O, GMW, and H_2O buffer. The collected experimental data in the case of BamABDE are reported in Fig. 4f (*see* **Note 19**).

3.8 Data Analysis

To analyse data for Neutron Reflectometry (NR), a theoretical model of the sample structure is compared to the collected experimental data using the refnx software suite [18]. This model includes a set number of parameters that are adjusted to improve the agreement between the theoretical NR curve and the experimental data. The model usually involves dividing the sample into layers with varying compositions. Each layer is assigned four parameters: thickness (t), scattering length density (ρ, Table 1), volume fraction (ϕ), and surface roughness (r). In this specific example, we will describe the basic steps for representing a BAM complex interacting with each other and the supported lipid bilayer as a stack of layers. By optimizing the model parameters to fit the experimental data, we can calculate the scattering length density profile (ρ), which provides information on the average distribution of molecules in the sample in the direction perpendicular to the substrate surface.

1. Start by analysing the data collected for Ni-NTA array on the gold surface. Use three uniform sublayers to describe the modified Ni-NTA group: one inner layer for Cr, one intermediate layer for the Au, followed by the outer layer for the Ni-NTA group. The Cr and Au layer present on the Si surface is not expected to change upon Ni-NTA modification. Keep the

Table 1
The theoretical SLD of materials in three buffer solutions

Material	$D_2O/\times 10^{-6}$ Å^{-2}	GMW/ $\times 10^{-6}$ Å^{-2}	$H_2O/\times 10^{-6}$ Å^{-2}
D_2O	6.36	–	–
GMW	–	4.63	–
H_2O	–	–	−0.56
Si	2.07	2.07	2.07
Cr	3.03	3.03	3.03
Au	4.63	4.63	4.63
His_8	4.5	4.16	3.13
Ni-NTA	2.80	2.80	2.80
β-Barrel BamA	3.28	2.98	2.09
POTRA3-5 BamA	3.20	2.87	1.90
POTRA1-2 BamA	3.06	2.75	1.83
POPC	0.26	0.26	0.26
BamB	3.08	2.78	1.90
BamC	3.12	2.81	1.87
BamD	3.18	2.86	1.92
BamE	3.09	2.78	1.84
DDM	0.64	0.64	0.64

Table 2
Characterization of the Ni-NTA array on the gold surface

Sublayers	t/ Å	$\rho/\times 10^{-6}$ Å^{-2}	Φ/ %	Water/ %	r/ Å
Cr	43	3.03	100	–	4
Au	146	4.66	91	9	8
Ni-NTA	12	3.66	43	57	6

structural parameters for Cr and Au fixed during the next steps of data analysis to reduce the number of free parameters in the model. In this case, the parameters for the Cr, Au, and Ni-NTA are shown in Table 2.

2. According to the studies on the conformation of the POTRA domains in BamA, the flexibility of POTRA1 and POTRA2 is constrained by several inter-domain interactions [19]. This creates a hinge between the POTRA domains and the β-barrel. There is also a constrained conformation for

Table 3
Characterization of the BamA molecules in DDM and after DDM removal on the Ni-NTA-modified surface

			Φ/ %			
Sublayer	**t/ Å**	**ρ/ × 10^{-6} Å^{-2}**	**BamA**	**DDM**	**Water**	**r/ Å**
In DDM						
His_8	6	6.2	–		93	4
β-Barrel	60	5.8	18	7	75	4
POTRA3-5	30	6.0	11	14	75	4
POTRA1-2	30	6.0	11	–	89	4
DDM removal						
His8	6	6.2	–	–	93	4
β-Barrel	60	5.8	18	–	82	4
POTRA3-5	30	6.0	11	–	89	4
POTRA1-2	30	6.0	11	–	89	4

POTRA3-4-5 domains due to a number of hydrophobic contacts and hydrogen-bonding interactions. The same study reported little inter-domain interaction between POTRA2 and 3, suggesting instead that this juncture is a hinge point allowing flexibility to the periplasmic domain of BamA [20]. Therefore, in this work, the periplasmic domain of BamA was treated as two rigid bodies: POTRA1-2 and POTRA3-5. Use four sublayers to divide BamA: a hydrated octa-histidine layer, the β-barrel domain of BamA layer (embedded in DDM), an aqueous layer including the POTRA3-5 domain, and another extended aqueous layer containing the POTRA1-2 domain of BamA. Simultaneously refine the three independent datasets collected from the isotopic contrasts using optical matrix method to obtain good fits resulting in real space SLD profiles where sublayers are assigned to different components of the array [21, 22]. The structural parameters obtained for the characterized BamA in DDM are reported in Table 3. (Note: the BamA volume fraction was fixed and the decrease of the SLD was counted as DDM filling the gap between BamA molecules).

3. With the structural information of the BamA in DDM in **step 2**, proceed with the analysis of the data collected after DDM removal. Compare the data collected with and without DDM. Visual inspection of the data trend can sometimes help designing a model for the data analysis. If the data trend for the curves collected with and without the DDM is identical, no effect is

induced on the structure of the immobilized BamA. In this specific case, as we only removed the solvent from the system, we build the same theoretical model for the DDM removal. The structural parameters obtained for the characterized BamA after DDM removal are reported in Table 3.

4. As a POPC bilayer was reconstituted around the immobilized BamA, use the same model as described in **step 3** to describe the supported lipid bilayer. If the data trend collected for the bilayer displays a series of Kiessig fringes in lower Q range with increased specular reflectivity, formation of hydrogenous lipid bilayer occurred on the gold surface. Fix the parameters for the thickness and volume fraction of BamA sublayers to reduce the number of free parameters in the model. The calculated SLD for POPC is 0.26×10^{-6} Å^{-2}, and introduction of POPC bilayer with small SLD value results in the reduction of SLD in the β-barrel domain in D_2O buffer from 5.80×10^{-6} Å^{-2} to 5.05×10^{-6} Å^{-2}, suggesting that the region was filled with POPC. The optimized structural parameters obtained for the introduction of BamA- POPC lipid bilayer are reported in Table 4.

Table 4
Characterization of BAM complex in the membrane layer

			Φ/ %					
Sublayer	***t***/ Å	ρ/ × 10^{-6} Å^{-2}	**BamA**	**BamB**	**BamDE**	**Bilayer**	**Water**	***r***/ Å
POPC reconstituted								
His_8	6	6.20	–	–		–	93	4
β-Barrel	60	5.05	18	–		14	68	4
POTRA3-5	30	4.16	11	–		34	55	4
POTRA1-2	30	6.00	11	–		–	89	4
BamB addition								
His_8	6	6.20	–	–		–	93	3
β-Barrel	60	5.10	18	–		14	68	3
POTRA3-5	30	4.96	11	11		10	68	3
POTRA1-2	30	5.95	11	–		–	89	3
BamCDE addition								
His_8	6	6.20	–	–	–	–	93	3
β-Barrel	60	5.05	18	–	–	14	68	3
POTRA3-5	40	5.08	11	11	10	5	63	3
POTRA1-2	30	6.00	11	–	–	–	89	3

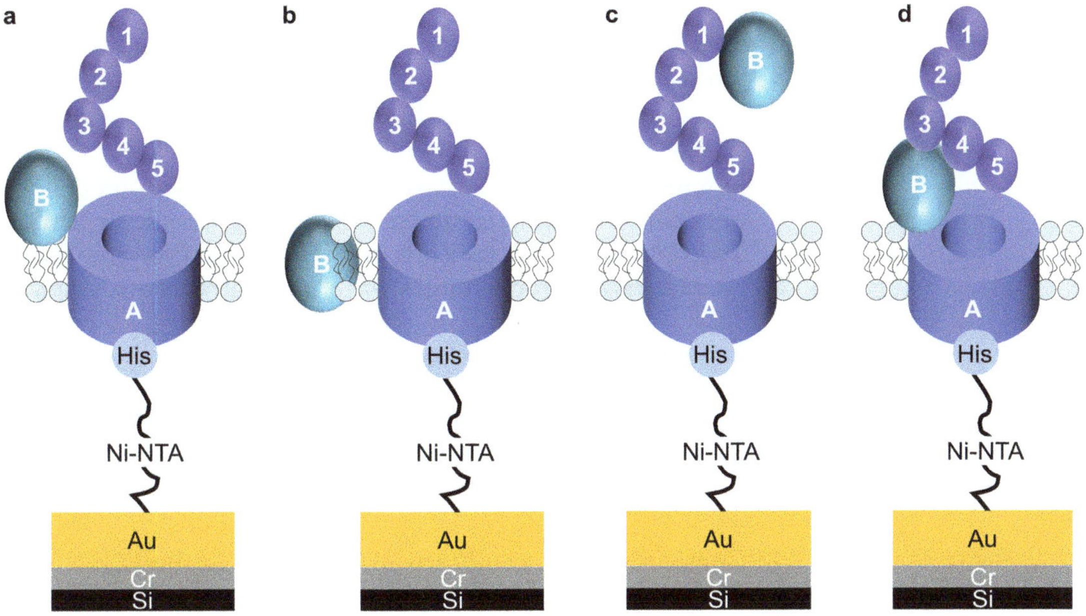

Fig. 5 Schematic representation of the different hypothesis for the BamB interaction with the BamA and/or POPC bilayer. (**a**) The protein molecule interacts with the POPC headgroups; (**b**) the protein molecule is able to cross the membrane and is hence found in the lipid tail region; (**c**) the protein molecules interact with POTRA1-2; (**d**) the protein molecules interact with POTRA3-5

5. Before analysing the data collected after adding BamB, make sure to compare it with the data collected without BamB. If the data trend for the curves collected with and without the protein is identical, no interaction occurred between the bilayer or BamA and BamB. Build different theoretical models based on reasonable hypothesis for the BamB-BamA/bilayer interaction. The following different scenarios can be considered in the case of BamB: (1) The protein molecule interacts with the POPC headgroups; (2) the protein molecule is able to cross the membrane and is hence found in the lipid tail region; (3) the protein molecule is present in the POTRA1–2 layer; (4) the protein molecule interacts with POTRA3-5. Figure 5 shows a schematic representation of the above-described potential locations of and protein interacting with a lipid membrane using the BamB as an example. Compare the different theoretical models to the experimental data and identify the one better fitting the experimental data. In the case of the BamB, model **d** produced the best agreement with the experimental data. According to this hypothesis, the model for the NR data analysis should consider four-layer model to describe the sample structure (i.e., layer 1, His_8; layer 2, β-barrel+POPC bilayer; layer 3, POTRA3-5+POPC bilayer+BamB; layer 4, POTRA1-2) and the additional layers to describe the Cr on the Si

substrate surface, Au and Ni-NTA layers (seven layers in total). The BamA volume fraction in layer 2–4 can be fixed to represent the same parameters compared to BamA only, as in this specific case, if BamB is hypothesized to fail to cause any conformational change to BamA.

6. Optimize the structural parameters of the best model for obtaining an accurate description of the sample structure. The optimized structural parameters obtained for the BamB interacting with the BamA/POPC lipid bilayer are reported in Table 4. In this case, we found that the BamB molecules located in POTRA3-5 layer without substantially affecting the overall BamA and membrane structure in layer 2, which is in good agreement with previous results [23, 24]. The total BamA thickness and the SLD for each sublayer remained almost unaffected if compared to the BamA before the BamB injection.
7. Considering that BamB was mainly found in the POTRA3-5 layer, the following different scenarios can be considered in the case of BamDE: (1) the protein molecules interact with POTRA1-2, (2) the protein molecules interact with POTRA3-5, and (3) the protein molecules interact with POTRA1-2 and POTRA3-5. Compare the different theoretical models to the experimental data and identify the one better reproducing the data. In the case of BamDE, scenario (2) produces the best agreement with the experimental data. The optimized structural parameters obtained for the BamDE interacting with the BamA are reported in Table 4. The thickness of POTRA3-5 layer extended an extra 10 Å with the addition of BamD and BamE. By fitting the SLD, we found that the POTRA3-5 layer was composed of 11% of BamA, 11% of BamB, 10% of BamDE, and 5% of lipid by volume fraction.

4 Notes

1. Always wear appropriate personal protective equipment such as gloves and eye protection when handling chemicals.
2. Add mercaptopropionic acid to isopropanol slowly and carefully, avoiding slashing. Stir the solution gently until the mercaptopropionic acid is completely dissolved and evenly distributed in the isopropanol. Add more isopropanol to bring the volume of the solution up to the desired final volume of 100 mL. Isopropanol is flammable, so keep it away from heat, sparks, open flames, and other sources of ignition.
3. Expressed and purified according to protocols reported elsewhere [16, 25]. Stored at 4 °C or −80 °C.

4. Dissolve 2.42 g of Tris (Tris(hydroxymethyl)aminomethane) and 8.77 g of NaCl in 800 mL of water, adjust the pH to 7.5 with HCl, and add more water to reach the desired final volume of 1 L. It is important to add the HCl slowly and carefully, while stirring the solution to ensure that it is well mixed. After adding the acid, wait for a few minutes and then measure the pH again to see if any further adjustments are needed. Remember to always add acid to water, never add water to acid, and to handle HCl with care as it is a strong acid.
5. D_2O is often used as a buffer in neutron reflectometry experiments because it has a different neutron scattering length density (SLD) than regular water (H_2O). This difference in SLD can be used to enhance the contrast between the sample and the buffer, making it easier to resolve the structure of the sample. Additionally, measurements with D_2O are characterized by a lower background due to the reduced incoherent scattering from protium nuclei. This results in a stronger signal from the sample and makes it easier to evaluate whether the sample is correctly aligned in the instrument. The critical edge shape and position can also be affected by sample misalignment, so starting the experiment with a measurement in D_2O can help to ensure accurate results.
6. Always follow the manufacturer's instructions and safety guidelines when operating an electron beam deposition system. Ensure that the electron beam deposition system is properly maintained and operated by trained personnel. The thickness for a gold layer is: 150–200 Å.
7. Handle concentrated Hellmanex solution with care as it is a strong alkaline solution that can cause skin and eye irritation. When diluting concentrated Hellmanex solution, always add the concentrated solution to water, never add water to the concentrated solution. Store the prepared 2% Hellmanex solution in a labelled container in a safe location away from incompatible materials.
8. The substrate surface is polished to remove any surface irregularities or defects that could affect the quality of the coating. A smooth surface allows for better adhesion and uniformity of the coating, which can improve the performance and reliability of the coated silicon blocks in their subsequent application. Additionally, a smooth surface can reduce scattering and improve the reflectivity of the silicon blocks in neutron reflectometry experiments.
9. In neutron reflectometry, the experimental data is typically compared to a theoretical model to extract information about the sample being studied. The theoretical model is defined by a set of parameters, such as the thickness and roughness of

different layers in the sample, which are adjusted to fit the experimental data. The agreement between the theoretical neutron reflectivity curve and the experimental data is quantified by a goodness-of-fit parameter. By using different contrast solutions, such as D_2O, H_2O, and GMW, one can obtain different contrast factors for the sample, which can help to reveal more information about its structure. By performing measurements with different contrast solutions and fitting the data to a theoretical model, one can obtain more accurate estimates of the model parameters and improve the agreement between the theoretical and experimental neutron reflectivity curves.

10. The fitted results of reflectivity data obtained from Ni-NTA at two differing solutions: 100% D_2O and 100% H_2O, were used to determine the relative contribution of the two components, Ni-NTA and water using a set of linear equations. The two individual components of a fitted layer will contribute to the SLD of this layer as shown in Eq. 1:

$$\rho = \rho_{\mathrm{Ni-NTA}}\,\Phi_{\mathrm{Ni-NTA}} + \rho_{\mathrm{water}}\,\Phi_{\mathrm{water}} \tag{1}$$

where ρ is the SLD of a given layer. $\rho_{\text{Ni-NTA}}$ and ρ_{water} are the individual SLDs of the Ni-NTA and solvent, respectively, and $\Phi_{\text{Ni-NTA}}$ and Φ_{water} (equals to $1 - \Phi_{\text{Ni-NTA}}$) are the volume fractions of these components within a particular layer.

11. For Ni-NTA assay, two isotopic contrasts, D_2O and H_2O, are enough to fit to the experimental data using three uniform sublayers: (i) chromium (Cr) layer on the silicon, (ii) the gold layer, and (iii) the NTA layer.
12. The detergent DDM is frequently used to isolate functional membrane proteins for biochemical assays and structure determination, having been empirically found to maintain activity of membrane proteins and protein complexes.
13. After the deposition of BamA, it is important to perform a quick visual inspection of the collected NR curves. This can provide valuable information about whether the BamA has formed successfully on the substrate surface. In the D_2O buffer contrast, the presence of BamA on the substrate surface should result in a higher reflectivity signal than the substrate alone. Failure to do so could result in missing important information about the success or failure of the experiment. If the BamA formation has failed and this issue is not addressed, further analysis could result in inaccurate or misleading data.

14. Handle chloroform with care as it is a volatile and potentially hazardous solvent. Work with chloroform in a well-ventilated area or under a fume hood to minimize inhalation exposure. Chloroform is also flammable, so keep it away from heat, sparks, open flames, and other sources of ignition.
15. Use freshly made POPC vesicles to ensure the integrity and stability of the vesicles. Over time, lipids can undergo hydrolysis or oxidation, which can affect the properties of the vesicles. Freshly prepared vesicles can help ensure that the vesicles are of the desired size, homogeneity, and composition.
16. Extrusion of multilamellar vesicle suspensions using membranes with a pore size >0.2 μm does not produce unilamellar vesicles. Vesicles produced with larger pore membranes will yield a polydisperse suspension of multilamellar vesicles. Polycarbonate membranes and filter supports are intended to be used for a single vesicle preparation and should not be reused. The lipid suspension should be kept above the phase transition temperature of the lipid during hydration and extrusion [26].
17. The interaction between the POPO lipid membrane and the BamA protein is a complex process that can be influenced by a number of factors, including the affinity of the protein for the specific lipid species present in the membrane. As such, the time required for this interaction to occur can vary significantly. In order to accurately determine the optimal incubation time for this interaction, it is recommended to use surface-sensitive techniques such as quartz crystal microbalance with dissipation monitoring (QCM-D) prior to performing a NR experiment. By utilizing these techniques, it is possible to more accurately define the conditions under which the interaction between the POPO lipid membrane and BamA will occur, considering the specific properties of the protein and the lipid membrane.
18. After the reconstitution of BamA, the partner subunits BamB, C, D, and E can be added freely to the above BamA surface to form a complex.
19. BamC, D, and E were prepared and added to the BamA protein. However, BamC was not able to bind to the BamA surface.

Acknowledgements

The work was supported by grants from Australian NHMRC. H.H.S. is an Australian NHMRC Career Development Research Fellow (GNT1106798).

References

1. Knowles TJ, Scott-Tucker A, Overduin M, Henderson IR (2009) Membrane protein architects: the role of the BAM complex in outer membrane protein assembly. Nat Rev Microbiol 7(3):206–214
2. Shen C et al (2023) Structural basis of BAM-mediated outer membrane β-barrel protein assembly. Nature 617:1–9
3. Hussain S, Bernstein HD (2018) The Bam complex catalyzes efficient insertion of bacterial outer membrane proteins into membrane vesicles of variable lipid composition. J Biol Chem 293(8):2959–2973
4. Tomasek D, Kahne D (2021) The assembly of β-barrel outer membrane proteins. Curr Opin Microbiol 60:16–23
5. Noinaj N et al (2013) Structural insight into the biogenesis of β-barrel membrane proteins. Nature 501(7467):385–390
6. Newstead S et al (2008) Insights into outer membrane protein crystallization. Mol Membr Biol 25(8):631–638
7. Kermani AA (2021) A guide to membrane protein X-ray crystallography. FEBS J 288(20):5788–5804
8. Ho C-M et al (2020) Bottom-up structural proteomics: cryoEM of protein complexes enriched from the cellular milieu. Nat Methods 17(1):79–85
9. Su C-C et al (2021) A 'build and retrieve' methodology to simultaneously solve cryo-EM structures of membrane proteins. Nat Methods 18(1):69–75
10. Wang HW, Wang JW (2017) How cryo-electron microscopy and X-ray crystallography complement each other. Protein Sci 26(1): 32–39
11. Renaud J-P et al (2018) Cryo-EM in drug discovery: achievements, limitations and prospects. Nat Rev Drug Discov 17(7):471–492
12. Hall SC et al (2021) Surface-tethered planar membranes containing the β-barrel assembly machinery: a platform for investigating bacterial outer membrane protein folding. Biophys J 120(23):5295–5308
13. Holt SA et al (2009) An ion-channel-containing model membrane: structural determination by magnetic contrast neutron reflectometry. Soft Matter 5(13):2576–2586
14. Le Brun AP et al (2008) Monitoring the assembly of antibody-binding membrane protein arrays using polarised neutron reflection. Eur Biophys J 37:639–645
15. James M et al (2011) The multipurpose time-of-flight neutron reflectometer "platypus" at Australia's opal reactor. Nucl Instrum Methods Phys Res A 632(1):112–123. https://doi.org/10.1016/j.nima.2010.12.075
16. Ding Y et al (2020) Characterization of BamA reconstituted into a solid-supported lipid bilayer as a platform for measuring dynamics during substrate protein assembly into the membrane. Biochim Biophys Acta Biomembr 1862(9):183317. https://doi.org/10.1016/j.bbamem.2020.183317
17. Chen X et al (2021) Substrate-dependent arrangements of the subunits of the BAM complex determined by neutron reflectometry. Biochim Biophys Acta (BBA)-Biomembranes 1863(9):183587
18. Holt SA, Oliver TE, Nelson AR (2022) Using refnx to model neutron reflectometry data from phospholipid bilayers. Methods Mol Biol 2402:179–197
19. Warner LR et al (2017) Flexibility in the periplasmic domain of BamA is important for function. Structure 25(1):94–106
20. Gatzeva-Topalova PZ, Walton TA, Sousa MC (2008) Crystal structure of YaeT: conformational flexibility and substrate recognition. Structure 16(12):1873–1881
21. Born M (1980) E. Wolf principles of optics. Pergamon Press 6(188):36–111
22. Clifton LA et al (2013) Asymmetric phospholipid: lipopolysaccharide bilayers; a gram-negative bacterial outer membrane mimic. J R Soc Interface 10(89):20130810. https://doi.org/10.1098/rsif.2013.0810
23. Chen Z et al (2016) Structural basis for the interaction of BamB with the POTRA3–4 domains of BamA. Acta Crystallogr D Biol Crystallogr 72(2):236–244
24. Jansen KB, Baker SL, Sousa MC (2015) Crystal structure of BamB bound to a periplasmic domain fragment of BamA, the central component of the β-barrel assembly machine. J Biol Chem 290(4):2126–2136
25. Shen HH et al (2014) Reconstitution of a nanomachine driving the assembly of proteins into bacterial outer membranes. Nat Commun 5:5078. https://doi.org/10.1038/ncomms6078
26. Avanti polar lipids, mini-extruder extrusion technique. https://avantilipids.com/divisions/equipment-products/mini-extruder-extrusion-technique

Chapter 18

Expression, Purification, and Cryo-EM Structural Analysis of an Outer Membrane Secretin Channel

Rebecca Conners, Mathew McLaren, Marjorie Russel, and Vicki A. M. Gold

Abstract

Secretin proteins form pores in the outer membranes of Gram-negative bacteria, and as such provide a means of transporting a wide variety of molecules out of or in to the cell. They are important components of several different bacterial secretion systems, surface filament assembly machineries, and virus assembly complexes. Despite accommodating a diverse assortment of molecules, including virulence factors, folded proteins, and whole viruses, the secretin family of proteins is highly conserved, particularly in their membrane-embedded β-barrel domain. We describe here a protocol for the expression, purification and cryo-EM structural determination of the pIV secretin from the Ff family of filamentous bacteriophages.

Key words Secretin, Membrane protein, Filamentous phage, Protein expression, Protein purification, Cryo-EM

1 Introduction

Secretins play an important role in the transport of a wide variety of molecules across cell membranes, and are key components of the Type II and Type III bacterial secretion systems (T2SS, T3SS), the type IV pilus assembly machinery, and the egress apparatus of filamentous bacteriophages [1, 2].

Secretins are found in the outer membranes of Gram-negative bacteria where they form cylindrical pores (Fig. 1a). The pore is a homo-oligomer of the secretin protein (which is present in 12–16 copies) and ranges in mass from 670 kDa to >1 MDa, depending on the species [3]. Despite transporting myriad types of molecules, the structures of secretins are highly conserved [4]. The highest structural conservation is seen at the C-terminus, within a double-walled β-barrel which is embedded in the outer membrane via a belt of hydrophobic residues and extends into the periplasm (Fig. 1a). A series of N-domains, differing in number and size between different

Raffaele Ieva (ed.), *Transmembrane β-Barrel Proteins: Methods and Protocols*, Methods in Molecular Biology, vol. 2778, https://doi.org/10.1007/978-1-0716-3734-0_18,

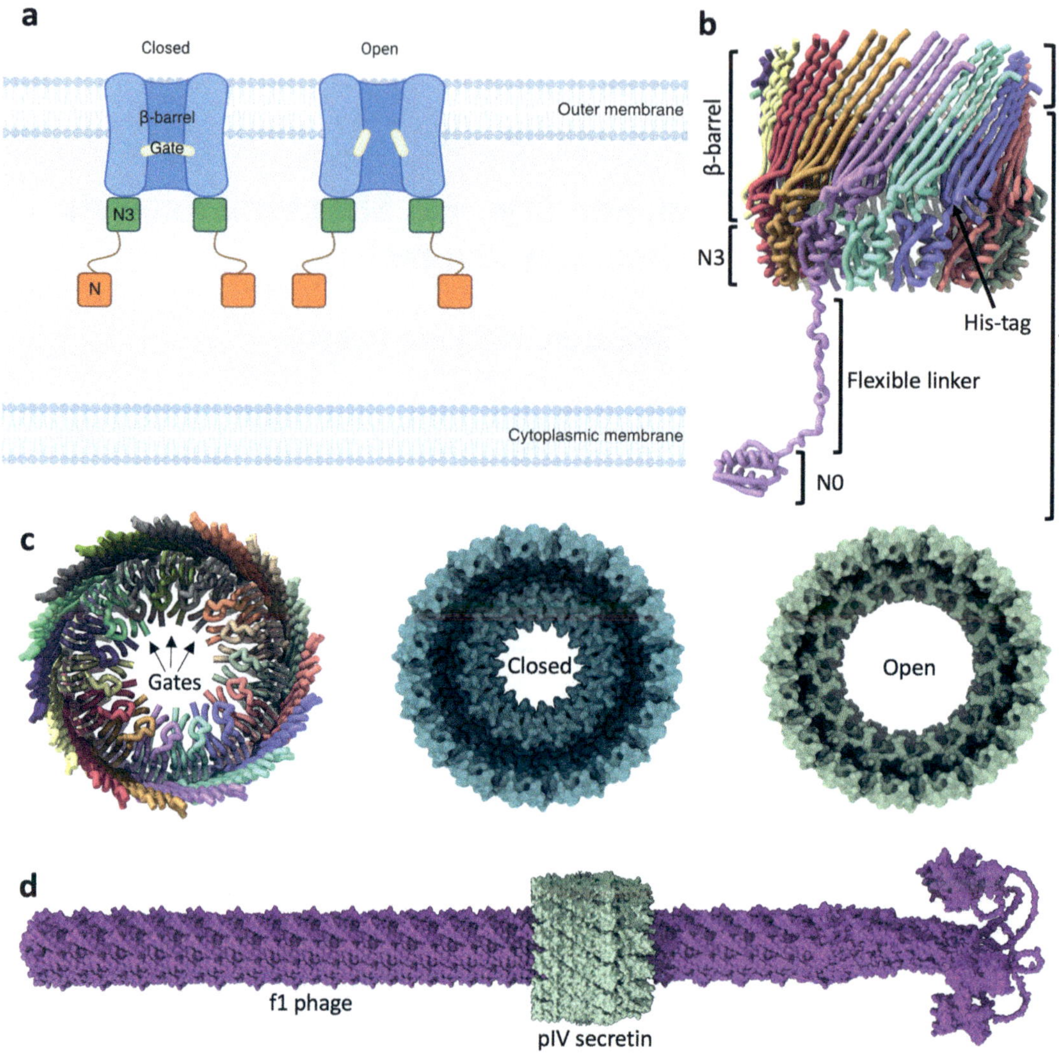

Fig. 1 Structure of the pIV secretin. (**a**) Schematic showing secretin pores embedded in the outer membrane and extending into the periplasm. The β-barrel is shown in blue, N3 domain in green, and additional N domains in orange. The gate is shown in cream in closed (left) and open (right) forms. Created with Biorender.com. (**b**) A side view of the pIV secretin structure labeled to show different domains and position in the cell, with individual chains of pIV colored differently. The flexible linker and N0 domain are shown for one pIV chain only and were predicted by AlphaFold [36] as they were not observed in our cryo-EM experiment. OM, outer membrane. (**c**) The structure of the pIV secretin is shown in top view and colored by chain (left), with loops extending into the center of the pore forming the gate (7OFH) [20]. The pIV secretin is shown in its closed (center, teal) and open (right, green) states, rendered as surfaces. The open state was modeled on the open state of a related protein, the Type III secretin InvG from *Salmonella typhimurium* (6PEE and 6Q15) [22]. (**d**) The open form of the pIV secretin (green) is large enough to accommodate an f1 phage particle (purple) [20, 37]. (Panel c is adapted from Fig. 4e in Conners et al. [20]; the article was licensed under a Creative Commons Attribution 4.0 International License, which permits use, sharing, adaptation, distribution, and reproduction in any medium or format (https:/creativecommons.org/licenses/by/4.0/))

species, are attached to this β-barrel via flexible linker regions. The periplasmic N-domains link the secretins to their partner protein complexes in the cytoplasmic membrane. The secretin pore has at least one "gate," located in the β-barrel, which has been observed in open and closed positions, so enabling a necessary level of control over the transport of molecules [5, 6] (Fig. 1a).

Our work has centered around a secretin from the Ff filamentous phage family (viruses f1, fd, and M13) [7], used in a variety of biotechnology applications including phage display and nanotechnology [8]. The bacteriophages infect Gram-negative bacteria *Escherichia coli* by binding to F-pili expressed on the cell surface, and do not cause lysis of their bacterial host [8]. Instead, the phage genome encodes an extrusion machinery that allows new virions to egress from their host bacteria without destroying them. With a genome of 6.4 kb, these simple viruses encode just eleven proteins, with three of them (pI, pXI, and pIV) making up the phage extrusion machinery [9]. Proteins pI (an ATPase) and pXI are located in the cytoplasmic membrane, and protein pIV (their secretin partner) is found in the outer membrane [10, 11]. After viral DNA has been replicated in the cytoplasm, it is packaged by the viral capsid proteins at the pI/pXI complex, and then the new viral particles are extruded across the outer membrane through pIV.

Work on purifying pIV for biochemical and structural studies began in the lab of Marjorie Russel at the Rockefeller University in the mid-nineties. N- or C-terminal His tags functioned poorly in phage assembly, so the wild-type protein was engineered to have a 9 residue His tag inserted at a predicted loop position (Fig. 1b), incurring an S318I mutation in the process that was found to stabilize the protein [12]. The first views of pIV were obtained through scanning transmission electron microscopy (STEM) [13]. These data indicated that pIV is cylindrical and has an internal pore wide enough to accommodate phage. Further improvements in the purification protocol followed [14], leading to the analysis of pIV by cryo-electron microscopy (cryo-EM) in 2003. This culminated in the first 3D views of the protein, a 22 Å resolution barrel-like complex with a large gated pore in the center [15]. Immunogold labeling predicted that the pIV multimer was comprised of 14 subunits [15]. In these previous studies [13, 15], much of the pIV preparation contained oligomers, where two pIV pores appeared to be joined at their C-termini.

Fast forward two decades and with major improvements in cryo-EM equipment and methodology [16], we re-initiated studies on pIV using the same His-tagged construct. Modern single particle cryo-EM is now recognized as a powerful structural biology technique used to determine the 3D structure of proteins, up to atomic resolution. Major challenges remain in the sample preparation steps and purification of membrane proteins can be particularly problematic due to their instability and low typical yields

[17]. There are several different methods for purifying membrane proteins [18], broadly divided into those that use detergents, and detergent-free protocols (such as nanodiscs, amphipols, and SMALPS [18]). We used the existing detergent-based pIV protocols as a starting point [14, 19], and made a few modifications to the protocol which improved the sample for high-resolution work. This included expression at a lower temperature (20 °C), the use of 5% CHAPS detergent for the initial solubilization of the membrane fraction, a final size exclusion chromatography step using a Sepharose S6 column, and the use of graphene oxide coated lacey carbon electron microscopy grids to enable sufficient particle density for data collection.

Our resulting 2.7 Å cryo-EM structure shows that the pIV secretin pore (Fig. 1b) is composed of 15 subunits which pack together to form a large ß-barrel composed of 60 ß-strands (4 ß-strands from each subunit) [20]. Within this ß-barrel sits a smaller inner barrel, also comprised of four ß-strands from each pIV subunit. This double-walled ß-barrel provides a scaffold for two extended loops from each pIV subunit, that project into the center of the pore, known as Gate 1 and Gate 2 [21]. By aligning our closed gate structure with the closed and open gate structures of the Type III secretin InvG from *Salmonella typhimurium* (6PEE and 6Q15 [22]) we modeled an open form and verified that phage particles could fit through the open pore (Fig. 1c, d). Attached to the ß-barrel by a short linker of 2 residues is the globular N3 domain, a three-stranded ß-sheet packed against two α-helices. The N3 domains are common to all secretins and are now considered part of the conserved C-terminal domain (i.e. the β-barrel and N3) [4]. Attached to the N3 domain is a 19 residue linker, predicted to be flexible, before a second globular domain, N0 (Fig. 1b). N0 and it's linker were not observed in our cryo-EM map due to flexibility; this movement of the N0 domain is a common occurrence in many of the solved secretin structures [4].

We describe here our protocol for expressing and purifying the pIV secretin membrane protein, followed by the techniques used for the determination of the protein structure by single particle cryo-EM.

2 Materials

Standard reagents were purchased from Sigma-Aldrich unless otherwise stated. The amounts of equipment and reagents required are shown for a 1 L culture of *Escherichia coli* cells that express the His-tagged pIV protein.

2.1 Expression of His-Tagged pIV Protein

1. Plasmid pPMR132(His) [13] (*see* **Note 1**).
2. Competent *E. coli* cells. We used electroporation-competent TG1 cells (Lucigen) (*see* **Note 2**).
3. Chloramphenicol stock solution: 50 mg/mL chloramphenicol in pure ethanol; filter sterilized.
4. Electroporator (we used Biorad Gene Pulser Xcell electroporator).
5. Super optimal broth with catabolite repression media (SOC): 2% (w/v) tryptone, 0.5% (w/v) yeast extract, 10 mM NaCl, 2.5 mM KCl, 10 mM $MgCl_2$, 10 mM $MgSO_4$, 20 mM glucose; sterilized by autoclaving.
6. One 2 L conical flask; sterilised by autoclaving.
7. Fortified broth media (FB): 2.5% (w/v) tryptone, 0.75% (w/v) yeast extract, 100 mM sodium chloride, 5 mM glucose, 50 mM Tris-HCl pH 7.6; sterilized by autoclaving.
8. FB-chloramphenicol agar plates: sterile FB medium supplemented with 1.5% (w/v) agar (*see* **Note 3**).
9. IPTG stock solution: 1 M Isopropyl β-D-1-thiogalactopyranoside (IPTG) in dH_2O, filter sterilized.

2.2 Small-Scale Comparison of Detergents for Optimal Solubilization of the pIV Secretin

1. TN buffer: 50 mM Tris-HCl pH 8, 100 mM NaCl, filter sterilized.
2. Protease inhibitor tablets e.g. cOmplete™, EDTA-free Protease Inhibitor Cocktail (Roche).
3. Sonicator, e.g., Vibra-cell (Sonics & Materials, Inc.).
4. DNase stock solution: 1 mg/mL DNase I in dH_2O, filter sterilized.
5. RNase stock solution: 10 mg/mL RNase in dH_2O, filter sterilized.
6. 1 M $MgCl_2$ in dH_2O, filter sterilized.
7. 5 M NaCl in dH_2O, filter sterilized.
8. 1 mL Dounce homogenizer.
9. Ultracentrifuge, e.g., Optima Max-TL Ultra with TLS55 rotor (Beckman Coulter).
10. Selection of detergents or detergent kit (*see* **Note 4**); here we used Popular Detergent Kit (Anatrace).
11. Extraction buffer for detergent screen: 50 mM Tris-HCl pH 7.6, 500 mM NaCl, 30 mM imidazole, 1–5% detergent of choice, filter sterilized.
12. 10% trichloroacetic acid (TCA), ice-cold.
13. Acetone.
14. 4% (w/v) sodium dodecyl sulphate (SDS) solution.

15. SDS-PAGE gels, SDS sample preparation buffer and standard protein gel running apparatus. We used Any kD mini-protean TGX gel (Bio-Rad).
16. Western transfer kits (e.g., Trans blot turbo transfer (Bio-Rad)) and standard Western blotting apparatus.
17. *E. coli* cells competitor lysate (not expressing pIV) (*see* **Note 5**).
18. 50 mM Tris-HCl pH 8.
19. 50 mM Tris-HCl pH 8 supplemented with 5 mM EDTA.
20. Lysozyme.
21. Tris-buffered saline (TBS): 20 mM Tris-HCl pH 8.0, 150 mM NaCl.
22. Binding buffer (1× TBS, 0.2% (w/v) BSA, plus 50 μL *E. coli* competitor lysate (giving a final protein concentration (w/v) of 0.66%)).
23. 17% (w/v) bovine serum albumin (BSA).
24. Non-fat dry milk.
25. Anti-pIV rabbit antibody.
26. Goat anti-rabbit IgG (H + L)-HRP conjugate (Bio-Rad).
27. Clarity™ Western ECL substrate (Bio-Rad).

2.3 Large-Scale Purification of pIV Protein Using CHAPS Detergent

1. TN buffer: 50 mM Tris-HCl pH 8, 100 mM NaCl, filter sterilized.
2. Protease inhibitor tablets: e.g., cOmplete™, EDTA-free Protease Inhibitor Cocktail, Roche.
3. Sonicator, e.g., Vibra-cell (Sonics & Materials, Inc.).
4. DNase stock solution: 1 mg/mL DNase I in dH_2O, filter sterilized.
5. RNase stock solution: 10 mg/mL RNase in dH_2O, filter sterilized.
6. 1 M $MgCl_2$ in dH_2O, filter sterilized.
7. 5 M NaCl in dH_2O, filter sterilized.
8. 10 mL Dounce homogenizer.
9. Ultracentrifuge, e.g., Optima LE-80 K Ultra with 70Ti rotor (Beckman Coulter).
10. CHAPS extraction buffer: 50 mM Tris-HCl pH 7.6, 500 mM NaCl, 30 mM imidazole, 5% (w/v) CHAPS (Melford), filter sterilized.
11. Ni^{2+} beads. We used Super Ni-NTA Affinity Resin (Generon).
12. Protein purification system: We used ÄKTA Prime or Pure models (GE Healthcare) (*see* **Note 6**).
13. Ni^{2+} start buffer (Buffer A): 50 mM Tris-HCl pH 7.6, 30 mM Imidazole, 500 mM NaCl, 1% (w/v) CHAPS, filter sterilized (*see* **Note 7**).

14. Ni^{2+} elution buffer (Buffer B): 50 mM Tris-HCl pH 7.6, 1 M Imidazole, 500 mM NaCl, 1% (w/v) CHAPS, filter sterilized.
15. EDTA stock solution: 500 mM EDTA, filter sterilized.
16. Superose 6 Increase 10/300 GL column (GE Healthcare).
17. Gel filtration buffer: 1% (w/v) CHAPS, 50 mM Tris-HCl pH 7.6, 500 mM NaCl, 0.5 mM EDTA, filter sterilized and degassed.
18. Vivaspin concentration device for 50 mL volume, 30 kDa molecular weight cut-off.
19. Nanodrop spectrophotometer (ThermoFisher Scientific).

2.4 Electron Microscopy

1. Carbon-coated (5 nm) copper grids (300 mesh) for negative stain.
2. Glow discharge cleaning system, e.g., PELCO easiGlow (Ted Pella Inc., Redding, USA).
3. 5% ammonium molybdate stain.
4. 1% uranyl acetate stain.
5. Graphene oxide-coated lacey carbon grids (300 mesh) for cryo-EM (Agar Scientific) (*see* **Note 8**).
6. High-precision tweezers, e.g., Dumont carbon steel (Plano, Wetzlar, Germany).
7. Whatman filter paper grade 1 for negative stain (GE Healthcare, Chalfont, UK).
8. Whatman filter paper grade 597 for cryo-EM (GE Healthcare, Chalfont, UK).
9. Grid storage boxes.
10. Plunge-freezing device, home-made or company produced, e.g., Vitrobot (ThermoFisher Scientific).
11. Liquid nitrogen.
12. Ethane gas (to fill a 10 mL container with liquified ethane).
13. Electron microscope with cryo-transfer holder for vitrified specimens, ideally operating at 300 keV, e.g., Titan Krios (ThermoFisher Scientific, Waltham, USA) or JEOL3200-FSC (JEOL, Tokyo, Japan) and fitted with a direct electron detector camera.
14. Computer software for data collection, e.g., Digital Micrograph (Gatan, Pleasanton, USA) or EPU (ThermoFisher Scientific) plus a variety of software for data processing, e.g., Relion [23], CryoSPARC [24], model building, e.g., Coot [25], ChimeraX [26], Isolde [27], ModelAngelo [28], and model refinement, e.g., Refmac [29], Phenix [30].

3 Methods

3.1 Expression of His-Tagged pIV Protein (See Note 9)

1. Transform plasmid pPMR132 into electrocompetent *E. coli* TG1 cells (*see* **Note 10**) using standard bacterial transformation methods. We used a Biorad Gene Pulser Xcell electroporator with a 0.1 cm gap electroporation cuvette (pulse of 1700 V (17 kV/cm field strength), resistance 200 Ω, capacitance 25 μF). We added 1 μL miniprepped plasmid DNA (263 ng/μL) to 50 μL competent *E. coli* TG1 cells, electroporated, and then added 950 μL SOC media (prewarmed to 37 °C).
2. Transfer the transformation mixture to a 15 mL Falcon tube and incubate for 1 h with shaking at 225 rpm at 37 °C.
3. Plate out 10, 50, and 250 μL of transformation mixture onto FB agar plates with 50 μg/mL chloramphenicol. Spread onto the plates with a sterile spreader and incubate overnight at 37 °C.
4. Pick a single colony with a sterile loop, and inoculate 10 mL FB broth with 50 μg/mL chloramphenicol. Grow overnight (12–18 h) at 37 °C with shaking at 225 rpm.
5. Use 1 mL of overnight culture to inoculate 1 L FB broth with 50 μg/mL chloramphenicol (in 2 L conical flasks) (*see* **Note 11**). Grow to an OD_{600} of 0.5–0.55 at 37 °C with shaking at 225 rpm.
6. Put the flasks at either 4 °C for 30 min or room temperature for 1 h to reduce the temperature of the cultures before induction. Induce protein expression by the addition of 1 mL of IPTG stock solution to each liter of culture (final concentration 1 mM IPTG). Incubate the cultures overnight (12–18 h) at 20 °C with shaking at 225 rpm (*see* **Note 12**).
7. Harvest the cells by centrifugation at 5000× *g* for 20 min at 4 °C.
8. Discard the supernatant and either proceed with the purification steps below or store the cell pellets at −20 °C.

3.2 Small-Scale Comparison of Detergents for Optimal Solubilization of the pIV Secretin

A small-scale prep was performed with a range of different detergents to compare their ability to solubilize the pIV protein. From this point on, all solutions should be ice-cold and the samples should be kept on ice.

1. Thaw a 1 L cell pellet.
2. Prechill ultracentrifuge and rotor.
3. Resuspend the pellet in 50 mL chilled TN buffer supplemented with one protease inhibitor tablet.
4. Sonicate with a medium tip, 70% amplitude, 20 s on/20 s off for 2 min.

5. Add DNase I and Rnase both to a final concentration of 5 μg/mL.
6. Add $MgCl_2$ to a final concentration of 10 mM.
7. Incubate for 1 h at 4 °C.
8. Add NaCl to a final concentration of 500 mM, which helps to release peripheral membrane proteins. Mix well.
9. Sonicate with a medium tip, 70% amplitude, 20 s on/20 s off for 1 min.
10. Add EDTA to a final concentration of 1 mM.
11. Centrifuge at 3260× *g* for 20 min at 4 °C to remove the cell debris. Keep the supernatant.
12. Divide the supernatant into equal parts depending on how many detergents are to be compared.
13. Centrifuge at 136,000× *g* for 1 h at 4 °C to isolate the total membrane fraction. Discard the supernatant.
14. Solubilize each pellet in 1 mL chilled extraction buffer containing different detergents (at 1–5% final concentration) using a 1 mL Dounce homogenizer.
15. Rotate at 4 °C overnight.
16. Prechill ultracentrifuge and rotor.
17. Centrifuge the samples at 136,000× *g* for 1 h at 4 °C to remove insoluble material. Keep the supernatant and discard the pellets.
18. Prepare the samples for SDS-PAGE analysis by performing a TCA precipitation (Subheading 3.2.1). Without TCA precipitation, pIV mostly remains in the stacking gel as boiling in SDS sample loading buffer is insufficient to denature the pIV multimer into monomers. Run a second, identical SDS-PAGE gel if Western blotting is required. Western blotting was required for pIV as the expression levels were too low to be able to compare resolubilization efficiency from the SDS-PAGE alone.

3.2.1 TCA Precipitation for SDS-PAGE and Western Blotting

1. Supplement each detergent-solubilized sample with TCA to a final concentration of 5% (w/v). Mix and let precipitate on ice for 10 min.
2. Centrifuge at 15,000× *g* for 10 min at 4 °C and remove supernatant.
3. Wash pellet with 1 mL acetone in a fume hood.
4. Centrifuge at 15,000× *g* for 10 min at 4 °C and remove supernatant.
5. Allow pellet to air dry in the fume hood.

6. Dissolve pellet in 20 μL 4% SDS.
7. Boil for 5–10 min.
8. Mix with equal amount of SDS sample loading buffer and load on SDS-PAGE gel.

3.2.2 Preparation of "Cold Competitor Lysate" for Western Blotting

To prevent the polyclonal anti-pIV antibody from binding nonspecifically to *E. coli* proteins, it was incubated with a lysozyme-treated "cold competitor" extract of *E. coli* cells (which did not express pIV) before blotting.

1. Grow 100 mL *E. coli* cells (*see* **Note 5**) to an OD_{600} of 0.8.
2. Harvest the cells by centrifugation at 5000× *g*, for 10 min at 4 °C.
3. Resuspend the pellet in 2.5 mL of 50 mM Tris-HCl pH 8 and pellet by centrifugation at 5000× *g* for 10 min at 4 °C.
4. Resuspend the pellet in 2.36 mL of 50 mM Tris-HCl pH 8, 5 mM EDTA.
5. Add lysozyme to a final concentration of 0.005% (w/v) and the EDTA-free protease inhibitor.
6. Incubate the reaction for 20 min at 37 °C.
7. Sonicate the sample (70% amplitude, 10 s on/10 s off, 3 min) and incubate at 37 °C for 1 h.
8. Take 50 μL aliquots, snap freeze in liquid nitrogen, and store at −80 °C.

3.2.3 Western Blotting

1. Preincubate pIV antiserum in 0.5 mL of binding buffer on a rotator for 1 h at room temperature. Dilute the mixture to the desired concentration for the Western blot with binding buffer.
2. For Western blotting, first separate pIV on an SDS-PAGE gel.
3. Electrophoretically transfer the proteins to a PVDF membrane.
4. Wash membrane 3 × 5 min in TBS.
5. Block membrane by incubating with TBS supplemented with 5% non-fat dry milk for 1–2 h at room temperature with gentle rocking.
6. Wash membrane 3 × 5 min with TBS.
7. Incubate membrane with the anti-pIV antibody prepared as in **Step 1** of Subheading 3.2.3 at 1:2000 dilution in TBS at 4 °C overnight with gentle rocking.
8. Wash membrane 5 × 10 min with TBS.
9. Incubate membrane with a secondary goat anti-rabbit IgG (H + L)-HRP conjugate at 1:3000 dilution for 1–2 h at room temperature with gentle rocking.
10. Wash membrane 3 × 5 min with TBS.
11. Detect using Clarity™ Western ECL substrate.

3.3 Large-Scale Purification of pIV Protein Using CHAPS Detergent

1. Thaw cell pellets.
2. Prechill ultracentrifuge and rotor.
3. Resuspend each pellet (from 1 L culture) in chilled 50 mL TN buffer freshly supplemented with EDTA-free Protease Inhibitor Cocktail.
4. Sonicate using a medium tip, 70% amplitude, 20 s on/20 s off for 2 min for each pellet.
5. Add to each tube DNase I and RNase, both to a final concentration of 5 μg/mL.
6. Add $MgCl_2$ to a final concentration of 10 mM.
7. Incubate for 1 h at 4 °C.
8. Add NaCl to a final concentration of 500 mM to release peripheral membrane proteins.
9. Sonicate using a medium tip, 70% amplitude, 20 s on/20 s off for 1 min for each pellet.
10. Centrifuge 3260× *g*, 20 min, 4 °C. Remove the cell debris and keep the supernatant.
11. Centrifuge at 136,000× *g*, 1 h, 4 °C. Discard the supernatant and keep the pellets.
12. Solubilise pellets in chilled extraction buffer. Use 10 mL extraction buffer per 1 L pellet, and a 10 mL Dounce homogeniser to aid solubilisation. Rotate at 4 °C overnight.
13. Prechill ultracentrifuge and rotor.
14. Centrifuge at 136,000× *g*, 1 h, 4 °C. Keep the supernatant and discard the pellets.
15. Prepare Ni^{2+} beads (we used 2 mL Ni^{2+} beads per 1 L culture) by washing 5 times with dH_2O and 2 times with extraction buffer. These can be prepared in advance and stored at 4 °C for 1–2 days.
16. Mix supernatant with prepared Ni^{2+} beads and rotate at 4 °C overnight.
17. Pour Ni^{2+} beads into an ÄKTA compatible column and pack.
18. Wash the column extensively (5–10 column volumes) with Buffer A until A_{280} trace is flat.
19. Elute pIV protein by applying 0.06 M, 0.37 M, and 1 M imidazole in steps and monitoring by A_{280} (*see* **Note 13**) (Fig. 2a).
20. Check fractions by SDS-PAGE analysis (Fig. 2a) and Western Blot (after TCA precipitation) and check protein concentrations with a Nanodrop.
21. Pool pIV containing fractions and concentrate in 50 mL Vivaspin concentrator with 30 kDa molecular weight cut-off.

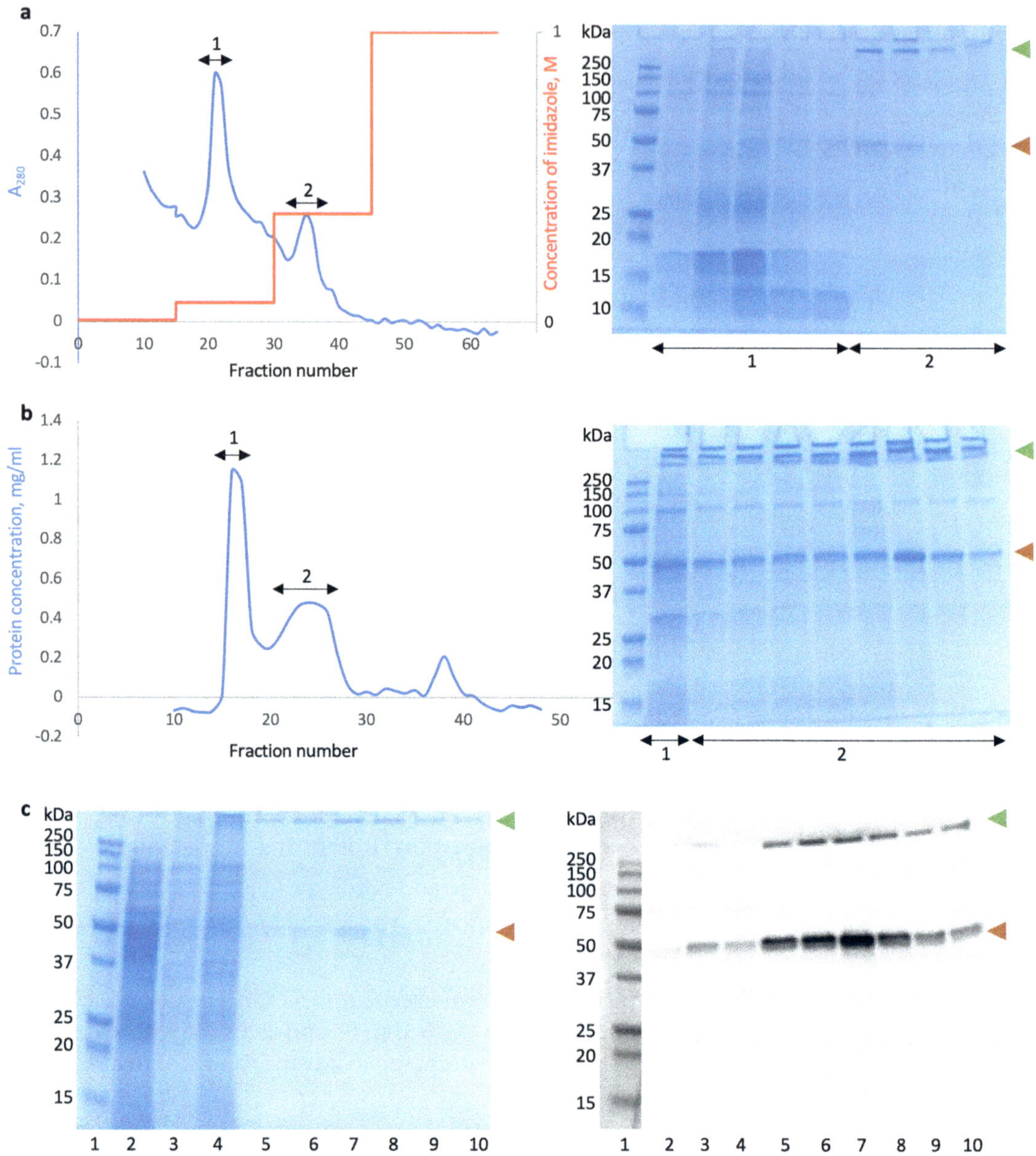

Fig. 2 Purification of the pIV secretin. (**a**) A_{280} trace and imidazole profile from the Ni^{2+} chromatography step (left) and corresponding SDS-PAGE analysis (right). The main peaks from the chromatography (1 and 2) were analyzed by SDS-PAGE. The pIV subunit is indicated by an orange arrowhead and the multimer by a green arrowhead. The presence of some multimeric pIV on the SDS-PAGE gel is due to the protein's resistance to solubilization by SDS. (**b**) Trace (left) and corresponding SDS-PAGE analysis (right) from the size exclusion chromatography step. pIV monomeric and multimeric fractions are indicated as in part a. (**c**) Whole purification scheme verified by SDS-PAGE (left) and Western blot analysis (right). Lane 1, Biorad Precision Plus All Blue Marker; lane 2, total soluble protein; lane 3, total membrane protein; lane 4, solubilized membrane protein; lanes 5–10, fractions from size exclusion chromatography. pIV monomeric and multimeric fractions are indicated as in part a. For the Western blot, Lane 1 (protein marker) was cut away and visualized with white light, hence the difference in coloring compared to the rest of the blot. (Panel c is taken from Supplementary Fig. 1a, b in Conners et al. [20]; the article was licensed under a Creative Commons Attribution 4.0 International License, which permits use, sharing, adaptation, distribution, and reproduction in any medium or format (https:/creativecommons.org/licenses/by/4.0/))

22. Equilibrate Superose 6 Increase 10/300 GL column in 1% (w/v) CHAPS, 25 mM Na HEPES pH 8.0, 500 mM NaCl, 0.5 mM EDTA. This size exclusion column separates molecules of 5000–5,000,000 Da in size.
23. Inject concentrated pIV protein (as concentrated as possible for optimal separation) and elute in 1% (w/v) CHAPS, 50 mM Tris pH 7.6, 500 mM NaCl, 0.5 mM EDTA. Monitor pIV elution by A_{280} trace (Fig. 2b).
24. Check fractions by SDS-PAGE analysis and Western Blot (after TCA precipitation) and check protein concentrations with a Nanodrop (Fig. 2b, c).
25. Concentrate pIV-containing fractions in 50 mL Vivaspin concentrator with 30 kDa molecular weight cut-off.

3.4 Electron Microscopy

Negative stain electron microscopy is a quick and useful technique to check the sample quality and concentration before moving on to the more expensive and time-consuming cryo-EM steps.

3.4.1 Negative Stain – Grid Preparation and Observation

1. Add 3 μL purified pIV onto a glow-discharged carbon-coated copper electron microscopy grid and leave for 1 min.
2. Wick away any excess liquid with filter paper.
3. Perform 3 washes of the grid in distilled H_2O by swiping the grid through three 30 μL drops of distilled H_2O on a parafilm-coated microscope slide (*see* **Note 14**).
4. Add 3 μL of stain (1% uranyl acetate or 5% ammonium molybdate) and leave for 20–60 s (*see* **Note 15**).
5. Wick away any excess liquid with filter paper and leave to dry.
6. Look at the grids on a transmission electron microscope and observe particle density, orientation and homogeneity. Once a satisfactory particle density is seen by negative stain (Fig. 3a), this should be increased tenfold (if possible) for an initial attempt at freezing cryo-EM grids.

3.4.2 Cryo-EM Grid Preparation

1. Add 3 μL purified pIV protein (~0.7 mg/mL) onto a graphene oxide-coated lacey carbon grid (without any prior glow discharge) and plunge freeze in liquid ethane (we used a Vitrobot Mark IV with settings of 4°, 100% relative humidity, wait 5, drain 0, blot force 0, blot time 4 s) (*see* **Note 16**).
2. Transfer grids from the liquid ethane into an electron microscopy grid box (under liquid nitrogen) and then transfer the grid box to long-term storage in liquid nitrogen.

3.4.3 Cryo-EM Data Collection

1. Collect micrographs on a 200 kV or (ideally) a 300 kV microscope (*see* **Note 17**).

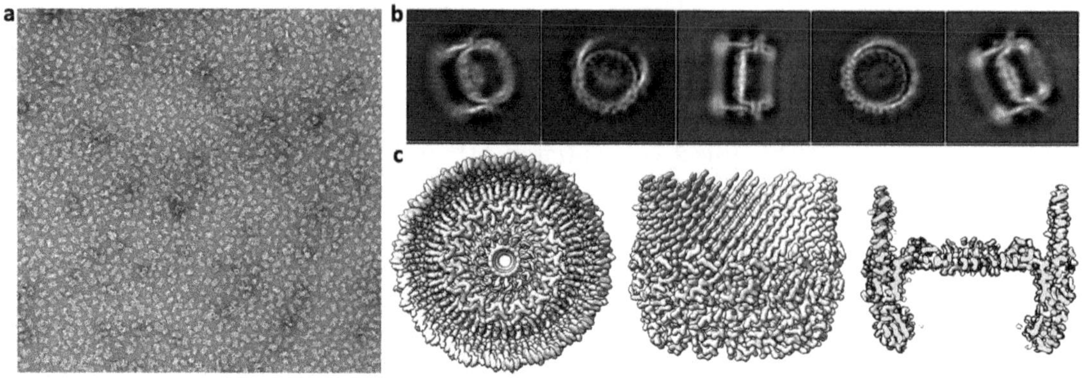

Fig. 3 Electron microscopy of the pIV secretin. (**a**) EM image of purified and negatively stained pIV. Particles are seen in top and side views covering the grid. (**b**) Representative 2D classes of the pIV secretin after cryo-EM. (**c**) Final 3D reconstruction shown from left to right as top view, side view, and side view sliced through the center. (Panel c was taken from Fig. 1b in Conners et al. [20]; the article was licensed under a Creative Commons Attribution 4.0 International License, which permits use, sharing, adaptation, distribution, and reproduction in any medium or format (https:/creativecommons.org/licenses/by/4.0/))

2. Collect images by using software that allows for automated data collection. Our data were collected using EPU software (ThermoFisher Scientific) with a pixel size of 1.072 Å (0.536 Å super-resolution) and a defocus range from −2.5 to −1.3 μm. A subset of data was collected in counting mode (dataset 1), and a second subset in super-resolution mode (dataset 2), and these were combined. Details of our data collection parameters are shown in Table 1.

3.4.4 Cryo-EM Image Processing

1. Carry out motion correction, CTF correction, and particle picking (we used Warp [31] software). 241,591 particles were picked from 7037 micrographs for dataset 1 and 330,389 from 14,336 micrographs for dataset 2. The pretrained model provided with Warp was used for the picking. Both sets of data were processed separately and combined in the latter stages of data processing.
2. Carry out rounds of 2D classification (Fig. 3b) and 3D refinement (we used Relion 3.1 software [23]) followed by CTF refinement and polishing. Our data produced a 2.7 Å resolution map for the pIV secretin (Fig. 3c).

3.4.5 Model Building and Refinement

If the resolution of the map is sufficient, an atomic model can be built into it. This can be performed manually, or by using software such as ModelAngelo [28], which uses a machine-learning approach to build protein and RNA/DNA sequence into maps of

Table 1
The statistics of data collection, model reconstruction, refinement, and validation

Data collection and processing	Dataset 1	Dataset 2
Acceleration voltage (kV)	300	300
Nominal magnification (x)	81 k	81 k
Pixel size (Å)	1.072	0.536 (super-res)
Frame rate (s^{-1})	11.93	11.93
Exposure time (s)	3.52	3.52
Total exposure (e^-/Å)	42.059	42.059
Particles		
Micrographs used for selection	7037	14,336
Defocus range (μm)	−2.5 to −1.3 (0.2 μm steps)	−2.5 to −1.3 (0.2 μm steps)
Total number of particles	241,591	330,389
In final 3D reconstruction	16,104	95,575
	Dataset 1 and dataset 2 combined	
Resolution		
"Gold-standard" at FSC 0.143 (Å)	2.7	
Map-sharpening B factor (Å^2)	0	
Model refinement		
Model composition	15 identical subunits	
Non-hydrogen atoms, per chain	1742	
Protein residues, per chain	231	
Ligands, per chain	2	
R.M.S. Z scores		
Bond lengths (Å)	0.48	
Bond angles (°)	0.72	
R.M.S.D		
Bond lengths (Å)	0.031	
Bond angles (°)	1.903	
Validation		
Ramachandran angles (%)		
Favoured	88	
Allowed	12	
Outliers	0	

(continued)

Table 1
(continued)

Data collection and processing	Dataset 1	Dataset 2
Clashscore	14	
Rotamer outliers (%)	2	
Resolution estimates (Å)		
Model resolution (0/0.143/0.5) – masked	2.5/2.6/2.7	
Model resolution (0/0.143/0.5) – unmasked	2.4/2.5/2.9	

These data were taken from Supplementary Table 1 in Conners et al. [20];

<3.5 Å resolution. The higher the resolution of the map, the higher the level of detail which can be modeled. At resolutions of ~15 Å the general protein shape can be seen, at ~10 Å protein folds can be observed, at ~5 Å the polypeptide backbone can be modelled, and at ~3 Å individual side chains can be placed [32].

1. If a structure of a homologous protein or domain is available, it can be manually placed in the map. In the case of pIV, we manually placed a homologous structure of the Type II secretin GspD from enteropathogenic *E. coli* (5W68 [33]) into our cryo-EM map using Chimera [26].
2. Make sequence changes to alter the homologous protein to your protein of interest with a model-building program, such as Coot [25]. If there are no homologous structures available, the polypeptide backbone can be built using a poly-Ala sequence, and then unambiguous regions of sequence that fit the observed density can be identified. Large, aromatic side chain residues (phenylalanine, tryptophan, and tyrosine) and glycine residues (no sidechain) can be used to help determine if the sequence is being correctly placed in the density.
3. Apply symmetry with software such as Chimera [26] – in the case of pIV, C15 symmetry was applied to the model to create the 15mer.
4. Refine the model using software such as Refmac [29] or Phenix [30].
5. After thorough manual checking of the fit of the model to the map, judge the quality of the model using software such as the validation tools in Coot [25], Molprobity [34], or the PDB validation server [35] (Table 1).

4 Notes

1. The wild-type pIV was engineered to have a 9 residue His tag (SAHHHHHHH) inserted at position 308. The protein also incurred an S318I mutation that stabilized it. N-terminal and C-terminal His tags were found to be unstable so the internal His tag was engineered in a predicted loop region at the base of the β-barrel [12] (Fig. 1b).
2. TG1 genotype: [F′ *traD36 proAB lacIqZ ΔM15] supE thi-1 Δ(lac-proAB) Δ(mcrB-hsdSM)5(rK - mK -)*.
3. After adding agar to the fortified broth media, dissolve it (and sterilize) by autoclaving. It will solidify at room temperature. To use: melt the agar in a microwave, add 50 μg/mL chloramphenicol (final concentration) once the molten agar has cooled to a temperature of 50–55 °C, and pour into sterile Petri dishes to set using aseptic technique.
4. A variety of detergent kits are available. Empirical testing is needed to identify the detergents that best solubilize a given membrane protein of interest.
5. A total lysate of *E. coli* that does not express pIV is used to clean up the polyclonal anti-pIV antiserum from non-specific antibodies. We used a *E. coli* K1508 strain (MC4100 [F^- $araD^-$ *Δlac* U169 $relA^-$ *thiA rpsL* (Str^R)] *ΔlamB106*). Any *E. coli* strain would be suitable for this step, e.g. TG1.
6. The ÄKTA Prime was used for the Ni^{2+} chromatography step and the ÄKTA Pure for size exclusion chromatography.
7. We used Bioultra ≥99.5% grade imidazole as it does not absorb at 280 nm and therefore makes small elution peaks easier to visualize on the ÄKTA A_{280} trace.
8. Several different types of electron microscopy grids are available. We tested different types for freezing and found that the pIV particles were usually observed on the carbon rather than in the holes. The lacey carbon grids with an additional graphene oxide layer gave us the best density of pIV particles within the holes.
9. Based on previously published methods [13, 14].
10. Bacterial cells were freshly transformed with plasmid for each prep – glycerol stocks were not used due to a significant increase in the loss of plasmid over storage time.
11. We typically grew 12 L FB per pIV prep due to low protein yield.
12. Small-scale expression trials were performed at different temperatures and IPTG concentrations to determine the conditions for maximum yield.

13. EDTA solution was pipetted into the ÄKTA collection tubes so that on elution the pIV protein would immediately be in a final concentration of 15 mM EDTA. Without this step, the pIV protein quickly degraded.
14. The washing of the electron microscopy grids in distilled water was critical to see the pIV particles clearly by negative stain, presumably due to the detergent in the sample.
15. Both uranyl acetate and ammonium molybdate are commonly used negative stain reagents with different properties and thus may enhance contrast of the sample differently. Testing both reagents allows for comparative analysis and judgement of which type of stain is better suited for the specific sample.
16. Many different types of electron microscopy grids and freezing conditions may need to be tested to obtain optimal grids in terms of particle concentration and distribution, and ice quality. Grids can be found with a multitude of different hole sizes and shapes, and a wide variety of additional support layers can be added to the grids such as coating them with layers of carbon or graphene oxide. This optimization stage is very sample dependent.
17. We used a 300 kV Titan Krios microscope (ThermoFisher Scientific) with a K3 direct electron detector (Gatan) at the Electron Bio-imaging Centre (eBIC) at Diamond Light Source, UK.

Acknowledgments

A Wellcome Trust Seed Award in Science (210363/Z/18/Z) and Leverhulme Trust Research Project Grant (RPG-2023-069) supported R.C. and a BBSRC responsive mode grant (BB/R008639/1) supported M.M, all awarded to V.G. We thank Prof. Jasna Rakonjac for sharing materials and for many helpful discussions. For the pIV structural work, we acknowledge Diamond Light Source for access and support of the cryo-EM facilities at the UK's national Electron Bio-imaging Centre (eBIC) [under proposal BI25452]. We acknowledge access and support of the GW4 Regional Facility for High-Resolution Electron Cryo-Microscopy, funded by the Wellcome Trust (202904/Z/16/Z and 206181/Z/17/Z) and BBSRC (BB/R000484/1). The deposited dataset was collected at eBIC, and the GW4 facility was used for sample screening. We are grateful to Ufuk Borucu of the GW4 Regional Facility for help with screening.

References

1. Green ER, Mecsas J (2016) Bacterial secretion systems: an overview. Microbiol Spectr 4:1–19
2. Silva YR de O, Contreras-Martel C, Macheboeuf P et al (2020) Bacterial secretins: mechanisms of assembly and membrane targeting. Protein Sci 29:893–904
3. Majewski DD, Worrall LJ, Strynadka NC (2018) Secretins revealed: structural insights into the giant gated outer membrane portals of bacteria. Curr Opin Struct Biol 51:61–72
4. Barbat B, Douzi B, Voulhoux R (2023) Structural lessons on bacterial secretins. Biochimie 205:110–116
5. Hu J, Worrall LJ, Hong C et al (2018) Cryo-EM analysis of the T3S injectisome reveals the structure of the needle and open secretin. Nat Commun 9:1–11
6. Yan Z, Yin M, Xu D et al (2017) Structural insights into the secretin translocation channel in the type II secretion system. Nat Struct Mol Biol 24:177–183
7. Marvin DA, Symmons MF, Straus SK (2014) Structure and assembly of filamentous bacteriophages. Prog Biophys Mol Biol 114:80–122
8. Rakonjac J, Gold VAM, León-Quezada RI et al (2023) Structure, biology, and applications of filamentous bacteriophages. Cold Spring Harb Protoc
9. Hay ID, Lithgow T (2019) Filamentous phages: masters of a microbial sharing economy. EMBO Rep 20:1–24
10. Feng J, Model P, Russel M (1999) A trans-envelope protein complex needed for filamentous phage assembly and export. Mol Microbiol 34:745–755
11. Russel M, Kazmierczak B (1993) Analysis of the structure and subcellular location of filamentous phage pIV. J Bacteriol 175:3998–4007
12. Linderoth NA, Model P, Russel M (1996) Essential role of a sodium dodecyl sulfate-resistant protein IV multimer in assembly-export of filamentous phage. J Bacteriol 178: 1962–1970
13. Linderoth NA, Simon MN, Russel M (1997) The filamentous phage pIV multimer visualized by scanning transmission electron microscopy. Science 278:1635–1638
14. Marciano DK, Russel M, Simon SM (1999) An aqueous channel for filamentous phage export. Science 284:1516–1519
15. Opalka N, Beckmann R, Boisset N et al (2003) Structure of the filamentous phage pIV multimer by cryo-electron microscopy. J Mol Biol 325:461–470
16. Chua EYD, Mendez JH, Rapp M et al (2022) Better, faster, cheaper: recent advances in cryo-electron microscopy. Annu Rev Biochem 91:1–32
17. Birch J, Cheruvara H, Gamage N et al (2020) Changes in membrane protein structural biology. Biology 9:401
18. Choy BC, Cater RJ, Mancia F et al (2021) A 10-year meta-analysis of membrane protein structural biology: detergents, membrane mimetics, and structure determination techniques. Biochim Biophys Acta Biomembr 1863: 1–9
19. Russel M, Linderoth NA, Šali A (1997) Filamentous phage assembly: variation on a protein export theme. Gene 192:23–32
20. Conners R, McLaren M, Łapińska U et al (2021) CryoEM structure of the outer membrane secretin channel pIV from the f1 filamentous bacteriophage. Nat Commun 12:6316
21. Spagnuolo J, Opalka N, Wen WX et al (2010) Identification of the gate regions in the primary structure of the secretin pIV. Mol Microbiol 76:133–150
22. Hu J, Worrall LJ, Vuckovic M et al (2019) T3S injectisome needle complex structures in four distinct states reveal the basis of membrane coupling and assembly. Nat Microbiol 4: 2010–2019
23. Zivanov J, Nakane T, Forsberg BO et al (2018) New tools for automated high-resolution cryo-EM structure determination in RELION-3. elife 7:e42166
24. Punjani A, Rubinstein JL, Fleet DJ et al (2017) CryoSPARC: algorithms for rapid unsupervised cryo-EM structure determination. Nat Methods 14:290–296
25. Emsley P, Cowtan K (2004) Coot: model-building tools for molecular graphics. Acta Crystallogr D Biol Crystallogr 60:2126–2132
26. Pettersen EF, Goddard TD, Huang CC et al (2004) UCSF Chimera – a visualization system for exploratory research and analysis. J Comput Chem 25:1605–1612
27. Croll TI (2018) ISOLDE: a physically realistic environment for model building into low-resolution electron-density maps. Acta Crystallogr D Struct Biol 74:519–530
28. Jamali K, Käll L, Zhang R et al (2023) Automated model building and protein identification in cryo-EM maps. bioRxiv

29. Murshudov GN, Vagin AA, Dodson EJ (1997) Refinement of macromolecular structures by the maximum-likelihood method. Acta Crystallogr D Biol Cryst 53:240–255
30. Liebschner D, Afonine PV, Baker ML et al (2019) Macromolecular structure determination using X-rays, neutrons and electrons: recent developments in Phenix. Acta Crystallogr D Struct Biol 75:861–877
31. Tegunov D, Cramer P (2019) Real-time cryo-electron microscopy data preprocessing with Warp. Nat Methods 16:1146–1152
32. Beckers M, Mann D, Sachse C (2021) Structural interpretation of cryo-EM image reconstructions. Prog Biophys Mol Biol 160:26–36
33. Hay ID, Belousoff MJ, Dunstan RA et al (2018) Structure and membrane topography of the vibrio-type secretin complex from the type 2 secretion system of enteropathogenic *Escherichia coli*. J Bacteriol 200:e00521-17
34. Williams CJ, Headd JJ, Moriarty NW et al (2018) MolProbity: more and better reference data for improved all-atom structure validation. Protein Sci 27:293–315
35. Burley SK, Berman HM, Bhikadiya C et al (2019) Protein data bank: the single global archive for 3D macromolecular structure data. Nucleic Acids Res 47:520–528
36. Jumper J, Evans R, Pritzel A et al (2021) Highly accurate protein structure prediction with AlphaFold. Nature 596:583–589
37. Conners R, León-Quezada RI, McLaren M et al (2023) Cryo-electron microscopy of the f1 filamentous phage reveals insights into viral infection and assembly. Nat Commun 14:1–15

Chapter 19

Recent Advances in Modeling Membrane β-Barrel Proteins Using Molecular Dynamics Simulations: From Their Lipid Environments to Their Assemblies

Anna L. Duncan, Ya Gao, Evert Haanappel, Wonpil Im, and Matthieu Chavent

Abstract

Spurred by advances in AI-driven modeling and experimental methods, molecular dynamics simulations are now acting as a platform to integrate these different approaches. This combination of methods is especially useful to understand β-barrel proteins from the molecular level, e.g., identifying specific interactions with lipids or small molecules, up to assemblies comprised of hundreds of proteins and thousands of lipids. In this minireview, we will discuss recent advances, mainly from the last 5 years, in modeling β-barrel proteins and their assemblies. These approaches require specific kinds of modeling and potentially different model resolutions that we will first describe in Subheading 1. We will then focus on different aspects of β-barrel protein modeling: how different types of molecules can diffuse through β-barrel proteins (Subheading 2); how lipids can interact with these proteins (Subheading 3); how β-barrel proteins can interact with membrane partners (Subheading 4) or periplasmic extensions and partners (Subheading 5) to form large assemblies.

Key words Molecular dynamics, Multiscale modeling, Outer membrane proteins, Protein-protein interactions, Protein-lipid interactions

Abbreviations

AA	All-atom
AI	Artificial intelligence
CG	Coarse grained
CIP	Ciprofloxacin
LOS	Lipooligosaccharide
LPS	Lipopolysaccharide
Lpt	Lipopolysaccharide transport
MD	Molecular dynamics
ML	Machine learning
MSM	Markov state model
NMR	Nuclear magnetic resonance

Raffaele Ieva (ed.), *Transmembrane β-Barrel Proteins: Methods and Protocols*, Methods in Molecular Biology, vol. 2778,
https://doi.org/10.1007/978-1-0716-3734-0_19,

OM	Outer membrane
OMP	Outer membrane protein
PG	Peptidoglycan
PNA	Peptide nucleic acid
TASS	Temperature accelerated sliced sampling

1 Introduction

β-barrel proteins are made of anti-parallel ß-sheets arranged in a cylindrical shape with a hydrophilic interior and a hydrophobic exterior. This confers interesting biophysical properties to this class of proteins, which can be modeled using molecular dynamics (MD) simulations at different scales, enabling us to capture and explain in particular the capability to span biological membranes and interact with protein partners, while allowing the diffusion of small molecules through their interior. Several decades of progresses have culminated in modeling in silico the complexity of biological membranes [1–3]. This is especially true for modeling bacterial membranes [4–7]. In this minireview, we will showcase recent advances, mainly obtained during the last 5 years, in modeling β-barrel proteins from understanding local interactions with small molecules, surrounding lipids, and protein partners to studying large protein assemblies. We will also show how methodological advances have helped to model β-barrel proteins and their interactions with intracellular partners extending the canonical 2D scope (i.e., in the membrane plane) to a fully 3D landscape.

2 Methodological Development to Model β-Barrel Proteins

Before presenting applications, it is important to first describe the methodological context. Methodological advances were driven by improvements both in computational power [8–10] and in experimental techniques capable of determining the structure of very large membrane protein assemblies [11–14]. It is currently possible to model both very large assemblies composed of millions of particles and very specific interactions between few atoms. To switch from one type of modeling to the other, it is often required to adapt the resolution of the model to the given size of the system [15]. Two widely used resolutions in MD simulations are atomistic (or all-atom, AA) and coarse-grained (CG) resolutions. The former allows modeling all the atoms and molecular interactions, while the latter groups several atoms (typically four heavy atoms, *see* **Notes 1** and **2**) into one bead, which considerably reduces the complexity of the model [16]. One of the main force fields (*see* **Notes 1** and **2**) used in CG-MD simulations is the so-called MARTINI force field

[17]. In comparison with CG-MD simulations, the diversity of the force fields for atomistic simulations is more pronounced [18] and researchers need to assess which force field is the most suited to their needs [19–22]. It is possible to switch from one type of resolution to the other using dedicated tools [23–25]. It is worth mentioning that artificial intelligence (AI) approaches are now clearly impacting the development of force fields for MD simulations [26, 27]. For interested readers who want to get started using MD simulations for their research, accessible tutorials and guides are available to model lipid membranes [28] and membrane proteins [29, 30].

Before performing the actual MD simulations, it is first necessary to build the simulation system, with all its molecular components (proteins, lipids, ligands, ions, and water molecules) properly placed. There now exist dedicated tools streamlining the process of model creation. One example is CHARMM-GUI *Membrane Builder*, a webserver allowing the creation of a complete membrane system both in AA and in CG resolution [31–34]. Notably, CHARMM-GUI Membrane Builder supports various lipopolysaccharide (LPS) models from many different Gram-negative bacteria [32]. To model membrane systems with the MARTINI CG force field, one can also use a Python script called *insane* (an acronym for INSert membrANE) to model either lipid membranes or membrane protein assemblies [35]. MemProtMD is a database containing thousands of membrane proteins, extracted from the Protein Data Bank (PDB), embedded into a model membrane [36]. Simulations and the subsequent analysis results can be conveniently accessed through a web browser, enabling users to explore interactive 3D visualizations of the assembled bilayer, as well as 2D visualizations depicting lipid contact data and membrane protein topology. To specifically model outer membrane proteins (OMP) from Gram-negative bacteria, Baltoumas and colleagues have developed an automated pipeline to insert OMP models into LPS-containing membranes [37].

Once the calculations for MD simulations are completed using one of the various simulation codes [38], one needs to visualize and analyze the results. This task has become increasingly difficult due to the size and the complexity of the models. Specific tools and methodologies are now available to render the visualization and the analysis of membrane systems within reach of the majority of researchers [28, 39].

3 Through the Pore: From Channels to Nanopores

In bacteria and mitochondria, β-barrel proteins situated in the OM constitute the passage point for numerous molecules. MD simulations at AA resolution allow the study of how these molecules can interact with protein loops and diffuse through the pore. Hereafter, we present a selection of recently modeled molecules, ranging from few-atom molecules (ions and small compounds) to oligonucleotides and peptides (*see* Fig. 1).

Extensive AA-MD simulations were used in combination with a Markov state model (MSM) to explore the dynamics of the L3 internal loop in the *E. coli* OmpF protein [40]. These simulations help to distinguish between open and closed states of OmpF. In this latter state, electrophysiology simulations revealed a significant reduction of ionic currents. AA-MD simulations were also used to study permeation paths of potassium and chloride ions in VDAC channels of *Saccharomyces cerevisiae* [41]. The ion permeation properties may be influenced by surrounding OM lipids. Using AA-MD simulations, Lee et al. have shown how the outer core and O-antigens of LPS may sterically occlude the channel entrance and decrease the diffusion constants of ions approaching the OccK5 protein (also known as OpdH) from *Pseudomonas aeruginosa* [42]. MD simulations were also used in combination with solid-state NMR to study the permeability of AlkL from *Pseudomonas putida*, a minimalistic OMP, for hydrophobic molecules such as carvone or octane [43]. In this work, the authors proposed the release of hydrophobic compounds in the membrane after a diffusion through extracellular loops. Contrary to hydrophilic compounds, here the hydrophobic compounds do not seem to traverse the pore. This lateral diffusion model was also proposed for the FadL channel of *P. putida* for the uptake of monoaromatic hydrocarbons (MAH) such as benzene or toluene [44].

It is possible to transport even larger hydrophilic compounds like amino acids through the pore of a β-barrel protein. Samsudin and Khalid performed steered MD and umbrella sampling simulations (*see* **Note 3**) to determine the permeation pathway of arginine through the OprD channel from *P. aeruginosa* [45]. MD simulations also suggested that the arginine is surrounded by a shell of water molecules during the translocation through the pore. The binding of the peptide substrate ARRA to *E. coli* OmpT was recently explored by combining AA-MD simulations and umbrella sampling [46]. The translocation of even larger peptides like protamine, a 32-amino-acid-long polycationic peptide, through the CymA channel from *Klebsiella oxytoca* has been studied [47]. MD simulations were also used to study the interaction of *E. coli* BtuB with a peptide nucleic acid (PNA) covalently linked to vitamin B_{12} [48]. PNA is a synthetic DNA analog with a peptide-like backbone that can strongly bind to nucleic acids essential for bacterial growth. This can thus constitute an interesting antibacterial strategy. The transport of vitamin B_{12} alone through *E. coli* BtuB was also explored by combining steered MD, umbrella sampling, and Gaussian force-simulated annealing [49].

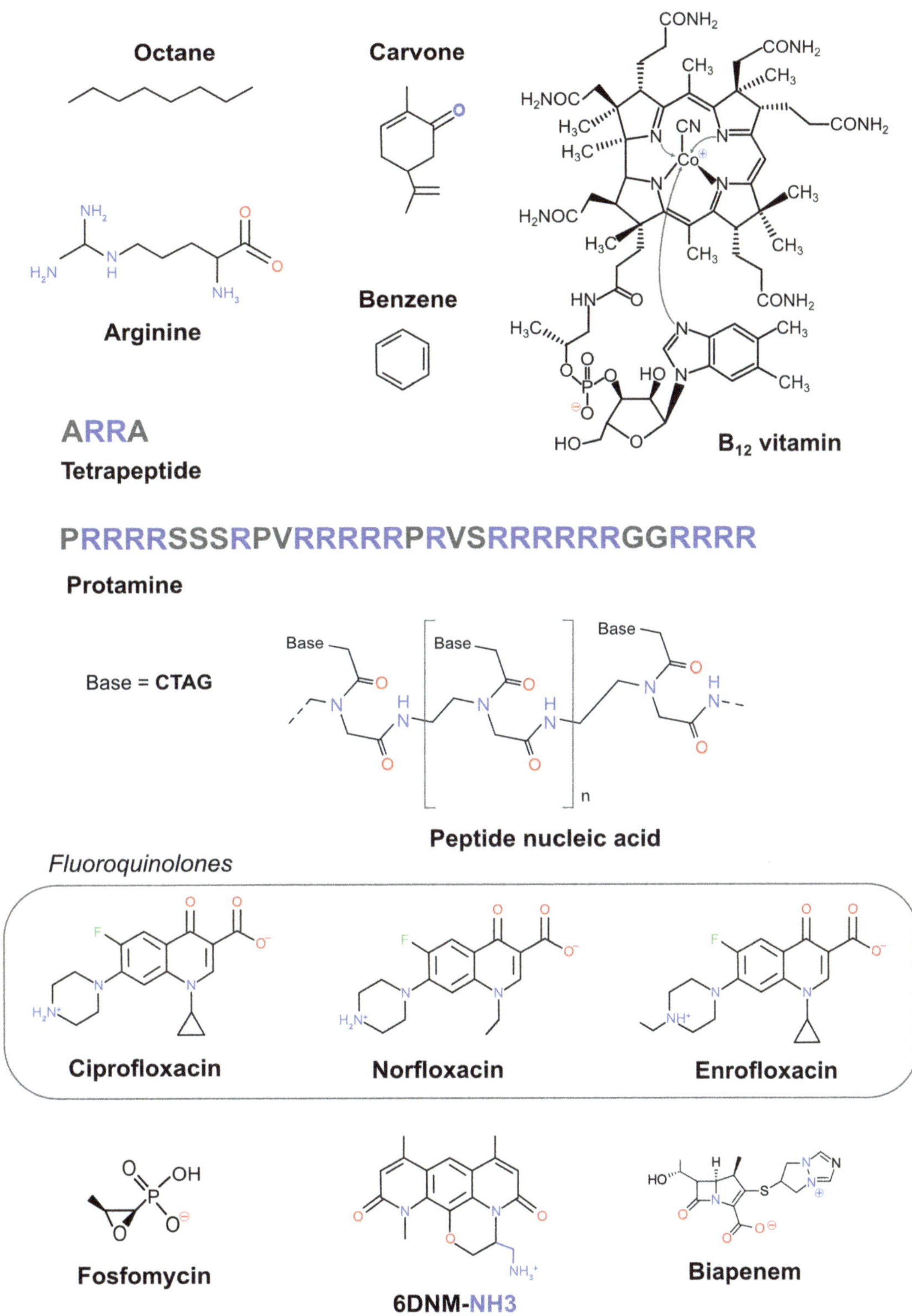

Fig. 1 Subset of small molecules and peptides, presented in Subheading 3, interacting with β-barrel proteins studied by multiscale molecular dynamics simulations

Due to their crucial role in the uptake of nutrients, β-barrel proteins are the target of numerous molecules that can potentially block this uptake and thereby act as antimicrobial drug candidates. MD simulations can give valuable insight into understanding how these molecules affect the structures and functions of β-barrel proteins. Fluoroquinolones are a class of broad-spectrum antibiotics. The permeation of ciprofloxacin (CIP), through OmpF, was studied using a temperature-accelerated sliced sampling (TASS) approach to characterize the two potential permeation pathways—the orientation of CIP inside the pore and its interactions with water molecules [50, 51]. The effect of divalent ions on the diffusivity of norfloxacin was investigated by AA-MD simulations and tested across several OM channels: OmpF and OmpC from *E. coli* and Omp35 and Omp36 from *Enterobacter aerogenes* [52]. Enrofloxacin and CIP permeation pathways across OmpC were also explored [53, 54]. Fosfomycin is a small phosphonic acid antibiotic discovered in *Streptomyces* strains. The fosfomycin permeability across the *E. coli* OmpF was investigated by AA-MD simulations including free energy and applied field techniques [55]. Fosfomycin translocation was also studied by AA-MD simulations for OrpO and OrpP from *P. aeruginosa* [56]. Recently an extensive set of AA-MD simulations and free energy calculations were performed to better understand how the introduction of a primary amine might enhance the permeation of antibiotics through OM channels [57]. The amine may enhance permeation by allowing the molecule to align its dipole with the electric field inside the porin's lumen, and it also establishes favorable electrostatic interactions with charged residues as the molecule moves through the pore. While these studies were all performed using AA-MD simulations, the recent developments of the MARTINI force field [17], especially aimed toward modeling small molecules [58], may allow researchers to use this less expensive approach to study antibiotic interactions with β-barrel proteins. Recent work explored how to use CG-MD simulations to investigate the interaction of the *P. aeruginosa* OccD3 porin with different carbapenems [59].

β-barrel proteins can be also engineered [60] and redesigned to serve as synthetic nanopores [61] to create, e.g., biomimetic membranes for water filtration [62] or to sequence nucleic acids [63] or proteins [64, 65]. MD simulations can help to better characterize these new molecules and their interactions with different solutes [61]. Two main classes of β-barrel proteins were studied using multiscale MD simulations: bacterial channels and pore-forming toxins (this is elaborated in Subheading 4). In the first category, one can cite the investigation of cyclodextrin and ions transport through the ΔCymA nanopore, a mutant of the protein without the 15 N-terminal residues, from *K. oxytoca* [66, 67]. MD simulations were also used to probe the stability of de novo design of a β-hairpin and its nanopore assembly in a membrane. This assembly formed a pore allowing the detection of a single polypeptide chain [68]. In the second category, the biophysical properties of several nanopores based on pore-forming toxin were investigated. The α-hemolysin was studied to assess its ability to recognize homopeptides [69] and its ionic transport and selectivity [70, 71]. For the aerolysin nanopore, translocation of different poly-arginine peptides [72], nucleotide discrimination [73], and detection of posttranslational modifications [74] were investigated. Ion conductance of both bacterial channels and pore-forming toxin nanopores was investigated using steric exclusion model and AA-MD simulations [75]. For interested readers, a recent perspective has summarized advances in understanding key issues in molecular simulations of antibiotic translocation and in the development of nanopore sensors [76].

4 β-Barrel Proteins in Their *Local* Environment

Lipids play fundamental roles in the folding [77, 78] and stability [79] of β-barrel proteins and in anchoring OMPs within the OM. MD simulations are uniquely capable of identifying and assessing specific protein-lipid interactions [80] as well as allowing the characterization of lipid binding sites [81].

Modeling of bacterial OM lipids has recently seen much progress, both at the atomistic scale [82, 83] and at CG resolution [33, 84, 85], enabling a wealth of new studies, in particular on OMP-LPS interactions (*see* Fig. 2a), which are challenging to study experimentally. AA-MD simulations have identified the role of calcium in specific OMP-LPS interactions and identified binding sites on *E. cloacae* OmpE36 that were in good agreement with those observed in crystal structures [90]. LPS has further been shown to have a unique interaction fingerprint with a diverse array of *E. coli* OMPs (OmpA, FhuA, OmpF, EstA, BtuB, and OmpX) [91]. Beyond the structural level, recent studies have shown that LPS plays a role in regulating OMP function. As an example, AA-MD simulations have shown that interactions of LPS with OprH affect the structure and dynamics of its extracellular loops [92]. As mentioned previously, AA-MD simulations of Occk5 demonstrated that LPS modulates ion transport through this OMP by hindering ion accessibility to the pore [42]. MD simulations in tandem with NMR spectroscopy showed that LPS interactions with the Ail protein change pathogen membrane properties, which confers enhanced resistance to the plague-causing bacterium *Yersinia pestis* [93]. MD simulations have also been used to uncover how LPS interactions with OmpD drive effective immunization with this OMP from *S. typhimurium* but not *S. enteritidis*, despite only the single amino acid difference between the OmpD homologs

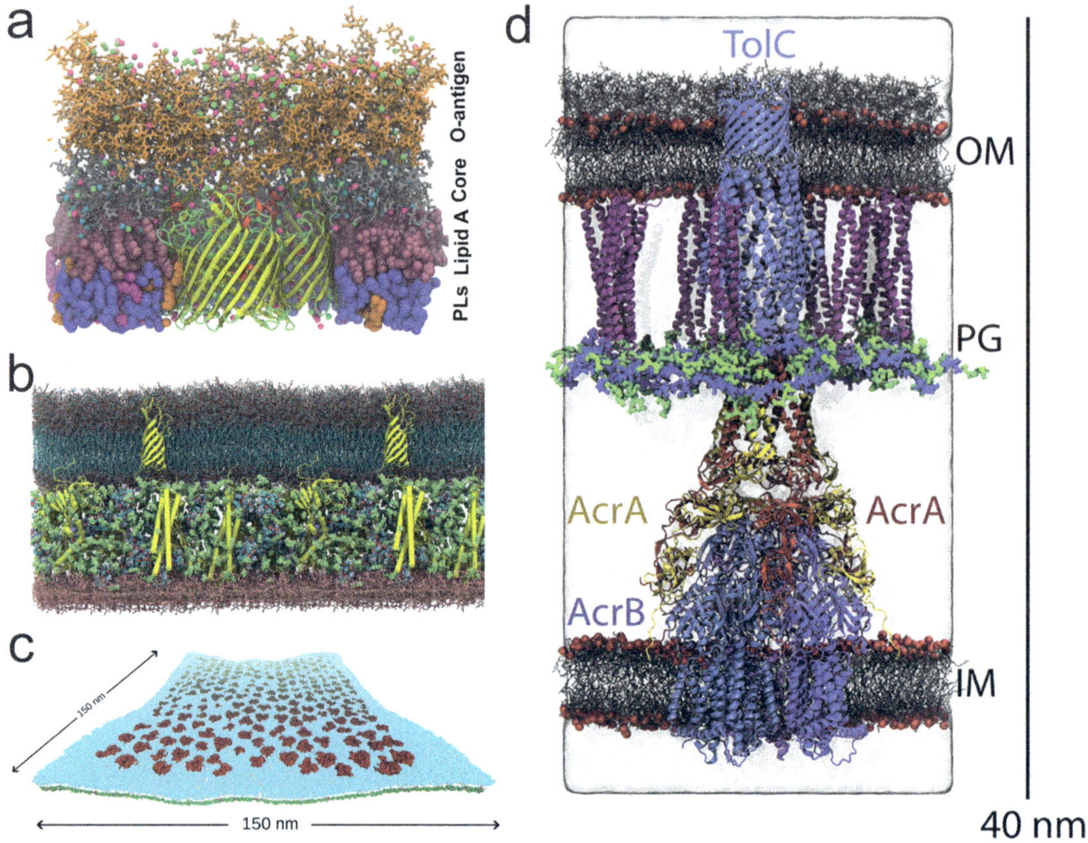

Fig. 2 Illustrative examples of systems presented in Subheadings 3, 4, and 5. (**a**) AA modeling of the OmpF protein surrounded by LPS [86]. (**b**) AA model of OM and periplasmic space containing proteins and osmolytes [87]. (**c**) CG modeling of an OMP island measuring 150 nm^2 along the membrane plane [88]. OMPs are shown in red and are embedded in an outer membrane model containing 100% RaLPS in the outer leaflet, and a 90: 5:5 ratio of POPE/POPG/cardiolipin in the inner leaflet. The RaLPS core region is shown in cyan, RaLPS phosphates are shown in yellow, and RaLPS acyl chains are shown in white. POPE, POPG, and cardiolipin lipids are shown in green, white, and pink, respectively. (**d**) AA modeling of the AcrAB-TolC multidrug efflux pump spanning the whole cell envelope [89] from the inner membrane (IM) through the peptidoglycan (PG) and up to the outer membrane (OM)

[94]. The interactions of surface-exposed loops with lipooligosaccharide (LOS) in a protein sequence dependent manner may alter the binding of antibodies to β-barrel proteins as seen for PorB from *N. meningitidis* [95]. MD simulations coupled with advanced mass spectrometry techniques have proposed how LPS is inserted by the LPS transport protein LptDE from *K. pneumoniae* [96] and helped to describe the mechanisms of deacylation of LPS by the β-barrel protein, LpxR, from *S. typhimurium* [97]. AA and CG MD simulations illustrated how thermodynamics drives membrane association of numerous lipoproteins, including the *E. coli* BAM complex, LptE, and CusC [98].

Multiscale simulations have also been used to show how other lipids affect β-barrel protein function. For instance, phosphatidylethanolamine of the mitochondrial OM influences the anion selectivity of VDAC and thereby regulates its function [99], while CG-MD simulations highlighted how ceramide lipids can bind to VDAC2 to trigger mitochondrial apoptosis, thus acting as a tumor-suppressing lipid [100]. Integrating MD and Brownian dynamics and electric field simulations with biochemical data has resulted in the ability to model protein complexes, such as the complex formed by the mitochondrial VDAC1 and hexokinase II, which would have been extremely challenging to achieve solely by experimental means, due to the interaction of hexokinase II with the membrane [101]. Phosphatidylglycerol binding to OmpF is sensitive to pH, and MD simulations mimicking different pH levels uncovered lipid interaction patterns suggesting that the lipid interactions attenuate *E. coli* OmpF channel closure [102].

5 Biogenesis and Large Assemblies of β-Barrel Proteins

β-barrel proteins are composed of a cylindrical arrangement of antiparallel β-sheets with hydrophobic residues tending to face the barrel's exterior. This structure allows specific features and constraints in term of protein folding, biogenesis, and interactions with membrane protein partners. Recently, de novo design was applied to engineer custom β-barrel proteins [103]. Multiscale (AA and CG) MD simulations are especially useful to gain insights into the biogenesis of these proteins, their folding, and the formation of larger membrane protein assemblies.

The biogenesis of OMPs in Gram-negative bacteria is mediated by the β-barrel assembly machinery (BAM) composed of five components called Bam A–E [104, 105]. The opening of BamA releasing newly folded β-barrel proteins into the OM, called lateral gating, was studied by AA-MD simulations [106]. Membrane thinning and distortion near this BamA's lateral gate were observed in simulations [107, 108]. The conformational plasticity of BAM was explored by Cryo-EM and MD simulations [109]. This study demonstrated that plasticity of the barrel domain of BamA is essential for the function of BAM. The BamA plasticity was also studied in the context of its interaction with its substrate EspP by Cryo-EM [110] and MD simulations [111] allowing a better characterization of the sequential conformational dynamics of BAM during the late stages of OMP assembly. Due to its central role of OMP biogenesis, BamA has been a target for the development of novel antibiotics. One of them, dynobactin A, identified by computational approaches, specifically targets lateral gating [112]. This BamA-antibiotic interaction was recently investigated by MD simulations [113]. Recently, a combination of cryo-electron microscopy, X-ray crystallography, native mass spectroscopy, in vivo experiments, and MD simulations was also used to decipher the association of Darobactin with BamA [114].

After the incorporation into the OM, β-barrel proteins can form large molecular assemblies with restricted diffusion, so-called islands [115]. CG-MD simulations suggested that island formation was driven by protein-protein interactions [115, 116]. Due to limitations of this CG modeling to depict realistic diffusion of LPS, these initial CG-MD simulations were performed without LPS molecules. Recently, however, these models were further extended to take into account the role of LPS in mediating protein-protein interactions [88] (*see* Fig. 2c). These large supramolecular assemblies of membrane proteins and lipids may affect the rigidity of the membrane as seen by CG-MD simulations [117]. Assemblies of other types of proteins, such as pore-forming proteins [118, 119], may even drastically affect the membrane organization. While individual pore-forming proteins are soluble, the assembly of multiple copies of these proteins at the membrane may lead to the formation of a large β-barrel pore. Assembly, interactions with lipids, and membrane pore formation by pore-forming proteins were investigated using both AA- and CG-MD simulations. Pneumolysin prepore interaction with lipids and intermediate steps leading to a complete pore were investigated by multiscale simulations [120]. Using a similar multiscale approach, the creation of a pore by gasdermin proteins has also been characterized [121–123].

6 Beyond the Membrane Plane

β-barrel proteins can also extend outside the membrane plane both via intra- and extracellular domains and interactions with protein partners [124, 125]. Here we will focus our attention on periplasmic domains. These may modulate β-barrel proteins functions, help transferring substrates, or favor interactions with other layers of the bacterial envelope such as the PG. The BamA subunit contains a large periplasmic domain composed of five globular polypeptide transport-associated (POTRA) motifs linked in tandem and numbered 1–5 from the N-terminal end [126]. Atomistic MD simulations have proposed a model of POTRA interacting with the phospholipids not only via hydrogen bonds but also via hydrophobic interactions engaged with tryptophan residues [127]. These interactions may favor different POTRA structural configurations by drifting at the membrane surface. Some of these conformations may be compatible with binding to BamB and BamD. CryoEM and AA-MD simulations of the BAM machinery (containing BamA +POTRA and BamB-E) in a reconstituted nanodiscs revealed membrane deformations [108]. The lipopolysaccharide transport (Lpt) machinery transfers LPS molecules into the OM [128]. The release of LPS molecules may also occur via a lateral gating mechanism as seen for BamA (*see* Subheading 4). AA-MD simulations have suggested that the periplasmic domain of LptDE is highly dynamic [129] and that the LPS substrate helps the opening of the lateral gate [96, 129].

The Khalid lab has used AA-MD simulations to study how the dimerization of OmpA protein through its cytoplasmic C-terminal domain may maintain a distance between the PG layer and the OM hence limiting the PG distortion [130]. AA-MD simulations also suggested that OmpA interactions with the PG layer are facilitated by tripartite contacts between Braun's lipoprotein, the PG layer, and the OmpA C-terminal domain [131]. The crowded OM and periplasm were also modeled using AA-MD simulations to study the travel of the antibiotic polymyxin B1 through the periplasm [87]. These simulations revealed that polymyxin B1 forms both transient and long-lived interactions with proteins, osmolytes, lipids of the OM, and the cell wall and is rarely uncomplexed when in the periplasm (*see* Fig. 2b).

Recent works have also started modeling how trans-envelope processes, such as mechanical stress sensing and metabolite efflux, are coordinated across the three envelope layers of Gram-negative bacterial cells, i.e., IM PG OM [89, 132] (*see* Fig. 2d).

7 Conclusion

With their β-barrel core, β-barrel proteins may be seen as less flexible than other types of proteins [133, 134]. However, based on recent advances coupling multiscale MD simulations with other experimental approaches, it is now clear that their intrinsic flexibility coupled to their diffusion in the membrane and their interactions with membrane peripheral partners play an important role in their function.

Here, we have given an overview of recent advances in MD simulations to decipher β-barrel protein dynamics at different scales: from the flexibility of extracellular loops and their interactions with lipids and small molecules to large assemblies of proteins and diffusion of molecules in different regions of the cell envelope. We are now moving toward very large and complex models to create digital twins of the bacterial cell envelope to develop new antibiotics or membranes composed of nanopores to design new biotechnological tools.

With recent advances in AI and experiments integrated to MD simulations, it will become increasingly feasible to model very complex biophysical mechanisms involving β-barrel proteins and shed new lights onto their function.

8 Notes

1. The term "force field" refers to a set of potential energy terms and specific implementation details, including parameter values, to calculate intra- and inter-molecular forces between atoms and particles (for CG systems, see below) [38]. Different force fields are optimized for specific types of molecules such as proteins [135], lipids [136] or DNA [137]. Thus, it is important to use the most appropriate force field for a given molecular system.
2. The goal of coarse grained (CG) mapping is to provide accelerated calculations by reducing the effective number of particles and using longer time scales, compared to AA resolution. In general, CG mapping requires a bottom-up strategy [138] by grouping several atoms together into a single bead and by assigning their overall chemical features (polarity, hydrophobicity, charge) and center of mass to that bead. For the MARTINI force field, the CG mapping originally consisted of grouping typically four heavy atoms (i.e., not including the hydrogen atoms) into one bead. With the development of the third version of MARTINI [139], it is now possible to select smaller bead types representing two to three atoms. Numerous tools exist that streamline CG mapping, such as PyCGTOOL [140], Auto-MARTINI [141], Swarm-CG [142, 143], CGCompiler [144], or MAD [145]. Recently, the use of machine learning (ML) has helped to automatically design new CG models for proteins [27, 146, 147].
3. The two approaches *known as steered molecular dynamics (SMD) simulations and umbrella sampling (US)* belong to the enhanced sampling methods category that allow sampling of larger portions of the configuration space of complex systems in a given amount of simulation time [148]. Steered molecular dynamics emulates atomic force microscopy experiments by introducing a fictitious 3D particle moving at constant velocity and connected to a molecule by a harmonic spring. It is often used to study folding/unfolding of proteins or ligand binding [149, 150]. Umbrella sampling allows exploration of one specific path by biasing the simulation along one (or more-dimensional) reaction coordinate to calculate energy barriers between different states [151].

Acknowledgments

We thank Pr S Khalid and Dr D Prakaash for providing pictures of the OM and periplasmic AA models (presented in Fig. 2b) and of the CG-model of OMP island (presented in Fig. 2c). We thank P Stansfeld for proofreading the manuscript and giving his feedback. MC is supported by the ITMO Cancer of Aviesan "Mathematiques et Informatique" on funds managed by Inserm. WI is supported by NSF (MCB-2111728).

References

1. Enkavi G, Javanainen M, Kulig W et al (2019) Multiscale simulations of biological membranes: the challenge to understand biological phenomena in a living substance. Chem Rev 119:5607–5774. https://doi.org/10.1021/acs.chemrev.8b00538
2. Corradi V, Sejdiu BI, Mesa-Galloso H et al (2019) Emerging diversity in lipid–protein interactions. Chem Rev 119:5775–5848. https://doi.org/10.1021/acs.chemrev.8b00451
3. Marrink SJ, Corradi V, Souza PCT et al (2019) Computational modeling of realistic cell membranes. Chem Rev 119:6184–6226. https://doi.org/10.1021/acs.chemrev.8b00460
4. Khalid S, Schroeder C, Bond PJ, Duncan AL (2022) What have molecular simulations contributed to understanding of Gram-negative bacterial cell envelopes? Microbiology 168:001165. https://doi.org/10.1099/mic.0.001165
5. Im W, Khalid S (2020) Molecular simulations of Gram-negative bacterial membranes come of age. Annu Rev Phys Chem 71:1–18. https://doi.org/10.1146/annurev-physchem-103019-033434
6. Khalid S, Berglund NA, Holdbrook DA et al (2015) The membranes of Gram-negative bacteria: progress in molecular modelling and simulation. Biochem Soc Trans 43:162–167. https://doi.org/10.1042/bst20140262
7. Patel DS, Qi Y, Im W (2017) Modeling and simulation of bacterial outer membranes and interactions with membrane proteins. Curr Opin Struct Biol 43:131–140. https://doi.org/10.1016/j.sbi.2017.01.003
8. Chang C, Deringer VL, Katti KS et al (2023) Simulations in the era of exascale computing. Nat Rev Mater 8:309–313. https://doi.org/10.1038/s41578-023-00540-6
9. Wieczór M, Genna V, Aranda J et al (2023) Pre-exascale HPC approaches for molecular dynamics simulations. Covid-19 research: a use case. Wiley Interdiscip Rev Comput Mol Sci 13:e1622. https://doi.org/10.1002/wcms.1622
10. Voelz VA, Pande VS, Bowman GR (2023) Folding@home: achievements from over 20 years of citizen science herald the exascale era. Biophys J 122:2852–2863. https://doi.org/10.1016/j.bpj.2023.03.028
11. Shi X, Chen M, Yu Z et al (2019) In situ structure and assembly of the multidrug efflux pump AcrAB-TolC. Nat Commun 10:2635. https://doi.org/10.1038/s41467-019-10512-6
12. Han F, Hu Y, Wu M et al (2023) Structures of Tetrahymena thermophila respiratory megacomplexes on the tubular mitochondrial cristae. Nat Commun 14:2542. https://doi.org/10.1038/s41467-023-38158-5
13. Tan J, Zhang X, Wang X et al (2021) Structural basis of assembly and torque transmission of the bacterial flagellar motor. Cell 184:2665–2679.e19. https://doi.org/10.1016/j.cell.2021.03.057
14. Sheedlo MJ, Chung JM, Sawhney N et al (2020) Cryo-EM reveals species-specific components within the Helicobacter pylori Cag type IV secretion system core complex. eLife 9:e59495. https://doi.org/10.7554/elife.59495
15. Chavent M, Duncan AL, Sansom MS (2016) Molecular dynamics simulations of membrane proteins and their interactions: from nanoscale to mesoscale. Curr Opin Struct Biol 40:8–16. https://doi.org/10.1016/j.sbi.2016.06.007
16. Kmiecik S, Gront D, Kolinski M et al (2016) Coarse-grained protein models and their applications. Chem Rev 116:7898–7936. https://doi.org/10.1021/acs.chemrev.6b00163

17. Marrink SJ, Monticelli L, Melo MN et al (2022) Two decades of Martini: better beads, broader scope. Wiley Interdiscip Rev Comput Mol Sci 13:42. https://doi.org/10.1002/wcms.1620
18. Nerenberg PS, Head-Gordon T (2018) New developments in force fields for biomolecular simulations. Curr Opin Struct Biol 49:129–138. https://doi.org/10.1016/j.sbi.2018.02.002
19. Sandoval-Perez A, Pluhackova K, Böckmann RA (2017) Critical comparison of biomembrane force fields: protein-lipid interactions at the membrane interface. J Chem Theory Comput 13:2310–2321. https://doi.org/10.1021/acs.jctc.7b00001
20. Pluhackova K, Kirsch SA, Han J et al (2016) A critical comparison of biomembrane force fields: structure and dynamics of model DMPC, POPC, and POPE bilayers. J Phys Chem B 120:3888–3903. https://doi.org/10.1021/acs.jpcb.6b01870
21. Fadda E (2022) Molecular simulations of complex carbohydrates and glycoconjugates. Curr Opin Chem Biol 69:102175. https://doi.org/10.1016/j.cbpa.2022.102175
22. Huang J, MacKerell AD (2018) Force field development and simulations of intrinsically disordered proteins. Curr Opin Struct Biol 48:40–48. https://doi.org/10.1016/j.sbi.2017.10.008
23. Vickery ON, Stansfeld PJ (2021) CG2AT2: an enhanced fragment-based approach for serial multi-scale molecular dynamics simulations. J Chem Theory Comput 17:6472–6482. https://doi.org/10.1021/acs.jctc.1c00295
24. Kroon PC, Grunewald F, Barnoud J et al (2023) Martinize2 and vermouth: unified framework for topology generation. eLife 12:RP90627. https://doi.org/10.7554/elife.90627
25. Wassenaar TA, Pluhackova K, Böckmann RA et al (2014) Going backward: a flexible geometric approach to reverse transformation from coarse grained to atomistic models. J Chem Theory Comput 10:676–690. https://doi.org/10.1021/ct400617g
26. Noé F, Tkatchenko A, Müller K-R, Clementi C (2020) Machine learning for molecular simulation. Annu Rev Phys Chem 71:1–30. https://doi.org/10.1146/annurev-physchem-042018-052331
27. Durumeric AEP, Charron NE, Templeton C et al (2023) Machine learned coarse-grained protein force-fields: are we there yet? Curr Opin Struct Biol 79:102533. https://doi.org/10.1016/j.sbi.2023.102533
28. Smith DJ, Klauda JB, Sodt AJ (2019) Simulation best practices for lipid membranes [Article v1.0]. Living J Comput Mol Sci 1:5966. https://doi.org/10.33011/livecoms.1.1.5966
29. Jefferies D, Khalid S (2021) Atomistic and coarse-grained simulations of membrane proteins: a practical guide. Methods 185:15–27. https://doi.org/10.1016/j.ymeth.2020.02.007
30. Lemkul J (2018) From proteins to perturbed Hamiltonians: a suite of tutorials for the GROMACS-2018 molecular simulation package [Article v1.0]. Living J Comput Mol Sci 1:5068. https://doi.org/10.33011/livecoms.1.1.5068
31. Feng S, Park S, Choi YK, Im W (2023) CHARMM-GUI membrane builder: past, current, and future developments and applications. J Chem Theory Comput 19:2161. https://doi.org/10.1021/acs.jctc.2c01246
32. Lee J, Patel DS, Ståhle J et al (2018) CHARMM-GUI membrane builder for complex biological membrane simulations with glycolipids and lipoglycans. J Chem Theory Comput 15:775–786. https://doi.org/10.1021/acs.jctc.8b01066
33. Hsu PC, Bruininks BMH, Jefferies D et al (2017) CHARMM-GUI Martini Maker for modeling and simulation of complex bacterial membranes with lipopolysaccharides. J Comput Chem 38:2354–2363
34. Qi Y, Ingólfsson HI, Cheng X et al (2015) CHARMM-GUI Martini Maker for coarse-grained simulations with the Martini force field. J Chem Theory Comput 11:4486–4494. https://doi.org/10.1021/acs.jctc.5b00513
35. Wassenaar TA, Ingólfsson HI, Böckmann RA et al (2015) Computational lipidomics with insane: a versatile tool for generating custom membranes for molecular simulations. J Chem Theory Comput 11:2144–2155. https://doi.org/10.1021/acs.jctc.5b00209
36. Newport TD, Sansom MSP, Stansfeld PJ (2018) The MemProtMD database: a resource for membrane-embedded protein structures and their lipid interactions. Nucleic Acids Res 47:D390–D397. https://doi.org/10.1093/nar/gky1047
37. Baltoumas FA, Hamodrakas SJ, Iconomidou VA (2019) The gram-negative outer membrane modeler: automated building of lipopolysaccharide-rich bacterial outer membranes in four force fields. J Comput Chem

40:1727–1734. https://doi.org/10.1002/jcc.25823

38. Talirz L, Ghiringhelli LM, Smit B (2021) Trends in atomistic simulation software usage [Article v1.0]. Living J Comput Mol Sci 3:1483. https://doi.org/10.33011/livecoms.3.1.1483
39. Corey RA, Baaden M, Chavent M (2023) A brief history of visualizing membrane systems in molecular dynamics simulations. Front Bioinform 3:1149744. https://doi.org/10.3389/fbinf.2023.1149744
40. Vasan AK, Haloi N, Ulrich RJ et al (2022) Role of internal loop dynamics in antibiotic permeability of outer membrane porins. Proc Natl Acad Sci 119:e2117009119. https://doi.org/10.1073/pnas.2117009119
41. Guardiani C, Magrì A, Karachitos A et al (2018) yVDAC2, the second mitochondrial porin isoform of Saccharomyces cerevisiae. Biochim Biophys Acta Bioenerg 1859:270–279. https://doi.org/10.1016/j.bbabio.2018.01.008
42. Lee J, Pothula KR, Kleinekathöfer U, Im W (2018) Simulation study of Occk5 functional properties in Pseudomonas aeruginosa outer membranes. J Phys Chem B 122:8185–8192. https://doi.org/10.1021/acs.jpcb.8b07109
43. Schubeis T, Marchand TL, Daday C et al (2020) A β-barrel for oil transport through lipid membranes: dynamic NMR structures of AlkL. Proc Natl Acad Sci 117:21014–21021. https://doi.org/10.1073/pnas.2002598117
44. Somboon K, Doble A, Bulmer D et al (2020) Uptake of monoaromatic hydrocarbons during biodegradation by FadL channel-mediated lateral diffusion. Nat Commun 11:6331. https://doi.org/10.1038/s41467-020-20126-y
45. Samsudin F, Khalid S (2019) Movement of arginine through OprD: the energetics of permeation and the role of lipopolysaccharide in directing arginine to the protein. J Phys Chem B 123:2824–2832. https://doi.org/10.1021/acs.jpcb.9b00063
46. Zhang Y, Baaden M (2023) Molecular insights into substrate binding of the outer membrane enzyme OmpT. Catalysts 13:214. https://doi.org/10.3390/catal13020214
47. Pangeni S, Prajapati JD, Bafna J et al (2021) Large-peptide permeation through a membrane channel: understanding protamine translocation through CymA from Klebsiella oxytoca**. Angew Chem Int Ed 60:8089–8094. https://doi.org/10.1002/anie.202016943
48. Pieńko T, Czarnecki J, Równicki M et al (2021) Vitamin B12-peptide nucleic acids use the BtuB receptor to pass through the Escherichia coli outer membrane. Biophys J 120:725–737. https://doi.org/10.1016/j.bpj.2021.01.004
49. Pieńko T, Trylska J (2020) Extracellular loops of BtuB facilitate transport of vitamin B12 through the outer membrane of E. coli. PLoS Comput Biol 16:e1008024. https://doi.org/10.1371/journal.pcbi.1008024
50. Acharya A, Prajapati JD, Kleinekathöfer U (2021) Improved sampling and free energy estimates for antibiotic permeation through bacterial porins. J Chem Theory Comput 17:4564–4577. https://doi.org/10.1021/acs.jctc.1c00369
51. Acharya A, Ghai I, Piselli C et al (2023) Conformational dynamics of loop L3 in OmpF: implications toward antibiotic translocation and voltage gating. J Chem Inf Model 63:910–927. https://doi.org/10.1021/acs.jcim.2c01108
52. Wang J, Prajapati JD, Kleinekathöfer U, Winterhalter M (2020) Dynamic interaction of fluoroquinolones with magnesium ions monitored using bacterial outer membrane nanopores. Chem Sci 11:10344–10353. https://doi.org/10.1039/d0sc03486j
53. Prajapati JD, Solano CJF, Winterhalter M, Kleinekathöfer U (2018) Enrofloxacin permeation pathways across the porin OmpC. J Phys Chem B 122:1417–1426. https://doi.org/10.1021/acs.jpcb.7b12568
54. Prajapati JD, Solano CJF, Winterhalter M, Kleinekathöfer U (2017) Characterization of ciprofloxacin permeation pathways across the porin OmpC using metadynamics and a string method. J Chem Theory Comput 13:4553–4566. https://doi.org/10.1021/acs.jctc.7b00467
55. Golla VK, Sans-Serramitjana E, Pothula KR et al (2019) Fosfomycin permeation through the outer membrane porin OmpF. Biophys J 116:258–269. https://doi.org/10.1016/j.bpj.2018.12.002
56. Golla VK, Piselli C, Kleinekathöfer U, Benz R (2022) Permeation of fosfomycin through the phosphate-specific channels OprP and OprO of Pseudomonas aeruginosa. J Phys Chem B 126:1388–1403. https://doi.org/10.1021/acs.jpcb.1c08696
57. Haloi N, Vasan AK, Geddes EJ et al (2021) Rationalizing the generation of broad spectrum antibiotics with the addition of a positive charge. Chem Sci 12:15028–15044. https://doi.org/10.1039/d1sc04445a

58. Alessandri R, Barnoud J, Gertsen AS et al (2022) Martini 3 coarse-grained force field: small molecules. Adv Theory Simul 5: 2100391. https://doi.org/10.1002/adts.202100391
59. Dai Y, Ma H, Wu M et al (2021) Development of the computational antibiotic screening platform (CLASP) to aid in the discovery of new antibiotics. Soft Matter 17:2725–2736. https://doi.org/10.1039/d0sm02035d
60. Kim DE, Jensen DR, Feldman D et al (2023) De novo design of small beta barrel proteins. Proc Natl Acad Sci 120:e2207974120. https://doi.org/10.1073/pnas.2207974120
61. Chowdhury R, Ren T, Shankla M et al (2018) PoreDesigner for tuning solute selectivity in a robust and highly permeable outer membrane pore. Nat Commun 9:3661. https://doi.org/10.1038/s41467-018-06097-1
62. Tu Y-M, Song W, Ren T et al (2020) Rapid fabrication of precise high-throughput filters from membrane protein nanosheets. Nat Mater 19:347–354. https://doi.org/10.1038/s41563-019-0577-z
63. Ayub M, Stoddart D, Bayley H (2015) Nucleobase recognition by truncated α-hemolysin pores. ACS Nano 9:7895–7903. https://doi.org/10.1021/nn5060317
64. Wei X, Penkauskas T, Reiner JE et al (2023) Engineering biological nanopore approaches toward protein sequencing. ACS Nano 17: 16369–16395. https://doi.org/10.1021/acsnano.3c05628
65. Asandei A, Muccio GD, Schiopu I et al (2020) Nanopore-based protein sequencing using biopores: current achievements and open challenges. Small Methods 4:1900595. https://doi.org/10.1002/smtd.201900595
66. Prajapati JD, Kleinekathöfer U (2020) Voltage-dependent transport of neutral solutes through nanopores: a molecular view. J Phys Chem B 124:10718–10731. https://doi.org/10.1021/acs.jpcb.0c08401
67. Bhamidimarri SP, Prajapati JD, van den Berg B et al (2016) Role of electroosmosis in the permeation of neutral molecules: CymA and cyclodextrin as an example. Biophys J 110: 600–611. https://doi.org/10.1016/j.bpj.2015.12.027
68. Shimizu K, Mijiddorj B, Usami M et al (2022) De novo design of a nanopore for single-molecule detection that incorporates a β-hairpin peptide. Nat Nanotechnol 17:67–75. https://doi.org/10.1038/s41565-021-01008-w
69. Muccio GD, Rossini AE, Marino DD et al (2019) Insights into protein sequencing with an α-hemolysin nanopore by atomistic simulations. Sci Rep 9:6440. https://doi.org/10.1038/s41598-019-42867-7
70. Basdevant N, Dessaux D, Ramirez R (2019) Ionic transport through a protein nanopore: a Coarse-Grained Molecular Dynamics Study. Sci Rep 9:15740. https://doi.org/10.1038/s41598-019-51942-y
71. Dessaux D, Mathé J, Ramirez R, Basdevant N (2022) Current rectification and ionic selectivity of α-hemolysin: coarse-grained molecular dynamics simulations. J Phys Chem B 126: 4189–4199. https://doi.org/10.1021/acs.jpcb.2c01028
72. Ouldali H, Sarthak K, Ensslen T et al (2020) Electrical recognition of the twenty proteinogenic amino acids using an aerolysin nanopore. Nat Biotechnol 38:176–181. https://doi.org/10.1038/s41587-019-0345-2
73. Li M-Y, Ying Y-L, Yu J et al (2021) Revisiting the origin of nanopore current blockage for volume difference sensing at the atomic level. JACS Au 1:967–976. https://doi.org/10.1021/jacsau.1c00109
74. Ensslen T, Sarthak K, Aksimentiev A, Behrends JC (2022) Resolving isomeric post-translational modifications using a biological nanopore as a sensor of molecular shape. J Am Chem Soc 144:16060–16068. https://doi.org/10.1021/jacs.2c06211
75. Wilson J, Sarthak K, Si W et al (2019) Rapid and accurate determination of nanopore ionic current using a steric exclusion model. ACS Sens 4:634–644. https://doi.org/10.1021/acssensors.8b01375
76. Acharya A, Prajapati JD, Kleinekathöfer U (2022) Atomistic simulation of molecules interacting with biological nanopores: from current understanding to future directions. J Phys Chem B 126:3995–4008. https://doi.org/10.1021/acs.jpcb.2c01173
77. Horne JE, Brockwell DJ, Radford SE (2020) Role of the lipid bilayer in outer membrane protein folding in Gram-negative bacteria. J Biol Chem 295:10340–10367. https://doi.org/10.1074/jbc.rev120.011473
78. Machin JM, Kalli AC, Ranson NA, Radford SE (2023) Protein–lipid charge interactions control the folding of outer membrane proteins into asymmetric membranes. Nat Chem 15:1754–1764. https://doi.org/10.1038/s41557-023-01319-6
79. Marx DC, Fleming KG (2021) Local bilayer hydrophobicity modulates membrane protein

stability. J Am Chem Soc 143:764–772. https://doi.org/10.1021/jacs.0c09412

80. Corradi V, Mendez-Villuendas E, Ingólfsson HI et al (2018) Lipid–protein interactions are unique fingerprints for membrane proteins. ACS Cent Sci 4:709–717. https://doi.org/10.1021/acscentsci.8b00143
81. Corey RA, Song W, Duncan AL et al (2021) Identification and assessment of cardiolipin interactions with E. coli inner membrane proteins. Sci Adv 7:eabh2217. https://doi.org/10.1126/sciadv.abh2217
82. Hsieh M-K, Klauda JB (2022) Leaflet asymmetry modeling in the lipid composition of Escherichia coli cytoplasmic membranes. J Phys Chem B 126:184–196. https://doi.org/10.1021/acs.jpcb.1c07332
83. Gao Y, Widmalm G, Im W (2023) Modeling and simulation of bacterial outer membranes with lipopolysaccharides and capsular polysaccharides. J Chem Inf Model 63:1592–1601. https://doi.org/10.1021/acs.jcim.3c00072
84. Jefferies D, Shearer J, Khalid S (2019) Role of O-antigen in response to mechanical stress of the E. coli outer membrane: insights from coarse-grained MD simulations. J Phys Chem B 123:3567–3575. https://doi.org/10.1021/acs.jpcb.8b12168
85. Vaiwala R, Ayappa KG (2023) Martini-3 coarse-grained models for the bacterial lipopolysaccharide outer membrane of Escherichia coli. J Chem Theory Comput. https://doi.org/10.1021/acs.jctc.3c00471
86. Patel DS, Re S, Wu EL et al (2016) Dynamics and interactions of OmpF and LPS: influence on pore accessibility and ion permeability. Biophys J 110:930–938. https://doi.org/10.1016/j.bpj.2016.01.002
87. Pedebos C, Smith IPS, Boags A, Khalid S (2021) The hitchhiker's guide to the periplasm: unexpected molecular interactions of polymyxin B1 in E. coli. Structure 29:444–456.e2. https://doi.org/10.1016/j.str.2021.01.009
88. Webby MN, Oluwole AO, Pedebos C et al (2022) Lipids mediate supramolecular outer membrane protein assembly in bacteria. Sci Adv 8:eadc9566. https://doi.org/10.1126/sciadv.adc9566
89. Gumbart JC, Ferreira JL, Hwang H et al (2021) Lpp positions peptidoglycan at the AcrA-TolC interface in the AcrAB-TolC multidrug efflux pump. Biophys J 120:3973–3982. https://doi.org/10.1016/j.bpj.2021.08.016
90. Kesireddy A, Pothula KR, Lee J et al (2019) Modeling of specific lipopolysaccharide binding sites on a Gram-negative porin. J Phys Chem B 123:5700–5708. https://doi.org/10.1021/acs.jpcb.9b03669
91. Shearer J, Jefferies D, Khalid S (2019) Outer membrane proteins OmpA, FhuA, OmpF, EstA, BtuB, and OmpX have unique lipopolysaccharide fingerprints. J Chem Theory Comput 15:2608–2619. https://doi.org/10.1021/acs.jctc.8b01059
92. Lee J, Patel DS, Kucharska I et al (2017) Refinement of OprH-LPS interactions by molecular simulations. Biophys J 112:346–355. https://doi.org/10.1016/j.bpj.2016.12.006
93. Singh C, Lee H, Tian Y et al (2020) Mutually constructive roles of Ail and LPS in Yersinia pestis serum survival. Mol Microbiol 114:510–520. https://doi.org/10.1111/mmi.14530
94. Domínguez-Medina CC, Pérez-Toledo M, Schager AE et al (2020) Outer membrane protein size and LPS O-antigen define protective antibody targeting to the Salmonella surface. Nat Commun 11:851. https://doi.org/10.1038/s41467-020-14655-9
95. Matthias KA, Strader MB, Nawar HF et al (2017) Heterogeneity in non-epitope loop sequence and outer membrane protein complexes alters antibody binding to the major porin protein PorB in serogroup B Neisseria meningitidis. Mol Microbiol 105:934–953. https://doi.org/10.1111/mmi.13747
96. Fiorentino F, Sauer JB, Qiu X et al (2021) Dynamics of an LPS translocon induced by substrate and an antimicrobial peptide. Nat Chem Biol 17:187–195. https://doi.org/10.1038/s41589-020-00694-2
97. Saunders GM, Macdonald HEB, Essex JW, Khalid S (2018) Prediction of the closed conformation and insights into the mechanism of the membrane enzyme LpxR. Biophys J 115:1445–1456. https://doi.org/10.1016/j.bpj.2018.09.002
98. Rao S, Bates GT, Matthews CR et al (2020) Characterizing membrane association and periplasmic transfer of bacterial lipoproteins through molecular dynamics simulations. Structure 28:475–487.e3. https://doi.org/10.1016/j.str.2020.01.012
99. Liefferinge FV, Krammer E-M, Sengupta D, Prévost M (2019) Lipid composition and salt concentration as regulatory factors of the anion selectivity of VDAC studied by coarse-grained molecular dynamics simulations. Chem Phys Lipids 220:66–76. https://doi.org/10.1016/j.chemphyslip.2018.11.002

100. Dadsena S, Bockelmann S, Mina JGM et al (2019) Ceramides bind VDAC2 to trigger mitochondrial apoptosis. Nat Commun 10: 1832. https://doi.org/10.1038/s41467-019-09654-4
101. Haloi N, Wen P-C, Cheng Q et al (2021) Structural basis of complex formation between mitochondrial anion channel VDAC1 and hexokinase-II. Commun Biol 4: 667. https://doi.org/10.1038/s42003-021-02205-y
102. Liko I, Degiacomi MT, Lee S et al (2018) Lipid binding attenuates channel closure of the outer membrane protein OmpF. Proc Natl Acad Sci 115:6691–6696. https://doi.org/10.1073/pnas.1721152115
103. Vorobieva AA, White P, Liang B et al (2021) De novo design of transmembrane β barrels. Science 371:eabc8182. https://doi.org/10.1126/science.abc8182
104. Wu R, Stephenson R, Gichaba A, Noinaj N (2020) The big BAM theory: an open and closed case? Biochim Biophys Acta Biomembr 1862:183062. https://doi.org/10.1016/j.bbamem.2019.183062
105. Noinaj N, Gumbart JC, Buchanan SK (2017) The β-barrel assembly machinery in motion. Nat Rev Microbiol 15:197–204. https://doi.org/10.1038/nrmicro.2016.191
106. Lundquist K, Bakelar J, Noinaj N, Gumbart JC (2018) C-terminal kink formation is required for lateral gating in BamA. Proc Natl Acad Sci 115:E7942–E7949. https://doi.org/10.1073/pnas.1722530115
107. Liu J, Gumbart JC (2020) Membrane thinning and lateral gating are consistent features of BamA across multiple species. PLoS Comput Biol 16:e1008355. https://doi.org/10.1371/journal.pcbi.1008355
108. Iadanza MG, Schiffrin B, White P et al (2020) Distortion of the bilayer and dynamics of the BAM complex in lipid nanodiscs. Commun Biol 3:766. https://doi.org/10.1038/s42003-020-01419-w
109. Wu R, Bakelar JW, Lundquist K et al (2021) Plasticity within the barrel domain of BamA mediates a hybrid-barrel mechanism by BAM. Nat Commun 12:7131. https://doi.org/10.1038/s41467-021-27449-4
110. Shen C, Chang S, Luo Q et al (2023) Structural basis of BAM-mediated outer membrane β-barrel protein assembly. Nature 617:185–193. https://doi.org/10.1038/s41586-023-05988-8
111. Kuo KM, Ryoo D, Lundquist K, Gumbart JC (2022) Modeling intermediates of BamA folding an outer membrane protein. Biophys J 121:3242–3252. https://doi.org/10.1016/j.bpj.2022.07.027
112. Miller RD, Iinishi A, Modaresi SM et al (2022) Computational identification of a systemic antibiotic for Gram-negative bacteria. Nat Microbiol 7:1661–1672. https://doi.org/10.1038/s41564-022-01227-4
113. Kuo KM, Liu J, Pavlova A, Gumbart JC (2023) Drug binding to BamA targets its lateral gate. J Phys Chem B 127:7509–7517. https://doi.org/10.1021/acs.jpcb.3c04501
114. Kaur H, Jakob RP, Marzinek JK et al (2021) The antibiotic darobactin mimics a β-strand to inhibit outer membrane insertase. Nature 593:125–129. https://doi.org/10.1038/s41586-021-03455-w
115. Rassam P, Copeland NA, Birkholz O et al (2015) Supramolecular assemblies underpin turnover of outer membrane proteins in bacteria. Nature 523:333–336. https://doi.org/10.1038/nature14461
116. Chavent M, Duncan AL, Rassam P et al (2018) How nanoscale protein interactions determine the mesoscale dynamic organisation of bacterial outer membrane proteins. Nat Commun 9:2846. https://doi.org/10.1038/s41467-018-05255-9
117. Fowler PW, Hélie J, Duncan A et al (2016) Membrane stiffness is modified by integral membrane proteins. Soft Matter 12:7792–7803. https://doi.org/10.1039/c6sm01186a
118. Peraro MD, van der Goot FG (2015) Pore-forming toxins: ancient, but never really out of fashion. Nat Rev Microbiol 14:77–92. https://doi.org/10.1038/nrmicro.2015.3
119. Broz P, Pelegrín P, Shao F (2020) The gasdermins, a protein family executing cell death and inflammation. Nat Rev Immunol 20: 143–157. https://doi.org/10.1038/s41577-019-0228-2
120. Vögele M, Bhaskara RM, Mulvihill E et al (2019) Membrane perforation by the pore-forming toxin pneumolysin. Proc Natl Acad Sci 116:13352–13357. https://doi.org/10.1073/pnas.1904304116
121. Schaefer SL, Hummer G (2022) Sublytic gasdermin-D pores captured in atomistic molecular simulations. eLife 11:e81432. https://doi.org/10.7554/elife.81432
122. Korn V, Pluhackova K (2022) Not sorcery after all: roles of multiple charged residues in membrane insertion of gasdermin-A3. Front Cell Dev Biol 10:958957. https://doi.org/10.3389/fcell.2022.958957
123. Mari SA, Pluhackova K, Pipercevic J et al (2022) Gasdermin-A3 pore formation

propagates along variable pathways. Nat Commun 13:2609. https://doi.org/10.1038/s41467-022-30232-8

124. Dorgan B, Liu Y, Wang S et al (2022) Structural model of a Porphyromonas gingivalis type IX secretion system shuttle complex. J Mol Biol 434:167871. https://doi.org/10.1016/j.jmb.2022.167871

125. White P, Joshi A, Rassam P et al (2017) Exploitation of an iron transporter for bacterial protein antibiotic import. Proc Natl Acad Sci 114:12051–12056. https://doi.org/10.1073/pnas.1713741114

126. Noinaj N, Kuszak AJ, Gumbart JC et al (2013) Structural insight into the biogenesis of β-barrel membrane proteins. Nature 501: 385–390. https://doi.org/10.1038/nature12521

127. Fleming PJ, Patel DS, Wu EL et al (2016) BamA POTRA domain interacts with a native lipid membrane surface. Biophys J 110:2698–2709. https://doi.org/10.1016/j.bpj.2016.05.010

128. Sperandeo P, Martorana AM, Polissi A (2017) The lipopolysaccharide transport (Lpt) machinery: a nonconventional transporter for lipopolysaccharide assembly at the outer membrane of Gram-negative bacteria. J Biol Chem 292:17981–17990. https://doi.org/10.1074/jbc.r117.802512

129. Lundquist KP, Gumbart JC (2020) Presence of substrate aids lateral gate separation in LptD. Biochim Biophys Acta Biomembr 1862:183025. https://doi.org/10.1016/j.bbamem.2019.07.013

130. Samsudin F, Ortiz-Suarez ML, Piggot TJ et al (2016) OmpA: a flexible clamp for bacterial cell wall attachment. Structure 24:2227–2235. https://doi.org/10.1016/j.str.2016.10.009

131. Samsudin F, Boags A, Piggot TJ, Khalid S (2017) Braun's lipoprotein facilitates OmpA interaction with the Escherichia coli cell wall. Biophys J 113:1496–1504. https://doi.org/10.1016/j.bpj.2017.08.011

132. Hwang H, Paracini N, Parks JM et al (2018) Distribution of mechanical stress in the Escherichia coli cell envelope. Biochim Biophys Acta Biomembr 1860:2566–2575. https://doi.org/10.1016/j.bbamem.2018.09.020

133. Pabis A, Risso VA, Sanchez-Ruiz JM, Kamerlin SC (2018) Cooperativity and flexibility in enzyme evolution. Curr Opin Struct Biol 48: 83–92. https://doi.org/10.1016/j.sbi.2017.10.020

134. Ramanathan A, Ma H, Parvatikar A, Chennubhotla SC (2021) Artificial intelligence techniques for integrative structural biology of intrinsically disordered proteins. Curr Opin Struct Biol 66:216–224. https://doi.org/10.1016/j.sbi.2020.12.001

135. Lopes PEM, Guvench O, MacKerell AD (2014) Molecular modeling of proteins. Methods Mol Biol 1215:47–71. https://doi.org/10.1007/978-1-4939-1465-4_3

136. Klauda JB (2021) Considerations of recent all-atom lipid force field development. J Phys Chem B 125:5676–5682. https://doi.org/10.1021/acs.jpcb.1c02417

137. Yoo J, Winogradoff D, Aksimentiev A (2020) Molecular dynamics simulations of DNA–DNA and DNA–protein interactions. Curr Opin Struct Biol 64:88–96. https://doi.org/10.1016/j.sbi.2020.06.007

138. Jin J, Pak AJ, Durumeric AEP et al (2022) Bottom-up coarse-graining: principles and perspectives. J Chem Theory Comput 18: 5759–5791. https://doi.org/10.1021/acs.jctc.2c00643

139. Souza PCT, Alessandri R, Barnoud J et al (2021) Martini 3: a general purpose force field for coarse-grained molecular dynamics. Nat Methods 18:382–388. https://doi.org/10.1038/s41592-021-01098-3

140. Graham JA, Essex JW, Khalid S (2017) PyCGTOOL: automated generation of coarse-grained molecular dynamics models from atomistic trajectories. J Chem Inf Model 57:650–656. https://doi.org/10.1021/acs.jcim.7b00096

141. Bereau T, Kremer K (2015) Automated parametrization of the coarse-grained Martini force field for small organic molecules. J Chem Theory Comput 11:2783–2791. https://doi.org/10.1021/acs.jctc.5b00056

142. Empereur-Mot C, Pesce L, Doni G et al (2020) Swarm-CG: automatic parametrization of bonded terms in MARTINI-based coarse-grained models of simple to complex molecules via fuzzy self-tuning particle swarm optimization. ACS Omega 5:32823–32843. https://doi.org/10.1021/acsomega.0c05469

143. Empereur-mot C, Pedersen KB, Capelli R et al (2023) Automatic optimization of lipid models in the Martini force field using SwarmCG. J Chem Inf Model 63:3827–3838. https://doi.org/10.1021/acs.jcim.3c00530

144. Stroh KS, Souza PCT, Monticelli L, Risselada HJ (2003) CGCompiler: Automated Coarse-Grained Molecule Parametrization via Noise-

Resistant Mixed-Variable Optimization. J Chem Theory Comput 19: 8384–8400
145. Hilpert C, Beranger L, Souza PCT et al (2023) Facilitating CG simulations with MAD: the MArtini Database Server. J Chem Inf Model 63:702–710. https://doi.org/10.1021/acs.jcim.2c01375
146. Majewski M, Pérez A, Thölke P et al (2023) Machine learning coarse-grained potentials of protein thermodynamics. Nat Commun 14: 5739. https://doi.org/10.1038/s41467-023-41343-1
147. Arts M, Satorras VG, Huang C-W et al (2023) Two for one: diffusion models and force fields for coarse-grained molecular dynamics. J Chem Theory Comput 19: 6151–6159. https://doi.org/10.1021/acs.jctc.3c00702
148. Hénin J, Lelièvre T, Shirts MR et al (2022) Enhanced sampling methods for molecular dynamics simulations [Article v1.0]. Living J Comput Mol Sci 4:1583. https://doi.org/10.33011/livecoms.4.1.1583
149. Isralewitz B, Gao M, Schulten K (2001) Steered molecular dynamics and mechanical functions of proteins. Curr Opin Struct Biol 11:224–230. https://doi.org/10.1016/s0959-440x(00)00194-9
150. Do P-C, Lee EH, Le L (2018) Steered molecular dynamics simulation in rational drug design. J Chem Inf Model 58:1473–1482. https://doi.org/10.1021/acs.jcim.8b00261
151. Kästner J (2011) Umbrella sampling. Wiley Interdiscip Rev Comput Mol Sci 1:932–942. https://doi.org/10.1002/wcms.66

Chapter 20

Structural Modeling of T9SS Outer Membrane Proteins and Their Complexes

Christian D. Lorenz, Michael A. Curtis, and James A. Garnett

Abstract

The type 9 secretion system (T9SS) is a recently discovered machinery that both transports cargo proteins across the Gram-negative bacterial outer membrane and attaches them to lipopolysaccharides on the extracellular surface. Outer membrane proteins (OMPs) are key components of the T9SS and are involved in both steps. In this chapter, we describe a method for the in silico modeling of T9SS OMPs and their complexes, and model validation. This is useful when the production of recombinant OMPs is difficult, and these protocols can also be applied to OMP complexes outside of the T9SS.

Key words T9SS, Protein secretion, Bacterial envelope, OMPs, Membrane-embedded β-barrel proteins, AlphaFold, AlphaFold multimer, MD simulations

1 Introduction

Bacteria have evolved highly sophisticated mechanisms to transport proteins across their outer membranes, which are tailored to the cargo's functionality. Bacterial secretion systems are prime examples of this sophistication [1], with the type 9 secretion system (T9SS), produced by members of the superphylum Fibrobacteres-Chlorobi-Bacteroidetes [2], being one of the largest and most complex of the 11 that have been identified to date. In *Porphyromonas gingivalis*, a Gram-negative periodontal pathogen, the T9SS is composed of at least 23 components and is responsible for the export of multiple virulence factors (e.g., the gingipain cysteine protease RgpB) to the bacterial surface, and their subsequent delivery into outer membrane vesicles (OMVs) [3–6]. In *Flavobacterium johnsoniae* and *Tannerella forsythia*, the T9SS also provides a mechanism for gliding motility [7] and S-layer biogenesis [8], respectively.

Raffaele Ieva (ed.), *Transmembrane β-Barrel Proteins: Methods and Protocols*, Methods in Molecular Biology, vol. 2778, https://doi.org/10.1007/978-1-0716-3734-0_20,

In *P. gingivalis*, the T9SS is composed of two major structural entities: a *translocon assembly* which drives delivery of cargo proteins from the periplasm across the outer membrane (OM) to the cell surface and an *attachment complex* which tethers the newly exported cargo to an anionic form of lipopolysaccharide (A-LPS) in the outer leaflet of the OM [9, 10]. The translocon assembly, which comprises the major structural proteins PorK, PorL, PorM, and PorN, forms a >1.4-MDa assembly spanning both the inner and outer membranes [11, 12] (Fig. 1). Twelve copies of a $PorL_5PorM_2$ subcomplex are arranged as a ring structure in the inner membrane, where PorL uses proton motive force to drive the rotation of PorM [13–16]. PorM bridges the periplasm and binds to PorK which together with PorN assembles to form a large ~50 nm ring on the periplasmic leaflet of the outer membrane [11, 12]. Eight copies of the translocon pore protein, Sov [12, 17], through which cargo proteins are secreted, are tethered to the PorK/PorN ring through the lipoprotein PorW [18].

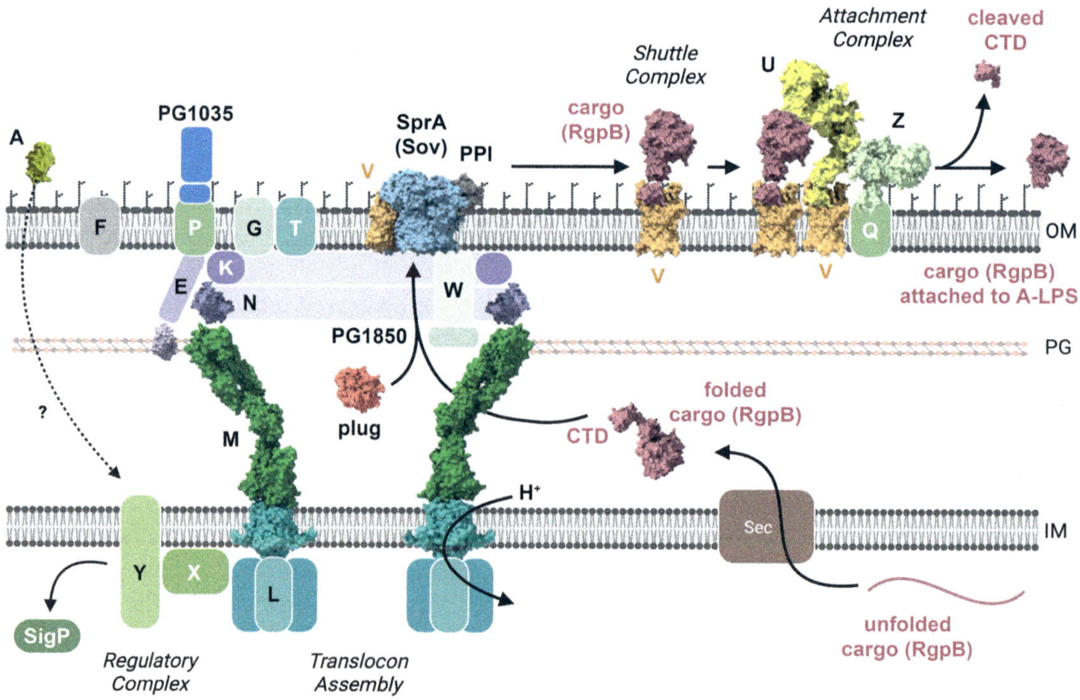

Fig. 1 Model of translocation and A-LPS surface attachment by the *P. gingivalis T9SS*. Structures displayed as surface representation indicate experimental structures or experimentally validated models. Schematics represent known components of the system where structural data is not available. OM: outer membrane; PG: peptidoglycan; IM: inner membrane. Cargo enters the periplasm where they fold and are then transported through the translocon complex and passed onto PorV, which dissociates and forms the shuttle complex. This is targeted to the attachment complex, which processed the cargo, claves the CTD, and covalently attaches it to A-LPS. Regulation of the system is through PorA, PorX, PorY, and SigP. (Created with BioRender.com)

Cargo proteins interact with the T9SS through a conserved Ig-like C-terminal domain (CTD) [2, 19–24]. After entry and folding in the periplasm via the Sec pathway, cargo proteins bind PorM and PorN on the periplasmic surface of the OM [11] before being directed through the Sov pore, which is associated with the β-barrel outer membrane protein (OMP) PorV [17] (Fig. 1). The CTD of the cargo protein within the Sov pore binds to the adjacent PorV, which triggers its release from Sov, thereby positioning the cargo in association with PorV on the external leaflet of the OM. Following release of the PorV:cargo complex from the translocon, Sov is sealed with a periplasmic plug protein, presumably to prevent cellular leakage [14]. The PorV:cargo complex is predicted to act as a shuttle by diffusing across the outer membrane to deliver the cargo to an attachment complex, composed of PorV in association with another OMP, PorQ [25], and PorU and PorZ, which contain CTDs that bind PorV and PorQ, respectively, anchoring them to the bacterial surface [26, 27]. PorZ is thought to deliver lipopolysaccharides (specifically A-LPS) to the sortase PorU, which removes the CTD from the cargo, and then covalently attaches it to A-LPS through a short carbohydrate linker [28–32] thereby securing the cargo to the outer leaflet of the outer membrane. Although other T9SS components have been identified, including several other β-barrel OMPs (i.e., PorF, PorG, PorP, PorT), their function in biogenesis and/or secretion is not yet clear [2, 25].

Over recent years, several experimentally derived T9SS structures have been published, which have significantly pushed forward our mechanistic understanding of type-9 dependent secretion [2, 12]. However, to date all OMP components have either been purified natively from source (i.e., cryo-electron microscopy structures of SprA:PPI:PorV and SprA:PPI:PorV complexes from *F. johnsoniae*; SprA is the Sov homologue) [17] or studied in situ (i.e., cryo-electron tomographic model of the translocon complex in *P. gingivalis*) [12]. Furthermore, we were unable to produce high enough yields of *P. gingivalis* PorV for structural studies using recombinant *Escherichia coli* expression approaches [33], and we speculate whether the folding of T9SS OMPs benefits from the presence of other T9SS components or the addition of other factors not available in *E. coli*. We therefore developed an approach where structural modeling of OMPs and OMP complexes is followed by molecular dynamics (MD) simulations for validation purposes and to probe for any major issues [33]. For complexes, interactions between proteins are then examined using an *E. coli* in cell binding assay, where mutations can be incorporated to further authenticate these models. As an example, we present our structural modeling of monomeric PorV and the PorV:RgpB-CTD shuttle complex and examination of wild-type binding between the two [33]. This is a versatile protocol and can be easily modified to study any other bacterial OMP or OMP complex where the production of recombinant OMP is limited.

2 Materials

2.1 Computational

2.1.1 In Silico Modeling

1. Amino acid sequences for RgpB (GenBank: USI98271.1) and PorV (GenBank: USI97876.1).
2. Computing resource for internet access to AlphaFold2 and AlphaFold2-multimer [34] run through the ColabFold notebook [35] (*see* **Note 1**).

2.1.2 MD Simulations and Analysis

1. Access to a high-performance computing resource (minimally a high-quality graphical processing unit). Here, we used the NMRBox server [36].
2. GROMACS molecular dynamics simulation package [37] (*see* **Note 2**).
3. C [38] and Python [39, 40] programming languages (*see* **Note 3**).
4. MDAnalysis python package [41–43] and CCP4 Programme Suite [44] (*see* **Note 3**).
5. Access to the CHARMM-GUI web server (which requires free registration for members of academic institutions) [45–49] (*see* **Note 4**).
6. Lipid composition of the bilayer environment for the proteins (*see* **Note 5**).

2.2 Bacterial

Prepare all solutions using ultrapure water (18 MΩ-cm at 25 °C) and analytical grade reagents. Prepare and store all reagents at room temperature (unless indicated otherwise).

2.2.1 PorV Expression

1. Purified wild-type RgpB C-terminal domain (RgpB-CTD) [50] (*see* **Note 6**).
2. Chemically competent *E. coli* BL21 (DE3) strain.
3. Wild-type PorV-FLAG plasmid (pOSS*porV-FLAG*) [33] (*see* **Note 7**) and pET28b (control) plasmid (Merck Millipore).
4. Water bath at 42 °C.
5. Autoclaved SOC media: 2% (w/v) tryptone, 0.5% (w/v) yeast extract, 10 mM NaCl, 2.5 mM KCl. Store at −20 °C.
6. Shaking incubator with shelf, at 37 °C.
7. Autoclaved LB media: 1% (w/v) tryptone, 0.5% (w/v) yeast extract, 86 mM NaCl.
8. Autoclaved LB-agar plates: 1% (w/v) tryptone, 0.5% (w/v) yeast extract, 86 mM NaCl, 15% (w/v) agar, 50 μg/mL kanamycin sulfate.
9. 50 mg/mL kanamycin sulfate in water.
10. Spectrophotometer set at 600 nm wavelength.

11. Clear polystyrene construction 1 mL cuvettes with 10 mm light path and suitable for use at 600 nm wavelength.
12. 1 M IPTG (isopropylthio-β-galactoside) in water.
13. Refrigerated centrifuge capable of speeds up to 3000 × *g*.

2.2.2 PorV:RgpB-CTD Surface Binding Assay

1. Autoclaved phosphate-buffered saline (PBS): 137 mM NaCl, 2.7 mM KCl, 10 mM Na_2HPO_4, 1.8 mM KH_2PO_4, pH 7.4.
2. Refrigerated centrifuge capable of speeds up to 1500 × *g*.
3. Monoclonal FLAG antibody, HRP-conjugated.
4. Platform rocker.
5. o-Phenylenediamine dihydrochloride (OPD) solution.
6. 96-well microplate.
7. Microplate reader.
8. Bovine serum albumin (BSA).
9. Autoclaved 10× Tris-buffered saline (TBS): 198 mM Tris–HCl, 1.50 M NaCl, pH 7.6.
10. TBS-Tween (TBST): 1× TBS, 0.1% (v/v) Tween20.
11. Blocking buffer: 3% (w/v) BSA in TBST.
12. Polyclonal rabbit RgpB-CTD antibody.
13. Swine anti-rabbit antibody, HRP-conjugated.
14. Enhanced chemiluminescence (ECL) substrate.
15. Chemiluminescence imaging system.

3 Methods

3.1 Computational

These experiments explain how OMPs and OMP complexes can be modeled and then validated with MD simulations. In addition, analysis is described for the identification of interfacial residues within modeled OMP complexes that can be mutated and tested using the bacterial assays. This is summarized in Fig. 2.

3.1.1 In Silico Modeling

1. Modeling of monomeric PorV and a 1:1 PorV:RgpB-CTD complex (*see* **Note 8**) was carried out with AlphaFold2 or AlphaFold2-multimer [34] using RgpB residues 672–736 and the intact mature sequence (minus signal sequence) of PorV, and run through the ColabFold notebook [35]. Sequence alignments and templates were generated through MMseqs2 [51] and HHsearch [52]. No prior template information was provided, and sequences used during modeling were both paired from the same species and unpaired from multiple sequence alignment (*see* **Note 9**).

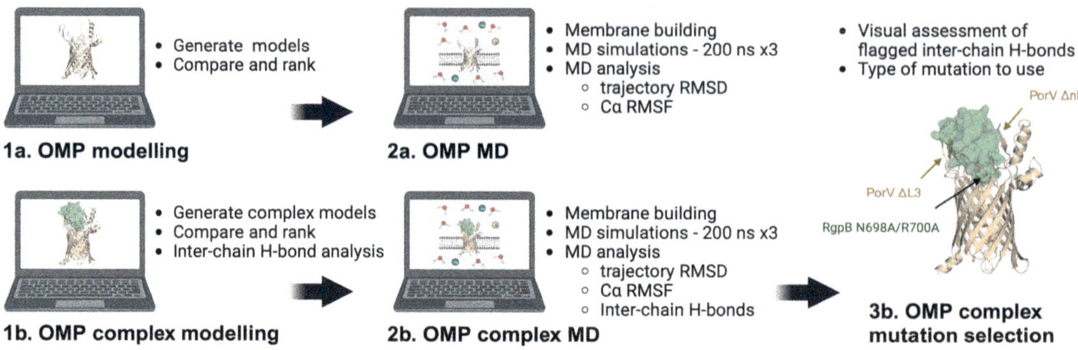

Fig. 2 Overview of the computational experiments. Flow diagram showing progression from in silico modeling, MD and through to the design of mutations. For (a) OMPs alone (e.g., PorV monomer), MD is used by itself to provide validity of the model, whereas for (b) OMP complexes (e.g., PorV:RgpB-CTD), additional analysis of inter-chain H-bonding provides a route for further model scrutiny with in cell binding assays. (Created with BioRender.com)

2. Model quality was assessed before proceeding to MD simulations [33]. As a model confidence cut-off, we only accept models with local Distance Difference Test (lDDT) scores for structured regions >90% and Predicted Aligned Error (PAE) scores <10 between C_α atoms in structured regions within the same chain. For OMP complexes, in addition to above, we only accepted models with an interface predicted template modeling score (ipTM) score > 0.75 [53], and PAE scores <10 between C_α atoms across the interface. The top-ranked model was then taken forward for MD simulations.

3.1.2 MD Simulations

1. Using the top-ranked AlphaFold2 models of monomeric PorV and the PorV:RgpB-CTD complex, the Positioning of Proteins and Membranes (PPM) server [54] was used to determine the optimal rotation and position of each within a lipid membrane.
2. We then fed the output from this server into the CHARMM-GUI membrane builder [45–49] to build the system that we wished to simulate. For these simulations, we constructed a lipid membrane that consisted of 100 lipids per leaflet, of which 80% were 1-palmitoyl-2-oleoyl-sn-glycero-3-phosphocholine (POPC) lipids and 20% 1-palmitoyl-2-oleoyl-sn-glycero-3-phosphoethanolamine (POPE) lipids. This composition was chosen as it is consistent with that used in the formation of nanodiscs as well as in simulations of β-barrel proteins from Gram-negative bacteria [55–57]. After creating this membrane, inserting the protein complex within it, and hydrating the membrane with approximately 2.5 nm of 150 mM NaCl aqueous solution on either side of the membrane, both initial systems had dimensions of 9 nm × 9 nm × 13 nm. We then

downloaded all of the simulation files required to perform the simulations using the GROMACS molecular dynamics engine [37].

3. Then using the NMRbox server [36], we carried out the series of simulations that are prescribed by the CHARMM-GUI server [49] in order to equilibrate the system at a temperature of 310 K and a pressure of 1 bar.
4. After equilibrating the system, we carried out three independent replica production simulations for 200 ns each. These replicas allow us to gather statistically relevant results.

3.1.3 MD Analysis

1. For both monomeric PorV and the ProV:RgpB-CTD complex, we first calculated the backbone Root Mean Squared Deviation (RMSD) over the course of the simulations for all replicas, using in-house Python scripts which employ the Python package, MDAnalysis [41–43], to ensure that the systems are stable and that the models are reliable. In addition, the Root Mean Square Fluctuation (RMSF) of C_α atoms was also analyzed to assess localized dynamics. As expected, all significant motion was observed in loop regions.
2. In order to understand the interactions that stabilize the complex formed by PorV and RgpB-CTD, we used the program CONTACT [44] to measure the intermolecular hydrogen bonds formed between the two proteins in the five top-ranking AlphaFold2 models of the complex. In this case, we defined a hydrogen bond as one in which the hydrogen bond donor and acceptor are within 3.5 Å of one another, and the donor-hydrogen-acceptor angle is between 90° and 120°.
3. To understand the interactions that stabilize the complexes during the three different independent replica simulations, we also measured the intermolecular hydrogen bonding using in-house Python scripts, which employ the Python package, MDAnalysis [41–43]. The same distance and angle criteria were used to identify that a given interaction was a hydrogen bond as in 1.
4. The occurrence of the various hydrogen bonds that were identified was then averaged across all the models in 1 and across the three replicas in 2. Then the frequency of each hydrogen bond occurring was calculated (between 0.0 and 1.0). Occupancies with values less than 0.08 were not included in any subsequent analysis. Based on this analysis, individual residues or loop regions that were considered important in contributing to the stability of the PorV:RgpB-CTD complex were selected for mutation and examined using our in cell assay described below.

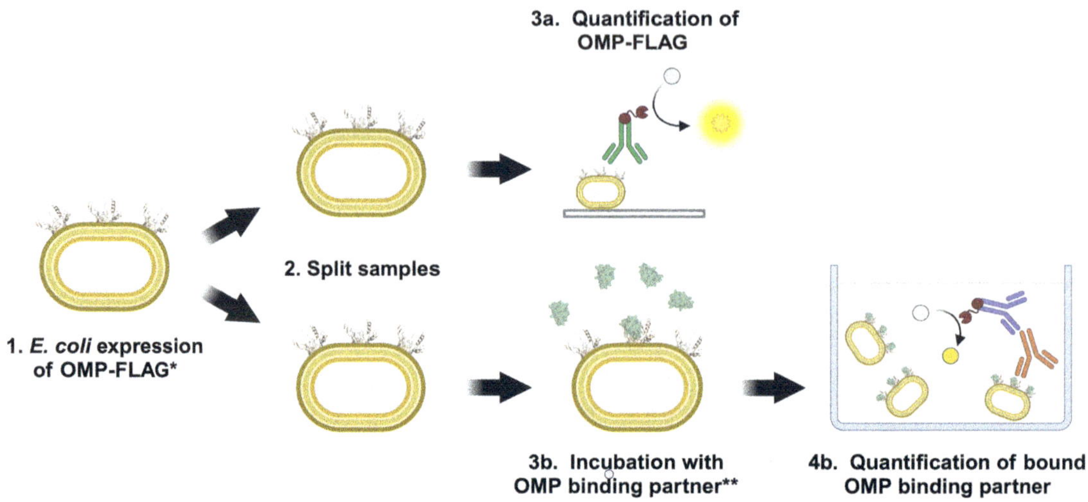

Fig. 3 Overview of the bacterial experiments. Flow diagram showing progression from OMP expression through to OMP quantification (e.g., PorV) and subsequent binding analysis of the binding partner (e.g., PorV-CTD). * In the presented method, only wild-type PorV is expressed, but constructs carrying mutations can also be expressed here, designed based on outputs from the computation experiments. ** Again, in the presented method, only wild-type RgpB-CTD is used to assess binding to wild-type PorV; however, RgpB constructs carrying point mutations or deletions can also be assessed here, designed based on outputs from the computation experiments. (Created with BioRender.com)

3.2 Bacterial

These experiments allow further authentication of OMP complex structures using an in cell assay that can assess the impact of site-specific mutations. While the experiments outlined here describe the examination of interactions between wild-type proteins, modified recombinant proteins and OMP constructs can also be used that incorporate the desired mutations identified from the above approach. This is summarized in Fig. 3. All procedures should be carried out at room temperature unless otherwise specified.

3.2.1 PorV Expression

1. Transfer two 20 μL aliquots of BL21 (DE3) cells from the −80 °C freezer to ice and defrost with the pOSS*porV-FLAG* and pET28b (control) plasmids for 5 min. Pipette ~10 ng of plasmid into the cells and incubate on ice for 30 min.
2. Transfer the cells to a 42 °C water bath for exactly 45 s and then onto ice for 2 min.
3. Add 200 μL of SOC media and then incubate in a shaking incubator (200 rpm) at 37 °C for 1 h.
4. Spread out 10–100 μL of the transformed cell suspensions onto LB-agar plates containing 50 μg/mL kanamycin under aseptic conditions and incubate statically at 37 °C overnight.
5. Select a colony and inoculate 5 mL LB media containing 50 μg/mL kanamycin. Incubate the cells in a shaking incubator (200 rpm) overnight at 37 °C.

6. Inoculate 50 mL of LB media containing 50 μg/mL kanamycin with 0.5 mL of overnight culture and incubate in a shaking incubator (200 rpm) at 37 °C (*see* **Note 10**). After 1 h, monitor bacterial growth by transferring a 1 mL sample to a cuvette and measure the optical density spectroscopically at 600 nm (OD_{600nm}). Once an OD_{600nm} of 0.6 has been reached, induce expression by adding IPTG to a final concentration of 0.5 mM, and then incubate with shaking for 4 h at 30 °C.
7. Measure the OD_{600nm} and volume of the cultures and then harvest the cells by centrifugation at 3000 × *g*, 4 °C for 15 min. Discard the supernatant and keep the cell pellets on ice.

3.2.2 PorV:RgpB-CTD Surface Binding Assay

1. Wash the cell pellets by resuspending them in 10 mL of PBS and then centrifuge at 1500 × *g*, 4 °C for 5 min.
2. Resuspend the cells in 2 mL of PBS and divide these into two 1 mL samples and keep on ice until needed.
3. To the first sample (from **step 2** of this section, *see* **Note 11**), add 1 μL of HRP-conjugated FLAG antibody (i.e., 1:1000 dilution) and place on a platform rocker at room temperature for 1 h.
4. Centrifuge the cells at 1500 × *g*, 4 °C for 5 min, wash three times with 1 mL PBS, and then resuspend in 1 mL OPD solution and mix at room temperature for 20 min.
5. Centrifuge the cells at 1500 × *g*, 4 °C for 5 min and then transfer 100 μL of each supernatant to a 96-well microplate and read the absorbance in a plate reader at 450 nm.
6. Significant expression is established using a Two-way ANOVA test comparing to the control plasmid.
7. Split the second sample (from **step 2** of this section, *see* **Note 12**) into two further 500 μL samples, incubate these with either 50 μM of purified RgpB-CTD or BSA (control) in PBS, and place on a platform rocker at room temperature for 1.5 h.
8. Centrifuge the cells at 1500 × *g*, 4 °C for 5 min, wash three times with 0.5 mL PBS, and then resuspend in 100 μL PBS.
9. Spot 5 μL of each sample onto a nitrocellulose membrane and allow to dry. Add blocking buffer and place on a platform rocker at room temperature for 2 h.
10. Wash the membrane once with TBST buffer and then add the primary RgpB-CTD antibody (1:1000 dilution) in TBST buffer and incubate on a platform rocker overnight at 4 °C.
11. Wash the membrane three times in TBST buffer and then add secondary swine anti-rabbit antibody (1:2500 dilution) in TBST buffer and incubate on a platform rocker at room temperature for 2 h.

12. Wash the membrane once in TBST buffer and then incubate for 5 min in ECL before detection by enhanced chemiluminescence.
13. ImageJ [58] can be used to quantify the highest intensity value of each dot. Values were adjusted based on OD_{450nm} values per 1 mL culture (*see* **Note 11**).

4 Notes

1. While we use AlphaFold2 and AlphaFold2-multimer [34] here, other approaches are available (e.g., RoseTTAFold [59]) and the outputs of different methods should be compared to assess, which represents the most accurate modeling [33].
2. While we use the GROMACS molecular dynamics simulation package [37], there are other codes that could be used instead including LAMMPS [60], AMBER [61], NAMD [62], OpenMM [63], and DL-POLY [64].
3. For our analysis of the simulations, we have chosen to use the CONTACT [44] and MDAnalysis [41–43] computer programs, which require to have the C and Python computer languages installed on your computer, respectively. As identifying a hydrogen bond within computer-based models of molecules is a very common analysis, there are a large variety of other tools that one might use (e.g., VMD [65]), which have other requirements.
4. We have used the CHARMM-GUI web server [45–49] to build our initial systems for the simulations we conducted. While CHARMM-GUI has a lot of functionality ranging from system preparation to the generation of the input files required to run the simulations, one could use other tools to generate systems of proteins embedded within a lipid environment including PACKMOL [66].
5. In this study, we used a two-component lipid membrane as a model of the lipid environment in which our proteins are found, it is possible to produce much more complex lipid environments (e.g., those found in [67, 68]), one just needs to input the desired number of each lipid component or the ratio of the various lipid compounds.
6. RgpB constructs were created as described previously [50]. In brief, wild-type RgpB C-terminal domain (RgpB-CTD; residues 662–736) was produced by PCR and cloned into pET46 Ek/LIC vector (Novagen). Engineered constructs with point mutations can also be assessed with this method, which we had synthesized commercially and cloned into the pET28b plasmid [33].

7. The plasmid pOSS*porV-FLAG* was created as described previously [33]. In brief, the intact mature PorV sequence (minus the *P. gingivalis* periplasmic signal sequence) was commercially synthesized with an *E. coli* OmpA signal sequence at its N-terminus for targeting to the outer membrane, and an internal dual FLAG sequence in the L9 loop for detection. This loop was used because our modeling suggested it was not involved in the formation of the PorV:RgpB-CTD complex. The position of the FLAG-tag should be assessed for each system being investigated. Engineered constructs with point mutations can also be assessed with this method, which again we had synthesized commercially.
8. The stoichiometry of the complex should be assessed biochemically/biophysically if enough material is available. Here, we used size exclusion chromatography and cross-linking approaches.
9. It may be necessary to modify other parameters if unrealistic complexes are output, for example, decreasing *max_msa*, increasing *num_seeds*, and/or using the *drop_outs* option.
10. All samples were grown in triplicate and experiments were repeated at least twice.
11. The first suspension of cells is used to quantify the level of PorV-FLAG that has been expressed and is displayed in the *E. coli* outer membrane. This is then used to normalize the level RgpB-CTD binding to the second sample.
12. The second suspension of cells is used to detect interactions with purified RgpB-CTD proteins.

References

1. Maphosa S, Moleleki LN, Motaung TE (2023) Bacterial secretion system functions: evidence of interactions and downstream implications. Microbiology (Reading) 169(4):001326. https://doi.org/10.1099/mic.0.001326
2. Gorasia DG, Veith PD, Reynolds EC (2020) The type IX secretion system: advances in structure, function and organisation. Microorganisms 8(8):1173. https://doi.org/10.3390/microorganisms8081173
3. Rangarajan M, Aduse-Opoku J, Hashim A et al (2017) LptO (PG0027) is required for lipid A 1-phosphatase activity in Porphyromonas gingivalis W50. J Bacteriol 199(11):e00751-16. https://doi.org/10.1128/JB.00751-16
4. Glew MD, Veith PD, Peng B et al (2012) PG0026 is the C-terminal signal peptidase of a novel secretion system of Porphyromonas gingivalis. J Biol Chem 287(29): 24605–24617. https://doi.org/10.1074/jbc.M112.369223
5. Chen YY, Peng B, Yang Q et al (2011) The outer membrane protein LptO is essential for the O-deacylation of LPS and the co-ordinated secretion and attachment of A-LPS and CTD proteins in Porphyromonas gingivalis. Mol Microbiol 79(5):1380–1401. https://doi.org/10.1111/j.1365-2958.2010.07530.x
6. Veith PD, Glew MD, Gorasia DG et al (2022) The type IX secretion system and its role in bacterial function and pathogenesis. J Dent Res 101(4):374–383. https://doi.org/10.1177/00220345211051599
7. McBride MJ, Zhu Y (2013) Gliding motility and Por secretion system genes are widespread among members of the phylum bacteroidetes. J Bacteriol 195(2):270–278. https://doi.org/10.1128/JB.01962-12

8. Tomek MB, Neumann L, Nimeth I et al (2014) The S-layer proteins of Tannerella forsythia are secreted via a type IX secretion system that is decoupled from protein O-glycosylation. Mol Oral Microbiol 29(6): 307–320. https://doi.org/10.1111/omi.12062
9. Shoji M, Ratnayake DB, Shi Y et al (2002) Construction and characterization of a nonpigmented mutant of Porphyromonas gingivalis: cell surface polysaccharide as an anchorage for gingipainsThe GenBank/EMBL/DDBJ accession number for the sequences reported in this paper is D64132. Microbiology 148(4): 1183–1191. https://doi.org/10.1099/00221287-148-4-1183
10. Rangarajan M, Aduse-Opoku J, Paramonov N et al (2008) Identification of a second lipopolysaccharide in Porphyromonas gingivalis W50. J Bacteriol 190(8):2920–2932. https://doi.org/10.1128/JB.01868-07
11. Vincent MS, Chabalier M, Cascales E (2018) A conserved motif of Porphyromonas type IX secretion effectors C-terminal secretion signal specifies interactions with the PorKLMN core complex. bioRxiv:483123. https://doi.org/10.1101/483123
12. Song L, Perpich JD, Wu C et al (2022) A unique bacterial secretion machinery with multiple secretion centers. Proc Natl Acad Sci U S A 119(18):e2119907119. https://doi.org/10.1073/pnas.2119907119
13. Vincent MS, Canestrari MJ, Leone P et al (2017) Characterization of the Porphyromonas gingivalis type IX secretion trans-envelope PorKLMNP core complex. J Biol Chem 292(8):3252–3261. https://doi.org/10.1074/jbc.M116.765081
14. Leone P, Roche J, Vincent MS et al (2018) Type IX secretion system PorM and gliding machinery GldM form arches spanning the periplasmic space. Nat Commun 9(1):429. https://doi.org/10.1038/s41467-017-02784-7
15. Sato K, Okada K, Nakayama K et al (2020) PorM, a core component of bacterial type IX secretion system, forms a dimer with a unique kinked-rod shape. Biochem Biophys Res Commun 532(1):114–119. https://doi.org/10.1016/j.bbrc.2020.08.018
16. Hennell James R, Deme JC, Kjær A et al (2021) Structure and mechanism of the proton-driven motor that powers type 9 secretion and gliding motility. Nat Microbiol 6(2): 221–233. https://doi.org/10.1038/s41564-020-00823-6
17. Lauber F, Deme JC, Lea SM et al (2018) Type 9 secretion system structures reveal a new protein transport mechanism. Nature 564(7734):77–82. https://doi.org/10.1038/s41586-018-0693-y
18. Gorasia DG, Lunar Silva I, Butler CA et al (2022) Protein interactome analysis of the type IX secretion system identifies PorW as the missing link between the PorK/N ring complex and the Sov translocon. Microbiol Spectr 10(1):e0160221. https://doi.org/10.1128/spectrum.01602-21
19. Lasica AM, Ksiazek M, Madej M et al (2017) The type IX secretion system (T9SS): highlights and recent insights into its structure and function. Front Cell Infect Microbiol 7: 215. https://doi.org/10.3389/fcimb.2017.00215
20. Shoji M, Sato K, Yukitake H et al (2011) Por secretion system-dependent secretion and glycosylation of Porphyromonas gingivalis hemin-binding protein 35. PLoS One 6(6):e21372. https://doi.org/10.1371/journal.pone.0021372
21. Seers CA, Slakeski N, Veith PD et al (2006) The RgpB C-terminal domain has a role in attachment of RgpB to the outer membrane and belongs to a novel C-terminal-domain family found in Porphyromonas gingivalis. J Bacteriol 188(17):6376–6386. https://doi.org/10.1128/JB.00731-06
22. Veith PD, Nor Muhammad NA, Dashper SG et al (2013) Protein substrates of a novel secretion system are numerous in the Bacteroidetes phylum and have in common a cleavable C-terminal secretion signal, extensive post-translational modification, and cell-surface attachment. J Proteome Res 12(10): 4449–4461. https://doi.org/10.1021/pr400487b
23. Kulkarni SS, Zhu Y, Brendel CJ et al (2017) Diverse C-terminal sequences involved in Flavobacterium johnsoniae protein secretion. J Bacteriol 199(12):e00884-16. https://doi.org/10.1128/JB.00884-16
24. Nguyen KA, Travis J, Potempa J (2007) Does the importance of the C-terminal residues in the maturation of RgpB from Porphyromonas gingivalis reveal a novel mechanism for protein export in a subgroup of Gram-negative bacteria? J Bacteriol 189(3):833–843. https://doi.org/10.1128/JB.01530-06
25. Veith PD, Gorasia DG, Reynolds EC (2021) Towards defining the outer membrane proteome of Porphyromonas gingivalis. Mol Oral Microbiol 36(1):25–36. https://doi.org/10.1111/omi.12320
26. Glew MD, Veith PD, Chen D et al (2017) PorV is an outer membrane shuttle protein for the type IX secretion system. Sci Rep 7(1):

8790. https://doi.org/10.1038/s41598-017-09412-w
27. Gorasia D, Chreifi G, Seers C et al (2020) In situ structure and organisation of the type IX secretion system. bioRxiv:2020.2005.2013.094771. https://doi.org/10.1101/2020.05.13.094771
28. Lasica AM, Goulas T, Mizgalska D et al (2016) Structural and functional probing of PorZ, an essential bacterial surface component of the type-IX secretion system of human oral-microbiomic Porphyromonas gingivalis. Sci Rep 6:37708. https://doi.org/10.1038/srep37708
29. Veith PD, Shoji M, O'Hair RAJ et al (2020) Type IX secretion system cargo proteins are glycosylated at the C terminus with a novel linking sugar of the Wbp/Vim pathway. mBio 11(5):e01497-20. https://doi.org/10.1128/mBio.01497-20
30. Mizgalska D, Goulas T, Rodriguez-Banqueri A et al (2021) Intermolecular latency regulates the essential C-terminal signal peptidase and sortase of the Porphyromonas gingivalis type-IX secretion system. Proc Natl Acad Sci U S A 118(40):e2103573118. https://doi.org/10.1073/pnas.2103573118
31. Madej M, Nowakowska Z, Ksiazek M et al (2021) PorZ, an essential component of the type IX secretion system of Porphyromonas gingivalis, delivers anionic lipopolysaccharide to the PorU sortase for transpeptidase processing of T9SS cargo proteins. mBio 12(1):e02262-20. https://doi.org/10.1128/mBio.02262-20
32. Gorasia DG, Veith PD, Chen D et al (2015) Porphyromonas gingivalis type IX secretion substrates are cleaved and modified by a sortase-like mechanism. PLoS Pathog 11(9):e1005152. https://doi.org/10.1371/journal.ppat.1005152
33. Dorgan B, Liu Y, Wang S et al (2022) Structural model of a Porphyromonas gingivalis type IX secretion system shuttle complex. J Mol Biol 434(23):167871. https://doi.org/10.1016/j.jmb.2022.167871
34. Bryant P, Pozzati G, Elofsson A (2022) Improved prediction of protein-protein interactions using AlphaFold2. Nat Commun 13(1):1265. https://doi.org/10.1038/s41467-022-28865-w
35. Mirdita M, Schütze K, Moriwaki Y et al (2021) ColabFold – making protein folding accessible to all. bioRxiv:2021.2008.2015.456425. https://doi.org/10.1101/2021.08.15.456425
36. Maciejewski MW, Schuyler AD, Gryk MR et al (2017) NMRbox: a resource for biomolecular NMR computation. Biophys J 112(8):1529–1534. https://doi.org/10.1016/j.bpj.2017.03.011
37. Abraham MJ, Murtola T, Schulz R et al (2015) GROMACS: high performance molecular simulations through multi-level parallelism from laptops to supercomputers. SoftwareX 1–2:19–25. https://doi.org/10.1016/j.softx.2015.06.001
38. Kernighan BW, Ritchie DM (1988) The C programming language. Prentice Hall, Englewood Cliffs
39. Rossum G (1995) Python reference manual. CWI (Centre for Mathematics and Computer Science), Amsterdam
40. Rossum GV, Drake FL (2009) Python 3 reference manual. CreateSpace, Scotts Valley
41. Gowers RJ, Linke M, Barnoud J et al (2016) MDAnalysis: a Python package for the rapid analysis of molecular dynamics simulations. In: Proceedings of the 15th Python in science conference (SciPy 2016), pp 98–105
42. Michaud-Agrawal N, Denning EJ, Woolf TB et al (2011) MDAnalysis: a toolkit for the analysis of molecular dynamics simulations. J Comput Chem 32(10):2319–2327. https://doi.org/10.1002/jcc.21787
43. Smith P, Ziolek RM, Gazzarrini E et al (2019) On the interaction of hyaluronic acid with synovial fluid lipid membranes. Phys Chem Chem Phys 21(19):9845–9857. https://doi.org/10.1039/c9cp01532a
44. Collaborative Computational Project Number (1994) The CCP4 suite: programs for protein crystallography. Acta Crystallogr D Biol Crystallogr 50(Pt 5):760–763. https://doi.org/10.1107/S0907444994003112
45. Jo S, Kim T, Iyer VG et al (2008) CHARMM-GUI: a web-based graphical user interface for CHARMM. J Comput Chem 29(11):1859–1865. https://doi.org/10.1002/jcc.20945
46. Jo S, Kim T, Im W (2007) Automated builder and database of protein/membrane complexes for molecular dynamics simulations. PLoS One 2(9):e880. https://doi.org/10.1371/journal.pone.0000880
47. Jo S, Lim JB, Klauda JB et al (2009) CHARMM-GUI Membrane Builder for mixed bilayers and its application to yeast membranes. Biophys J 97(1):50–58. https://doi.org/10.1016/j.bpj.2009.04.013
48. Wu EL, Cheng X, Jo S et al (2014) CHARMM-GUI Membrane Builder toward realistic biological membrane simulations. J Comput Chem 35(27):1997–2004. https://doi.org/10.1002/jcc.23702

49. Lee J, Cheng X, Swails JM et al (2016) CHARMM-GUI input generator for NAMD, GROMACS, AMBER, OpenMM, and CHARMM/OpenMM simulations using the CHARMM36 additive force field. J Chem Theory Comput 12(1):405–413. https://doi.org/10.1021/acs.jctc.5b00935
50. Corsini PM, Wang S, Rehman S et al (2022) Molecular and cellular insight into Escherichia coli SslE and its role during biofilm maturation. NPJ Biofilms Microbiomes 8(1):9. https://doi.org/10.1038/s41522-022-00272-5
51. Steinegger M, Soding J (2017) MMseqs2 enables sensitive protein sequence searching for the analysis of massive data sets. Nat Biotechnol 35(11):1026–1028. https://doi.org/10.1038/nbt.3988
52. Steinegger M, Meier M, Mirdita M et al (2019) HH-suite3 for fast remote homology detection and deep protein annotation. BMC Bioinformatics 20(1):473. https://doi.org/10.1186/s12859-019-3019-7
53. Yin R, Feng BY, Varshney A et al (2022) Benchmarking AlphaFold for protein complex modeling reveals accuracy determinants. Protein Sci 31(8):e4379. https://doi.org/10.1002/pro.4379
54. Lomize MA, Pogozheva ID, Joo H et al (2012) OPM database and PPM web server: resources for positioning of proteins in membranes. Nucleic Acids Res 40(Database issue): D370–D376. https://doi.org/10.1093/nar/gkr703
55. Piggot TJ, Holdbrook DA, Khalid S (2013) Conformational dynamics and membrane interactions of the E. coli outer membrane protein FecA: a molecular dynamics simulation study. Biochim Biophys Acta 1828(2): 284–293. https://doi.org/10.1016/j.bbamem.2012.08.021
56. Hicks R (2017) Preparation of membrane models of Gram-negative bacteria and their interaction with antimicrobial peptides studied by CD and NMR. Methods Mol Biol 1548: 231–245. https://doi.org/10.1007/978-1-4939-6737-7_16
57. Boags A, Hsu PC, Samsudin F et al (2017) Progress in molecular dynamics simulations of Gram-negative bacterial cell envelopes. J Phys Chem Lett 8(11):2513–2518. https://doi.org/10.1021/acs.jpclett.7b00473
58. Schindelin J, Arganda-Carreras I, Frise E et al (2012) Fiji: an open-source platform for biological-image analysis. Nat Methods 9(7): 676–682. https://doi.org/10.1038/nmeth.2019
59. Baek M, DiMaio F, Anishchenko I et al (2021) Accurate prediction of protein structures and interactions using a three-track neural network. Science 373(6557):871–876. https://doi.org/10.1126/science.abj8754
60. Thompson AP, Aktulga HM, Berger R et al (2022) LAMMPS – a flexible simulation tool for particle-based materials modeling at the atomic, meso, and continuum scales. Comput Phys Commun 271:108171. https://doi.org/10.1016/j.cpc.2021.108171
61. Case DA, Belfon K, Ben-Shalom IY, Brozell SR, Cerutti DS, Cheatham TE et al (2023) Amber 2023. University of California, San Francisco
62. Phillips JC, Hardy DJ, Maia JDC et al (2020) Scalable molecular dynamics on CPU and GPU architectures with NAMD. J Chem Phys 153(4):044130. https://doi.org/10.1063/5.0014475
63. Eastman P, Swails J, Chodera JD et al (2017) OpenMM 7: rapid development of high performance algorithms for molecular dynamics. PLoS Comput Biol 13(7):e1005659. https://doi.org/10.1371/journal.pcbi.1005659
64. Todorov IT, Smith W, Trachenko K et al (2006) DL_POLY_3: new dimensions in molecular dynamics simulations via massive parallelism. J Mater Chem 16(20): 1911–1918. https://doi.org/10.1039/B517931A
65. Humphrey W, Dalke A, Schulten K (1996) VMD: visual molecular dynamics. J Mol Graph 14(1):33–38, 27–38. https://doi.org/10.1016/0263-7855(96)00018-5
66. Martinez L, Andrade R, Birgin EG et al (2009) PACKMOL: a package for building initial configurations for molecular dynamics simulations. J Comput Chem 30(13):2157–2164. https://doi.org/10.1002/jcc.21224
67. Yee SM, Gillams RJ, McLain SE et al (2021) Effects of lipid heterogeneity on model human brain lipid membranes. Soft Matter 17(1): 126–135. https://doi.org/10.1039/D0SM01766C
68. Pogozheva ID, Armstrong GA, Kong L et al (2022) Comparative molecular dynamics simulation studies of realistic eukaryotic, prokaryotic, and archaeal membranes. J Chem Inf Model 62(4):1036–1051. https://doi.org/10.1021/acs.jcim.1c01514

Chapter 21

Rationale in Custom Design of Transmembrane β-Barrel Pores

Anastassia A. Vorobieva

Abstract

Biological nanopores incorporated into synthetic membranes are widely used for single-molecule analytical applications such as DNA sequencing. The ability to engineer custom membrane proteins with a pore would allow the generation of a multitude of nanopores for the sensing/sequencing of small molecules and (bio)polymers. The de novo design of transmembrane β-barrel pores has recently enabled the generation of nanopores with custom size, shape, and properties. In this chapter, I describe the rationale of transmembrane β-barrel design and computational methods to assemble the backbones, design sequences, and select the designs for experimental validation.

Key words Protein design, Protein folding, Pores, Transmembrane β-barrels, Rosetta

1 Introduction

Biological nanopores are emerging as powerful tools for the single-molecule analysis of complex samples and for the sequencing of biopolymers. Nanopore sequencing technologies rely on these tiny biological transmembrane channels to sequence DNA [1] or detect epigenetic markers (e.g., DNA methylation [2]). As the DNA molecule is translocated through the pore under an electric current, it causes partial obstruction of the channel. The intensity and dwell time of these current blockades are deconvolved to obtain the nucleotide sequence [3]. Nanopores have also been used to sequence synthetic polymers [4], fingerprint peptides [5], and study protein folding pathways [6] and enzyme turn-over [7] at the single-molecule level. To be suitable for analytical applications, nanopores must form channels with highly reproducible conductance properties and must be stable over time. Therefore, most proteins used in the nanopore industry are transmembrane β-barrels (TMBs). Formed from one β-sheet closing on itself to

Raffaele Ieva (ed.), *Transmembrane β-Barrel Proteins: Methods and Protocols*, Methods in Molecular Biology, vol. 2778,
https://doi.org/10.1007/978-1-0716-3734-0_21,

form a pore, they offer rigid scaffolds for stable and reproducible transport of molecules across the membrane.

To generate a detectable current change, the analyte of interest must be temporarily retained in the transmembrane channels through specific or nonspecific interactions [8]. The size, the shape, and the chemical properties of the transmembrane channel are therefore key parameters that define analyte capture, detection, and specificity. Fine-tuning these parameters in re-purposed naturally occurring TMB nanopores remains challenging, and the number of engineerable natural nanopores is limited. Unprecedented control over nanopore structure can instead be achieved by designing the proteins from first principles (de novo). De novo design starts with the explicit assembly of a protein backbone with the desired shape, which is used to guide the design of a compatible amino acid sequence [9]. No a priori knowledge of natural sequences is required, and this approach can, in principle, be used to design synthetic TMBs with completely novel properties and membrane preference. The de novo design of proteins folding into lipid membranes builds on quite different principles from those of proteins folding in water. For TMB nanopores, the challenge consists in (i) assembling backbones with native-like properties (designable) [10] and (ii) generating amino acid sequences incorporating negative design against strong secondary structure propensity [11].

This chapter describes the full computational pipeline for the de novo design of TMB nanopores with Rosetta. A strategy to assemble and customize the structure of TMB backbones by arranging the position of structural motifs is described, together with principles to design TMB sequences. Using this pipeline, TMB nanopores with 8–14 β-strands [11] have been designed [12] with an experimental success rate of 10–50%. It is important to note that many computational methods for protein design exist. Rosetta was selected here because it builds on a molecular model described with a combination of physical parameters. It provides a flexible and intuitive interface to implement otherwise challenging negative design principles. However, the comprehensive description of the design rationale and principles herein should allow the implementation of TMB design in other, present or future, computational models.

2 Materials

Scripts to run the demo are provided in the GitHub repository https://github.com/vorobieva/demo_TMB_design. To run the computational pipeline, we recommend having access to clusters with 200–350 CPU cores available for the calculations. The

Rosetta molecular modeling suite [13] is a molecular modeling software that can be applied to a multitude of design problems. The software is free to use and can be obtained by visiting https://www.rosettacommons.org/software/license-and-download. It can be compiled according to instructions in the documentation (https://www.rosettacommons.org/docs/latest/). The pipeline presented in this chapter is built on the Rosetta scripts interface. Once the software is compiled, a rosetta_script executable should be present in the Rosetta/main/source/bin directory. The protocol also requires Pyrosetta for data analysis, which can be obtained from https://www.pyrosetta.org/downloads. We recommend installing Pyrosetta in a python3 virtual environment together with matplotlib, pandas, and seaborn. Setting up a second virtual environment (Biopython and SciPy packages) is recommended for running the generate_blueprint.py script. Other software requirements include RaptorX [14] property prediction (http://raptorx.uchicago.edu/download/), Tango aggregation [15] prediction (http://tango.crg.es/), and the ColabFold [16] AlphaFold [17] notebook.

3 Methods

3.1 Blueprint Design of the β-Barrel Backbones

3.1.1 Define Backbone Architecture

The design starts with defining the desired geometric properties of the β-barrel backbones. β-barrels are characterized by the total number of strands (n) and the shear number (S), which is a numerical expression of the connectivity between the β-strands (Fig. 1a, b). It is the total number of residue shifts (register shifts) between the first and last strands (*see* **Note 1**). Together, n and S define the theoretical outside diameter (the real size of the pore varies with the shape and the sidechains packing), the staggering angle of the β-strands to the transmembrane axis (θ, Eq. 1), and the arrangement of sidechains in the pore (number of Cβ-stripes, which equals to ½ S) [18]. Table 1 can be used to select the n and S parameters based on the desired geometric properties.

3.1.2 Transmembrane Span

The transmembrane span of the designed TMB (Z) needs to match the hydrophobic thickness of the membrane (Fig. 1b). It is calculated using Eq. 2, where d is the average distance between two residues along a β-strands (3.3 Å), z is the number of residues per β-strand, and θ is the strands stagger angle. For de novo design, Z is defined based on the desired membrane thickness, and Eq. 2 can be used to calculate the number of residues per β-strand (z) necessary to span Z. The angle θ is calculated with Eq. 1 based on n, S, and D (the average distance between two β-strands) [18]. Most natural TMBs folding in the outer membrane of *E. coli* have a hydrophobic transmembrane span Z of approximately 24 Å [20].

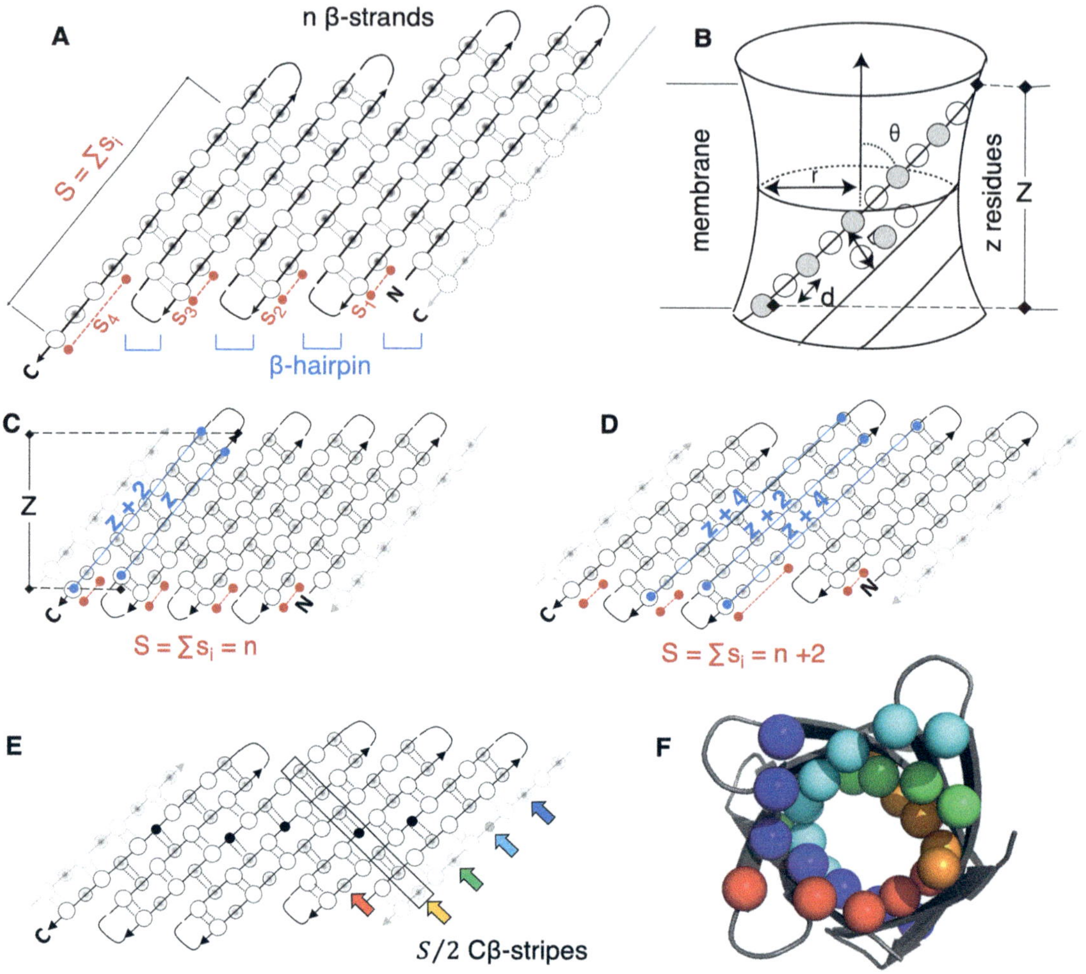

Fig. 1 Geometric parameters to consider for the assembly of designable TMB backbones. (**a**, **b**) Definition of the number of strands, the shear number, and the transmembrane span. (**c**–**f**) Stepwise design of the TMB blueprint. Cβ-stripes are highlighted with colored arrows (**e**) and mapped on a 3D β-barrel structure (**f**)

$$\tan(\theta) = \frac{Sd}{nD} \tag{1}$$

$$Z = zd\ \cos(\theta) \tag{2}$$

3.1.3 Blueprint Generation

The parameters calculated above are used to generate a Rosetta blueprint representation of the backbone. First, the lengths of individual strands must be adjusted around the average calculated span z according to the geometric and β-strand connectivity constraints inherent to the β-barrel fold. The β-strands in TMBs are almost exclusively antiparallel and organized into pairs—the β-hairpins (Fig. 1a). To account for the staggered β-hairpins, the lengths of the odd and even β-strands in the blueprint are set to z and $z + 2$ residues, respectively (Fig. 1c).

Table 1
Number of Cβ-strips, outside radius (*r*) of the β-barrel, and staggering angle of the β-strands to the membrane axis (*θ*) as a function of the number of strands (*n*) and the shear number (*S*). The values were calculated as described in [19]

Number of strands (*n*)	Shear number (*S*)	*r* (Å)	*θ* (°)	Number of core Cβ-strips
8	8	7.3	36.3	4
8	10	8.0	42.5	5
10	12	9.7	41.3	6
12	12	10.8	36.3	6
12	14	11.4	40.5	7
12	16	12.2	44.4	8
12	18	12.9	47.7	9
12	24	15.4	55.7	12
14	14	12.5	36.3	7
14	16	13.2	40.0	8
16	16	14.3	36.3	8
16	18	15.0	39.5	9
16	20	15.6	42.5	10
16	22	16.4	45.2	11
18	18	16.1	36.3	9
18	20	16.7	39.2	10
18	22	17.4	41.9	11
19	20	17.3	37.7	10
22	24	20.2	38.7	12
24	26	22.0	38.5	13
26	30	24.5	40.2	15
36	18	27.5	20.1	9
60	60	53.3	36.3	30

For TMBs with regular β-strands arrangement ($S = n$), each β-hairpin is shifted relative to the previous one by two residues (Fig. 1c). If $S > n$, the register shifts will be unequally distributed between the hairpins (Fig. 1d). An even-numbered β-strand is then chosen to incorporate the additional register shift(s). We recommend choosing a β-strand close to the middle of the β-sheet. Its length and those of the next two β-strands are increased by two

```
1   A   L   .
0   V   EB  R
0   V   EB  R
0   V   EB  R
0   V   EB  R
0   V   EB  R
0   V   EB  R
0   V   EB  R
0   V   EB  R
0   V   EB  R
0   V   EB  R
0   V   LA  R
0   V   LA  R
0   V   LG  R
0   V   EB  R
```

Fig. 2 Example of a short blueprint file

residues (total length: $z + 2$ and $z + 4$ residues for odd and even β-strands, respectively) (Fig. 1d). The result is formatted into a string describing the final lengths of β-strands and loops (the β-strands with the additional register shifts are underlined):

$\mathbf{E_1 10\text{-}L3\text{-}E_2 12\text{-}L2\text{-}E_3 10\text{-}L3\text{-}\underline{E_4 14}\text{-}L2\text{-}\underline{E_5 12}\text{-}L3\text{-}\underline{E_6 14}\text{-}L2\text{-}E_7 10\text{-}L3\text{-}E_8 12}$

Such string is used as input for the generate_blueprint.py script, which produces a blueprint file and the corresponding Rosetta constraints file. Each line in the blueprint describes the structure property of an individual residue in the polypeptide (Fig. 2). For de novo design, all residues are remodeled ("R," fourth column) but the first one – the single alanine residue provided in the input .pdb file. Valine ("V," second column) is used at every position as a placeholder sidechain. The third row describes the desired structure properties. The secondary structure of the residue is defined with a single-letter code (("L")oop, ("E") xtended, ("H")elix). The second (optional) single-letter code ("A," "B," "E," "G," "O") refers to the torsion angles bin of the backbone atoms [21]. It provides a second level of control over the protein structure and the opportunity to introduce irregular structure features. The generate_blueprint.py script will also provide a cst file containing Rosetta constraints describing the backbone-backbone hydrogen bond interactions present in the β-barrel.

A stepwise guide to run the generate_blueprint.py script.

1. Open your command line interface.
2. Load the *generate* virtual environment:

```
#bash script-------------------------------------

-----------
>source
/path_to_virual_envs/generate_blueprint/bin/activate
      #-----------------------------------------
-----------------
```

3. Create a new directory.

```
#bash script-------------------------------------

-----------
>mkdir new_design_directory
>cd new_design_directory
      #-----------------------------------------
----------------
```

4. Modify the *generate_blueprint.py* script to match the desired blueprint and run the script.

Generate_blueprint.py script modification:

```
topol = "E[23]L[3]E[24]L[2]E[22]L[3]E[24]L[2]E[22]L[3]E[26]L[2]E[24]L[3]E[26]L[2]E[22]L[3]E[24]L[2]E[22]L[3]E[24]L[1]" # Copy here the string describing the blueprint, designed as explained above
common_bulges = {'E2': [24], 'E4': [24], 'E6': [26], 'E8': [26], 'E10': [24], 'E12': [24]} #By default one β-bulge residue is added as the last residue of each even-numbered strand
cap = 2 #For transmembrane β-barrels, keep the default value "2".
topology = [('E1','E2'),('E2','E3'),('E3','E4'),('E4','E5'),('E5','E6'),('E6','E7'),('E7','E8'),('E8','E9'),('E9','E10'),('E10','E11'),('E11','E12'),('E12','E1')] #Description of the desired β-strand pairing. Pairing is sequential; finally the last and the first strands are paired.
#bash script-------------------------------------

-----------
>python generate_blueprint.py -xml input.xml -refpdb input.pdb -prefix b1
```

```
#Specifies the xml and pdb inputs. The prefix is an
arbitrary string.
#------------------------------------------
--------------------
```

Expected Result The script generates two text files. The *bp* file contains a blueprint representation of the desired β-barrel backbone (*see* **Note 2**). The *cst* file contains the constraints describing the hydrogen bond connectivity between the β-strands and their geometry.

3.1.4 Sculpting the β-Barrel with Torsional Irregularities

Glycine residues (kinks) in an extended conformation (positive φ angle) are necessary to relieve strain in the curved β-sheet and to generate designable TMB backbones. Currently, their position is defined by editing the corresponding lines in the blueprint output by the generate_blueprint.py script. (V)aline is replaced with (G)-lycine, and the third column is changed to "EE" (β-strand residue with positive φ angle). As a general principle, one kink is incorporated per Cβ-strip (row of sidechains lining the pore in the direction of the hydrogen bonds between β-strands, Fig. 1e). To design an initial solution, we recommend generating 3D backbones, identifying the *S*/2 Cβ-strips facing the pore and placing one kink in the middle of each Cβ-strip (Fig. 1e, f). If the transmembrane span increases, more than one kink per Cβ-strip might be necessary (*see* **Note 3**). The distribution of kinks shapes the pore, since kink positions will define locally bended "corners" in the β-barrel (*see* **Note 3**).

3.1.5 Explicit Assembly of β-Barrel Backbones

The backbones are assembled with BlueprintBDR application [22] using the Rosetta scripts interface. The script takes three inputs: the blueprint (*bp*) and the constraints (*cst*) files generated previously, and an input pdb containing a single alanine amino acid (the starting point of the simulation). The backbone assembly protocol alternates between collecting sequence-free protein fragments from an internal Rosetta database, randomly combining them to fold a coarse-grained representation of the protein (centroid model) and minimizing with the constraints contained in the *cst* file to form all expected β-sheet hydrogen bonds (Fig. 3). The process is repeated for 2000–5000 trajectories (more trajectories might be necessary to build larger TMB pores, *see* **Note 4**).

1. Running the Rosetta script

```
#bash script-------------------------------------

-----------
>/path_to_Rosetta/main/source/bin/rosetta_scripts
```

```
<ROSETTASCRIPTS>
<SCOREFXNS>
<ScoreFunction name="SFXN1" weights="fldsgn_cen_omega02.wts" />
#The weights have been developed to minimize artifacts while minimizing the coarse-
grained backbone model.
#They can be retrieved from /path_to_Rosetta/main/database/scoring/weights/
</SCOREFXNS>

<FILTERS>
<ScoreType name="cen_total" scorefxn="SFXN1" score_type="total_score"
threshold="1000000" />
<ScoreType name="vdw" scorefxn="SFXN1" score_type="vdw" threshold="1000000" />
<ScoreType name="cen_rama" scorefxn="SFXN1" score_type="rama" threshold="1000000" />
#These filters report the centroid scores in the score file output.
</FILTERS>

<TASKOPERATIONS>
</TASKOPERATIONS>

<MOVERS>
<Dssp name="dssp"/>
<SwitchResidueTypeSetMover name="fullatom" set="fa_standard"/>
<SwitchResidueTypeSetMover name="cent" set="centroid"/>
<BluePrintBDR name="bdr1" scorefxn="SFXN1" use_abego_bias="1" blueprint="TMB_blueprint"
constraint_file="hbond_constraints"/>
<ConstraintSetMover name="addcst1" add_constraints="1" cst_file="cst"/>
<MinMover name="min1" scorefxn="SFXN1" chi="1" bb="1"
type="dfpmin_armijo_nonmonotone_atol" tolerance="0.0001"/>

<ParsedProtocol name="cenmin1" >
<Add mover_name="cent" />
<Add mover_name="addcst1" />
<Add mover_name="min1" />
<Add mover_name="fullatom" />
</ParsedProtocol>
<ParsedProtocol name="bdr1ss" >
<Add mover_name="bdr1" />
<Add mover_name="cenmin1" />
<Add mover_name="dssp" />
</ParsedProtocol>
</MOVERS>

<PROTOCOLS>
<Add mover_name="bdr1ss" />
<Add mover_name="fullatom" />
<Add filter_name="cen_total" />
<Add filter_name="vdw" />
<Add filter_name="cen_rama" />
</PROTOCOLS>
</ROSETTASCRIPTS>
```

Fig. 3 Description of the blueprint.xml script. Rosetta scripts documentation can be found here: https:/www.rosettacommons.org/docs/latest/scripting_documentation/Scripting-Documentation

```
-parser:protocol blueprint.xml -s input.pdb
-nstruct 100 -rama_map /path_to_Rosetta/main/data-
base/scoring/score_functions/rama/Rama_XPG_3le-
vel.txt
#The script is often run in parallel in several
directories; 100 structures (nstruct) are generated
per directory for a total of 2000-5000 models. The
```

```
-rama_map specifies smoother Ramachandran statis-
tics to smoothen local minima.
#----------------------------------------
--------------------
```

2. Selecting good backbone for the sequence design stage

   The backbone analysis code can be found in the *backbone-s_analysis.ipynb* Jupyter notebook. Run the first six code blocks to load the model scores into one dataframe and to visualize the distributions of Van der Waals (*vdw*) and torsional scores (*omega*, *cen_rama*). Backbones that do not contain clashes (*vdw* < 1.0) and with torsional scores (*omega*, *cen_rama*) better than 50% of the average distribution are selected for the next design step. In the example below, the average *omega* score of the backbone ensemble is 21.0 and the average *cen_rama* score is 5.0.

```
#python script------------------------------------
-------------
>df = pd.read_csv(score_file, sep='\s+', header=1)
#Load the score into one pandas dataframe
>dpicked = df.loc[(dsc['vdw'] < 1.0) & (dsc['ome-
ga'] < 21.0) & (dsc['cen_rama'] < 5.0)]
#select best-scored backbones based on the score
distributions in the ensemble
>for model in dpicked.description:
>     source = str(folder_id) + "/" + model + ".pdb"
>     target = "best_200/" + model + ".pdb"
>     copyfile(source, target)
#Copy the selected backbones into a new folder named
"best_200"
#----------------------------------------
--------------------
```

Expected Results Around 250 high-quality protein backbones representing different conformations of the same blueprint.

3.2 Placing the Mortise-Tenon Folding Motifs

The mortise-tenon is a YGD/E motif spanning 2–3 β-strands, and it is critical for the correct assembly of the β-barrel (Fig. 4a) [23]. The tyrosine and glycine are positioned opposite to each other on two hydrogen bonded β-strands and facing the pore of the TMB. The aromatic ring of the tyrosine is positioned in a planar manner directly above the two hydrogen bonds formed with the glycine, and it closely interacts with the grove resulting from the missing glycine sidechain. The conformation of the tyrosine

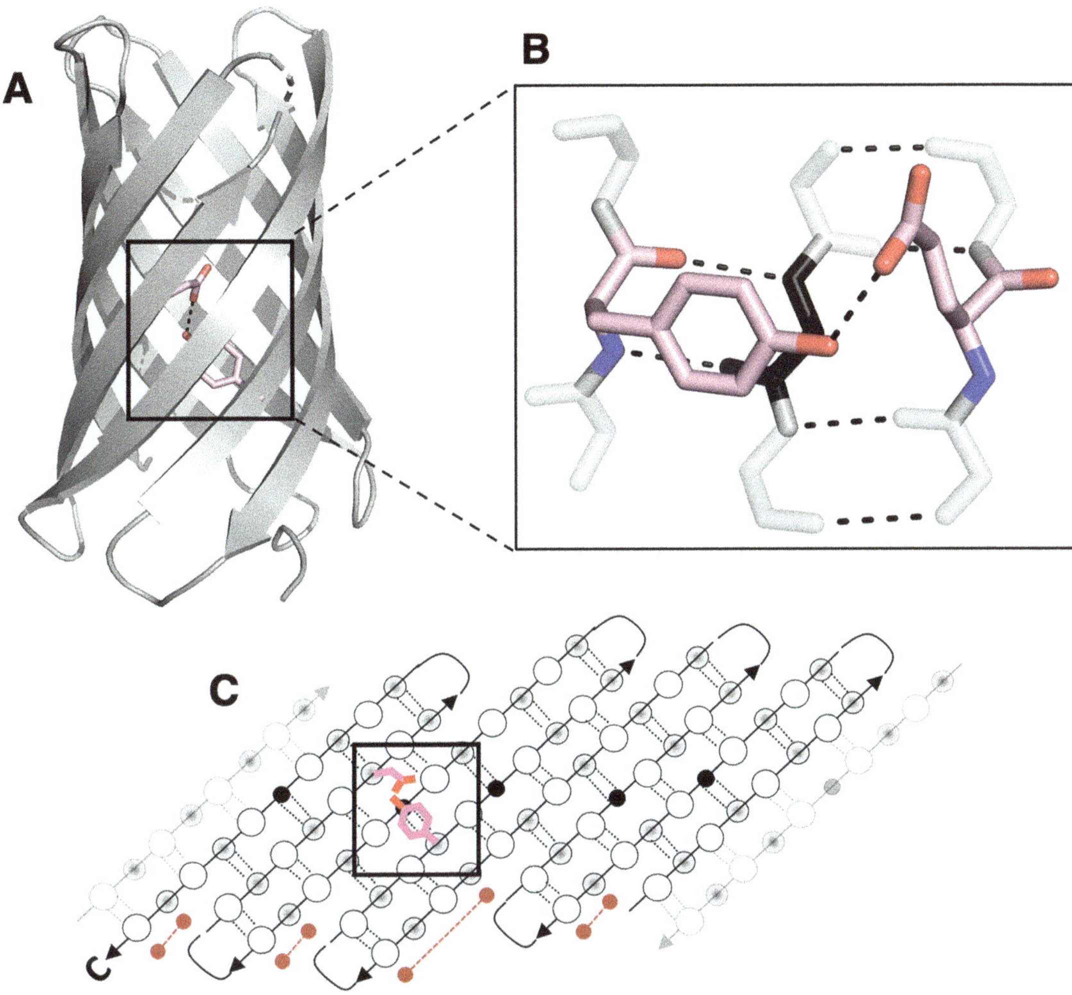

Fig. 4 Molecular definition of the mortise-tenon (YGD/E) folding motif (**a**, **b**) and recommended placement in the β-barrel blueprint (**c**)

sidechain is further stabilized by a hydrogen bond to a negatively charged aspartate or glutamate residue (Fig. 4b). As a general principle, we recommend introducing one motif on the β-hairpin that shifts by more than two residues relative to its neighbor in the blueprint (Fig. 4c).

The motifs are designed in three steps. First, the coarse-grained representations of the backbones are refined with atomistic details (*refine.xml* script). The tyrosines of the mortise-tenon motifs are introduced in the models in the correct rotameric state using a resfile and torsional constraints. Second, Rosetta HBNet is used to exhaustively search possible positions for the D/E residue in the motif that would enable strong hydrogen bond interaction with the tyrosine. Third, a script (get_all_motifs.py) generates all possible combinations of motifs (if more than one mortise-tenon motifs were designed in the TMB).

1. Atomistic refinement of the backbones.

```
#bash script-----------------------------------------

-----------
#Create a new directory to run the refinement
>mkdir new_refinement_directory
#Copy input backbones created previously and make
one new directory per model
>cp path_to_backbones/*.pdb .
>for pdb in *.pdb ; do mkdir ${pdb%.pdb}; done
#Run the refinement refine.xml script for one input
new_backbone.pdb
>cd new_backbone/ && /path_to_Rosetta/main/source/
bin/rosetta_scripts -parser:protocol ../refine.xml
-s ../new_backbone.pdb -nstruct 5 -rama_prepro_s-
teep -beta_nov15
#Run the refinement of all inputs in parallel using
distributed computing
#-----------------------------------------
----------------
```

2. Analyze and select refined backbones.

   The backbone analysis and selection code can be found at the end of the *backbones_analysis.ipynb* Jupyter notebook. The best-scored models are selected based on Rosetta *total_score* (25th percentile of the designs ensemble), torsional scores (*omega*, *rama_prepro*, 50th percentile), and β-sheet hydrogen bonds (*hbond_lr_bb*, 90th percentile).

```
#python script-------------------------------------
-------------
>df = pd.concat([pd.read_csv(f, sep='\s+', head-
er=1) for f in glob.glob('best_200/*_input_*/
score.sc')], ignore_index = True)
#Load the score into one pandas dataframe
>best = df.loc[(df['total_score'] < -370) & (df
['omega'] < 13.5) & (df['rama_prepro'] < -3.6) &
(df['hbond_lr_bb'] < -85)]
#select best-scored backbones based on the score
distributions in the ensemble
>for model in best.description:
>    folder_id = "_".join(model.split("_")[0:4])
>     source = "best_200/" + str(folder_id) + "/" +
model + ".pdb"
>    target = "hbnet/" + model + ".pdb"
>    copyfile(source, target)
```

```
#Copy the selected refined models to a new directory
called "hbnet"
#-----------------------------------------
---------------------
```

3. Run HBNet to design the full YGD/E motifs.

   The hbnet.xml script is modified to properly identify the positions of the tyrosine. The tyrosine positions are listed in the *<RESIDUE_SELECTORS>* section and identified based on their residue indices in the PDB models. In the example below, the tyrosines are placed in positions 10 and 68.

```
#Rosetta script-----------------------------------
--------------
<RESIDUE_SELECTORS>
    <Index name="tyr10" resnums="10"/>
    <Index name="tyr68" resnums="68"/>
</RESIDUE_SELECTORS>
#-----------------------------------------
----------------------
```

   Next, create a new directory for each model and run the hbnet.xml script.

```
#bash script--------------------------------------

-----------
>cd hbnet/
>for pdb in *.pdb ; do mkdir ${pdb%.pdb}; done
>for file in *.pdb ; do echo ${pdb%.pdb} >> task.
list; done
>while read line ; do cd $line/ && /path_to_rosetta/
main/source/bin/rosetta_scripts -s ../$line.pdb
-beta_nov16 -parser:protocol ../hbnet.xml && cd ..
; done < task.list
#-----------------------------------------
-------------------
```

4. Combine individual motifs with the get_all_motif.py script.

   Open the *get_all_motif.py* script and modify it to correctly reference the positions of the tyrosines in the amino acid sequence (*Y_10*, *Y_68* in this example). The script will combine the motifs in position 10 and 68 (designed independently by HBNet) into one model, write the output model into a new directory (e.g., *../design_core/*), and modify an amino acid identity constraints file *../all.resfile* for the subsequent sequence design stage.

```
#bash script---------------------------------------
-----------
>mkdir ../design_core/
>python get_all_motifs.py
#------------------------------------------
--------------------
```

Expected Results 1000–2500 models that will be used as input for the next step (design of TMB sequences). In the new directory *design_core/*, each model pdb is copied into a new subdirectory *model/* together with a new version of the *resfile* and a *cst* file describing the Asp or Gly residues designed with HBNet and their interactions with the tyrosines in the motif.

3.3 Design of TMB Amino Acid Sequences

Combinatorial design of TMB sequences consist in designing the residues lining the solvent-accessible pore with mostly polar amino acids, and the membrane-facing residues with mostly hydrophobic amino acids. The pore-facing YGD/E motifs are used as seeds to design a network of polar interactions, which is reenforced by several cycles of optimization and selection. A major challenge for TMB design is the necessity to apply negative design (incorporating sequence/structure frustration) to delay the folding of the β-strands in water. The goal is to achieve a delicate balance between a stable native state and partially destabilized local secondary structures. If the β-sheet propensity of the sequence is too strong, the polypeptide will aggregate in water before it has a chance to reach a membrane to fold. Such sequences often fail to express completely in *E. coli*, likely because of the toxicity of the aggregated/misfolded water-soluble species [11]. On the other hand, if the native state becomes too destabilized, the polypeptide will not fold as intended in detergent micelles or lipid membranes.

We apply two strategies to reduce the β-strand propensity of the sequence. First, a small number of hydrophobic amino acids are introduced among the polar pore-facing residues and polar amino acids (Ser, Thr) on the lipid-facing surface. The goal is to disrupt the alternation pattern of polar and hydrophobic sidechains that characterizes strong β-sheets. Second, Ala and Gly amino acids are designed with high frequency in the pore and on the surface of the β-barrel to increase the disorder property of the unfolded state.

Instead of using a physical membrane model for design calculations (which is challenging to build for membrane proteins featuring a transmembrane pore), we use a resfile to constrain possible amino acid identities per position. In our pipeline, the pore-facing and membrane-facing residues are designed independently. The *resfile* used to design lipid-exposed surface residues also defines specific aromatic motifs (Tyr, Trp) located at the lipid/water boundaries. An example resfile is provided in Fig. 5. It is recommended to closely follow the example structure, including the N-terminus and alternating loop1 and loop2 consensus sequences. The number and length of β-strands, as well as the positions of the glycine kinks, must be matched to the desired blueprint.

1. Design of the pore-facing residues

   Design calculations typically run in parallel on several input models using distributed computing. The *design_core_all.xml* script requires TMB-specific scoring function weights (*see* **Notes 5** and **6**).

```
#bash script---------------------------------------

-----------
>/path_to_rosetta/mais/source/bin/rosetta_scripts
-s pdb_input.pdb -parser:protocol design_core_all.
xml -nstruct 5
        #-------------------------------------------
--------------------
```

2. Analysis of combinatorial pore-facing residues design

   The analysis code can be found in the *select_best_motif.ipynb* jupyter notebook. The Rosetta scores are loaded into one Pandas Dataframe and visualized as population-wide distributions. Next, the stability of the designed YGD/E motifs is evaluated based on residue-level Rosetta scores. Finally, the best-scored models are selected based on the computed stability and hydrogen bond energy of the Asp or Glu residue in the mortise-tenon motif (50th percentiles), torsional energies (*omega*, *rama_prepro*, 50th percentiles), and β-sheet hydrogen bonds (*hbond_lr_bb*, 90th percentile). A new directory *../design_surface/* is created to copy the selected designs.

```
#python script-------------------------------------
-----------
>best = df.loc[(df['mean_hbond_DE']<-1.8) & (df
['mean_energy_DE']<-10) & (df['omega']<=18) & (df
['rama_prepro']<=-5) & (df['hbond_lr_bb']
<=-127.5)]
```

```
NATAA

start

1 A NOTAA C                     #N-terminus
2 A PIKAA NTEDPGRKQS            #N-terminus
3 A NOTAA C                     #N-terminus
4 A NOTAA C                     #N-terminus
5 A PIKAA G                     #N-terminus
6 A PIKAA STDN                  #N-terminus
7 A PIKAA VLI                   #surface
8 A PIKAA ADEFHKLMNQRSTVW       #core
9 A PIKAA VLA                   #surface
10 A PIKAA G                    #core
11 A PIKAA VLA                  #surface
12 A PIKAA ADEFHKLMNQRSTVW      #core
13 A PIKAA VLA                  #surface
14 A PIKAA ADEFHKLMNQRSTVW      #core
15 A PIKAA Y                    #surface
16 A PIKAA N                    #loop1
17 A PIKAA T                    #loop1
18 A PIKAA D                    #loop1
19 A PIKAA N                    #loop1
20 A PIKAA T                    #loop1
21 A PIKAA VLI                  #surface
22 A PIKAA ADEFHKLMNQRSTVW      #core
23 A PIKAA VLA                  #surface
24 A PIKAA G                    #core
25 A PIKAA VLA                  #surface
26 A PIKAA Y                    #core
27 A PIKAA VLA                  #surface
28 A PIKAA ADEFHKLMNQRSTVW      #core
29 A PIKAA Y                    #surface
30 A PIKAA ADEFHKLMNQRSTVW      #core
31 A PIKAA VLI                  #surface
32 A PIKAA STD                  #loop2
33 A PIKAA P                    #loop2
34 A PIKAA DEHTY                #loop2
35 A PIKAA VLI                  #surface
36 A PIKAA ADEFHKLMNQRSTVW      #core
37 A PIKAA VLA                  #surface
38 A PIKAA ADEFHKLMNQRSTVW      #core
39 A PIKAA VLA                  #surface
40 A PIKAA ADEFHKLMNQRSTVW      #core
41 A PIKAA Y                    #surface
42 A PIKAA ADEFHKLMNQRSTVW      #core
43 A PIKAA W                    #surface
44 A PIKAA N                    #loop1
45 A PIKAA N                    #loop1
46 A PIKAA S                    #loop1
47 A PIKAA S                    #loop1
48 A PIKAA L                    #loop1
```

Fig. 5 Example resfile to design TMB sequences

```
Dihedral N 29 CA 29 CB 29 CG 29 CIRCULARHARMONIC -3.0 0.20
```

Fig. 6 Example of a constraint on an aromatic residue rotamer

```
>for model in best.description:
>     folder_id = "_".join(model.split("_")[0:4]) +
"_" + model.split("_")[4][3:]
>     source = folder_id + "/" + model + ".pdb"
>     target = "../design_surface/" + model + ".pdb"
>     copyfile(source, target)
#The selected models are copied in the new folded
../design_surface/
#------------------------------------------
--------------------
```

3. Design of the membrane-exposed surface residues

   The *design_core_all.xml* script requires TMB-specific scoring function weights (*see* **Notes 5** and **6**) as well as a *resfile* specifying only surface residues and a cst_aromatics file constraining the rotameric state of the aromatic residues designed at the water/lipid interface (in the example below, the residue 29 is an aromatic amino acid (Fig. 6)).

   Running the design_surface.xml script:

```
# b a s h
script-----------------------------------------
-----------
#This is how the folder structure can be prepared
for distributed computing.
>cd design_surface/
>for pdb in *.pdb ; do mkdir ${pdb%.pdb}; done
>for file in *.pdb ; do echo ${pdb%.pdb} >> task.
list; done
#Command to run the design_surface.xml script
Cd new_design && /path_to_rosetta/main/source/bin/
rosetta_scripts -s new_design.pdb -parser:protocol
design_surface.xml -nstruct 10
#-------------------------------------------------
-------------
```

4. Analysis of combinatorial surface-exposed residues design

   The surface design step generates ten models per input; the surface residues are designed while the residues facing the pore are only repacked. The ten models are analyzed to identify inputs with stable enough networks of hydrogen bonds around the YGD/E motifs to remain conserved after repacking. The code in the *analyze_round2_surf.py* script calculates the frequency of hydrogen bonds retention around the motifs and selects inputs with more than 70% retention. For each stable input, the lowest-energy design is further selected.

```
#bash script---------------------------------------

-----------
#First source the pyrosetta environment
>source /path_to_venvs/pyrosetta/bin/activate
>python analyze_round2_surf.py
#-------------------------------------
--------------------
```

Expected Result The design and filtering iterations should result in 500–1000 designed TMB sequences.

3.4 Final Selection of Designs for Experimental Characterization

The final selection consists in verifying the sequence/structure compatibility of the designs using AlphaFold2 [17] structure prediction to select designed sequences with enough residual structure information to encode a β-barrel despite negative design. On the opposite side of the sequence/structure compatibility spectrum, sequences designed without incorporating negative design have high secondary structure propensity and are therefore readily folded by AlphaFold2 although they often fail to express experimentally. To achieve just the right balance between secondary and tertiary structures, the sequences are prescreened against high β-sheet and aggregation propensities using classic bioinformatics tools (RaptorX and Tango).

1. Running RaptorX and Tango on several sequences.

   The code below generates fasta-format amino acid sequences from a PDB structure. The sequences are used as input for RaptorX [14] and Tango [15].

```
#bash script---------------------------------------

-----------
#Obtain amino acid sequences in fasta format from
PDB models
>for file in *.pdb; do python pdb2fasta.py $file A >
${file%.pdb}.fasta ; done
#Run RaptorX prediction
>for f in *.fasta ; do /path_to_RaptorX/oneline_-
command.sh $f 1 0 ; done
#Generate a Tango input file and run Tango
>n=1; for file in *.fasta ; do echo "$n N N 7.0
```

```
298 0.02 $(echo $seq | awk -F: 'NR==2 {print $1}'
${file})" >> tango.list ; echo "${file%.fasta} $n"
>> tango_index.txt ; n=$((n+1)) ; done
>/path_to_tango/tango_x86_64_release -inputfile=-
tango.list
#-----------------------------------------
--------------------
```

2. Analyze predictions.

   The outputs from the previous step are analyzed with the *analyze_sequence_properties.ipynb* Jupyter notebook. The final designs are pre-filtered based on their β-sheet (35–50%), α-helix (=0%), and aggregation (<1400) propensities. The aggregation propensity cut-off is indicative and can be adjusted to select a reasonable number to test with AlphaFold2 (~100).

```
#python script-----------------------------------
-----------
#Collect all physical property descriptors per de-
sign into one Dataframe.
>df.loc[name] = [sequence,ss,identity,hydro,hy-
dro_core,hydro_surf,hydro_diff,F_freq,aro_freq,
beta_sheet,coil,helix,aggregation,result]
#Select the designs based on the physical proper-
ties descriptors
>df_picked = df.loc[(df_12_16["beta_sheet"]<50.0)&
(df_12_16["Aggregation propensity"]<1400)&
(df_12_16["helix"]==0.0)&(df_12_16["beta_sheet"]
>35.0)]
#-----------------------------------------------
------------
```

3. Run AlphaFold2 predictions.

   AlphaFold2 is run through the Colab interface using the AlphaFold2_batch.ipynb notebook (https://colab.research.google.com/github/sokrypton/ColabFold/blob/main/batch/AlphaFold2_batch.ipynb) [16]. The msa_mode option is set to *single_sequence* and the number of recycles increased to 48. The best designs are identified based on the plDDT model accuracy prediction (>0.8) [24].

Expected Result At least 10–15 designs meet the full validation criteria. The exact number depends on blueprint complexity, which increases with the size of the β-barrel and with the difference between the number of β-strands *n* and the shear number *S*. After experimental characterization of the designs, we typically observe a success rate from 10% to up to 50%.

4 Notes

1. The shear number S of a β-barrel can be challenging to calculate based on the structure using methods described in the literature. We propose a method that is robust to large S values and to the presence of structural irregularities in the β-barrel. The structure is first visualized in Pymol (Molecular Graphics System, Schrödinger, LLC) with the *flat_sheet* option set to *False* (Settings > Cartoon > Flat Sheets) and with backbone-backbone hydrogen bonds. Starting from a position on the first β-strand, follow the hydrogen bonds to the residue directly connected to it on β-strand 2, and then move one position up the sequence on that same β-strand. Follow this pattern (one move along the hydrogen bonds, one move along the β-strand) until strand 1 is reached again. Count the number of residues between your start and end positions on β-strand 1. The shear number S is the sum of that number and the number of β-strands n (Fig. 7).
2. TMBs often feature another torsional irregularity: the β-bulge. It consists in the insertion of one β-sheet residue in α-helical conformation ("EA" in the blueprint) into a stretch of regular β-sheet residues ("EB"). The β-bulges that are deemed necessary to assemble designable TMB backbones are present in the final blueprint and placed by the *generate_blueprint.py* script.
3. With practice, the designer will be able to adjust the position of the glycine kinks in the blueprint to define the shape of the β-barrel. As a general principle, the number of sidechains lined

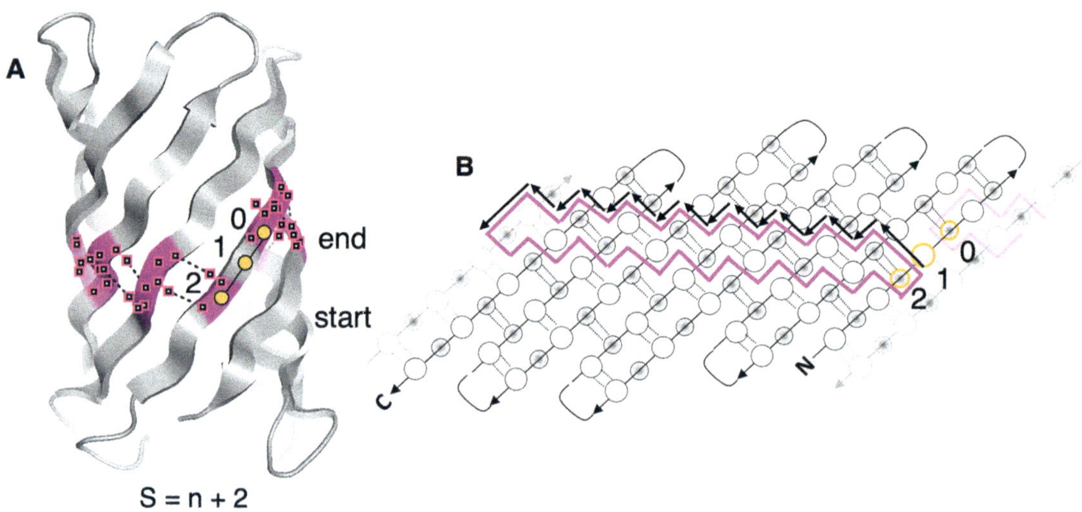

Fig. 7 Guidelines to calculate the shear number S based on a 3D β-barrel structure (**a**) and represented in a 2D blueprint (**b**)

up in a Cβ-strip cannot exceed the number of residues (*a*) spanning ¼ of the total circumference of the β-barrel. If the number of residues in the Cβ-strip exceeds *a* residues, the pattern is interrupted with a glycine kink. The radius *r* of the barrel (used to calculate the circumference) is obtained with Eq. 3. The distance *A* spanned by *a* residues lined up in a Cβ-strip is calculated with Eq. 4.

$$r = \sqrt{(Sd)^2 + (nD)^2} / (2n \sin(\pi/n)) \quad (3)$$

$$A = aD \sin(\theta) \quad (4)$$

4. The efficiency of TMB backbone assembly decreases with the complexity and the size of the blueprint. While sampling 5000 backbone assembly trajectories is enough to design 8-strand TMBs, larger pores of 12–14 strands might require up to 11,000–15,000 trajectories to collect 200–250 backbones passing the selection criteria.
5. The de novo design of TMBs requires a modified version of the Rosetta reference energy function (*ref2015.wts* [25]) to design the right balance of polar/hydrophobic residues on the surface and in the pore, and the relevant frequencies of alanine and glycine residues. We recommend testing the surface and pore design scripts on a small number of backbones and checking the distributions of amino acids and biophysical properties of the output. To this end, the Jupyter notebook *amino_acid_-properties.ipynb* is provided.
6. The *adapt_TMB_weights.py* script is used to adjust the energy function weights if necessary. The script prepares all the input to run short design simulations to sample a range of weights for the *fa_sol*, *fa_elec* scores and for the reference energies of alanine and glycine.

References

1. Van der Verren SE, Van Gerven N, Jonckheere W et al (2020) A dual-constriction biological nanopore resolves homonucleotide sequences with high fidelity. Nat Biotechnol 38:1415–1420
2. Yuen ZW-S, Srivastava A, Daniel R et al (2021) Systematic benchmarking of tools for CpG methylation detection from nanopore sequencing. Nat Commun 12:3438
3. Wang Y, Zhao Y, Bollas A et al (2021) Nanopore sequencing technology, bioinformatics and applications. Nat Biotechnol 39:1348–1365. https://doi.org/10.1038/s41587-021-01108-x
4. Cao C, Krapp LF, Al Ouahabi A et al (2020) Aerolysin nanopores decode digital information stored in tailored macromolecular analytes. Sci Adv 6:eabc2661
5. Huang G, Willems K, Soskine M et al (2017) Electro-osmotic capture and ionic discrimination of peptide and protein biomarkers with FraC nanopores. Nat Commun 8:935
6. Schmid S, Stömmer P, Dietz H et al (2021) Nanopore electro-osmotic trap for the label-

free study of single proteins and their conformations. Nat Nanotechnol 16:1244–1250

7. Galenkamp NS, Maglia G (2022) Single-molecule sampling of dihydrofolate reductase shows kinetic pauses and an endosteric effect linked to catalysis. ACS Catal 12:1228–1236. https://doi.org/10.1021/acscatal.1c04388
8. Robertson JWF, Ghimire ML, Reiner JE (2021) Nanopore sensing: a physical-chemical approach. Biochim Biophys Acta Biomembr 1863:183644
9. Huang P-S, Boyken SE, Baker D (2016) The coming of age of de novo protein design. Nature 537:320–327
10. Dou J, Vorobieva AA, Sheffler W et al (2018) De novo design of a fluorescence-activating β-barrel. Nature 561:485–491
11. Vorobieva AA, White P, Liang B et al (2021) De novo design of transmembrane β barrels. Science 371:eabc8182
12. Berhanu S, Majumder S, Müntener T et al (2023) Sculpting conducting nanopore size and shape through de novo protein design. bioRxiv. https://doi.org/10.1101/2023.12.20.572500
13. Leman JK, Weitzner BD, Lewis SM et al (2020) Macromolecular modeling and design in Rosetta: recent methods and frameworks. Nat Methods 17:665–680
14. Wang S, Li W, Liu S et al (2016) RaptorX-Property: a web server for protein structure property prediction. Nucleic Acids Res 44:W430–W435
15. Fernandez-Escamilla A-M, Rousseau F, Schymkowitz J et al (2004) Prediction of sequence-dependent and mutational effects on the aggregation of peptides and proteins. Nat Biotechnol 22:1302–1306
16. Mirdita M, Schütze K, Moriwaki Y et al (2022) ColabFold: making protein folding accessible to all. Nat Methods 19:679–682
17. Jumper J, Evans R, Pritzel A et al (2021) Highly accurate protein structure prediction with AlphaFold. Nature 596:583–589
18. Murzin AG, Lesk AM, Chothia C (1994) Principles determining the structure of beta-sheet barrels in proteins. I. A theoretical analysis. J Mol Biol 236:1369–1381
19. McLachlan AD (1979) Gene duplications in the structural evolution of chymotrypsin. J Mol Biol 128:49–79
20. Lomize MA, Pogozheva ID, Joo H et al (2012) OPM database and PPM web server: resources for positioning of proteins in membranes. Nucleic Acids Res 40:D370–D376
21. Lin Y-R, Koga N, Tatsumi-Koga R et al (2015) Control over overall shape and size in de novo designed proteins. Proc Natl Acad Sci U S A 112:E5478–E5485
22. Koga N, Tatsumi-Koga R, Liu G et al (2012) Principles for designing ideal protein structures. Nature 491:222–227
23. Michalik M, Orwick-Rydmark M, Habeck M et al (2017) An evolutionarily conserved glycine-tyrosine motif forms a folding core in outer membrane proteins. PLoS One 12:e0182016
24. Hermosilla AM, Berner C, Ovchinnikov S et al (2023) Validation of de novo designed water-soluble and transmembrane proteins by in silico folding and melting. bioRxiv. https://doi.org/10.1101/2023.06.06.543955
25. Park H, Bradley P, Greisen P Jr et al (2016) Simultaneous optimization of biomolecular energy functions on features from small molecules and macromolecules. J Chem Theory Comput 12:6201–6212

Chapter 22

Stress-Based Screening for Compounds That Inhibit β-Barrel Outer Membrane Protein Assembly in Gram-Negative Bacteria

Laurence Cleenewerk, Joen Luirink, and Peter van Ulsen

Abstract

Biogenesis of the outer membrane (OM) of Gram-negative bacteria involves two processes essential for growth, that is, the insertion of β-barrel outer membrane proteins (OMPs) by the Bam complex and the assembly of the LPS-containing outer leaflet of the OM by the LptD/E complex from the Lpt pathway. These processes have only recently gained attention as targets for antimicrobial drugs. Our laboratory has developed a simple screening tool to identify compounds that target processes that disrupt the biogenesis of the cell envelope, among which the activity of the Bam complex. The tool is based on the observation that such a disruption triggers cell envelope stress response systems, such as the σ^E, Rcs, and Cpx responses. In essence, specific stress-responsive promoters are fused to a gene encoding a bright fluorescent protein to serve as a panel of easy-to-monitor stress reporter plasmids. Using these plasmids, compounds triggering these stress systems and, therefore, putatively disrupting the biogenesis of the cell envelope can be identified by the nature and kinetics of the induced stress responses. We describe here the use of the stress reporter plasmids in high-throughput phenotypic screening using multi-well plates.

Key words Antimicrobial drugs, Gram-negative, Outer membrane, Stress responses, Phenotypic screening, β-barrel membrane proteins, Membrane integrity, Cell envelope

1 Introduction

There is an urgent need for novel antimicrobial drugs due to the rise in bacterial infections by multidrug-resistant pathogens. Apart from *Mycobacterium tuberculosis*, the World Health Organization prioritized the Gram-negative species *Acinetobacter baumannii*, *Pseudomonas aeruginosa* as well as the *Enterobacteriaceae*, which include pathogenic *Escherichia coli*, for development of new treatments [1]. Gram-negative bacteria are typified by a cell envelope that is hard to penetrate for many drugs [2]. It consists of a cytoplasmic membrane, a periplasmic space that contains the peptidoglycan layer, and an outer membrane (OM) connected to the

Raffaele Ieva (ed.), *Transmembrane β-Barrel Proteins: Methods and Protocols*, Methods in Molecular Biology, vol. 2778,
https://doi.org/10.1007/978-1-0716-3734-0_22,

peptidoglycan. This basic architecture may be further extended by a polysaccharide capsule. The OM itself consists of a lipid bilayer with phospholipids in the inner leaflet and lipopolysaccharides (LPS) in the outer leaflet in which integral β-barrel membrane proteins (OMPs) are inserted and lipoproteins are anchored. In most Gram-negative species, the OM-based complexes required to assemble LPS and β-barrel OMPs are essential, and their malfunctioning results in an increased permeability of the OM. The integrity of the cell envelope is important for the survival of Gram-negative bacteria in their respective niches. Consequently, multiple stress-response systems exist that monitor this integrity and respond to breaches in membrane permeability and obstructions in the biogenesis of β-barrel proteins and LPS molecules [3–5].

The β-barrel assembly machinery (Bam) complex mediates the folding of β-barrel OMPs and their insertion into the OM [6]. It consists of the essential integral OMP BamA and four OM-associated lipoproteins BamB-E, of which BamD is also essential (*see* Chapter 1). It is noteworthy that impaired expression of the Bam subunits renders the bacteria more permeable to toxic drugs [7]. Drugs targeting the Bam complex are not in use in the clinic as this machinery has only recently been identified as a highly interesting target [8, 9]. Interfering with its function could either directly kill the bacteria or potentiate the access of established antibacterial drugs with intracellular targets. Importantly, the complex includes periplasmic and surface-exposed parts making it relatively easy to access by drugs.

Our laboratory has developed simple screening tools to identify compounds that target essential processes in the biogenesis of the OM, one among them being the activity of the Bam complex [10–12]. The basic approach is based on the observation that interfering in these vital processes triggers one or several of the existing cell envelope stress systems. We constructed a panel of stress reporter plasmids by fusing the promoter of a gene that belongs to the regulon of a specific extracellular stress system to the gene that encodes the bright fluorescent protein NeonGreen (NG) [10] (Table 1). The panel of plasmids allows for monitoring a specific stress response by measuring the fluorescence of treated bacteria in culture in a 96-well plate format, which could be adapted to a 384-well [12] and even a 1536-well format containing cultures of only a few microliters (unpublished). The cell envelope stress systems monitored are the σ^E, Rcs, and Cpx cell envelope stress responses monitoring periplasmic accumulation of OMPs, membrane insertion of OMPs, and accumulation of periplasmic proteins and lipoproteins, respectively [3–5] (Table 1). A plasmid carrying the *groES* promoter reports on the heat-shock response, which is mainly cytoplasmic and therefore serves as a counter-selective control. The nature of the induced responses and their timing, early or

Table 1
Plasmids used for compound screening

Plasmid	Reference	Stress system	Trigger	Promoter coordinates[a]
pUA66-*rpoE*-NG	[10]	σ^E	Cell envelope—periplasmic accumulation of unfolded OMPs	2,710,068–2,710,152 (reversed orientation)
pUA66-*rprA*-NG	[11]	Rcs	Cell envelope—impaired membrane integration of OMPs	1,770,230–1,770,371
pUA66-*cpxP*-NG	[11]	Cpx	Cell envelope—accumulation of periplasmic proteins and lipoproteins	4,105,706–4,105,806
pUA66-*groES*-NG	[10]	Heat shock	Cytoplasmic—accumulation of unfolded proteins	4,370,550–4,370,665
pEH3-Hbp	[13]	n.a.[b]	n.a.	n.a.; *hbp* under control of IPTG-inducible P_{trc}

[a]Coordinates apply to *E. coli* K-12 strain MG1655 (NCBI Reference Sequence NC_000913.3)
[b]*n.a.* not applicable

later in the growth experiment, were found to roughly report on the type of cell envelope process that is targeted by a specific compound, be it the biogenesis of LPS, lipoproteins, peptidoglycan, or β-barrel OMPs [11] (Table 2).

We have applied the panel of stress reporter plasmids to identify three compounds that induce the σ^E and Rcs stress responses and inhibit the biogenesis of β-barrel OMPs. In two initial screening experiments [10, 12], we used the σ^E stress reporter plasmid with the intention to identify compounds that inhibit the secretion of the autotransporter Hemoglobin-binding protease (Hbp). Autotransporters require the Bam complex for their passage across the OM and include an OMP-like β-barrel domain that interacts with this complex [14]. A slight obstruction of this interaction or the translocation step results in the periplasmic accumulation of misfolded and protease-sensitive forms of Hbp, which in turn triggers the σ^E cell envelope stress response [10, 12]. Further analysis indicated that the three identified compounds did affect not only Hbp secretion but also the biogenesis of β-barrel OMPs (Fig. 1). For one compound, direct interference with OMP insertion by the Bam complex could be unequivocally demonstrated by using proteoliposomes with reconstituted Bam complexes [12] (*see also* Chaps. 5 and 6 for similar methods). These results led us to conclude that, apparently, expressing the autotransporter Hbp sensitizes the cell envelope stress systems, which helped the selection of compounds with a generic effect on cell envelope biogenesis, such

Table 2
Signature of stress responses targeting different cellular processes[a]

Cellular process	Stress system[b] Rcs[c] (P_{rprA})	σ^E (P_{rpoE})	Cpx (P_{cpxP})	Heat-shock (P_{groES})	Example compound
OMP biogenesis (Bam)	Later	Induced	None	None	VUF15259, MRL-494
LPS biogenesis (LptD)	Early	Not all compounds	Not all compounds	None	Polymixin B nonapeptide
Lipoprotein biogenesis	Later	None	Induced	None	Globomycin
Peptidoglycan synthesis	Later	None	None	None	Ampicillin
Cytoplasmic protein production	None	None	None	Not all compounds	Chloramphenicol

[a]Based upon [11]
[b]The promoter used for the reporter plasmid is given in brackets
[c]The Rcs system has been shown to respond early, that is, detectable after 30 min, peaking at 2 h, whereas other responses were later, that is, detectable after 90 min and either peaking after 3 h or showing steady increase

as inhibition of the Bam complex. Importantly, when incubated with cells not expressing *hbp*, the compounds also triggered the Rcs and σ^E stress systems, albeit to a lesser extent [11, 12].

The screens used the σ^E reporter plasmid as a first indicator of stress on cells that moderately expressed *hbp* from an IPTG-inducible plasmid [10, 12] (Fig. 1). However, the strategy can be easily adapted, for example by applying another stress reporter plasmid for the primary compound screen, by expression of a different cell envelope protein to skew and sensitize specific stress responses, or even through directly monitoring stress induction without any sensitizing trigger. Here, we will discuss the requirements for experiments using stress reporter plasmids in a 96-well format, but the upscaling to higher throughput formats using 384-well [12] or even 1536-well plates (unpublished) is feasible.

The procedure is summarized in Fig. 2. Cultures of *E. coli* K-12 laboratory strains carrying the pEH3-Hbp and pUA66-*rpoE-NG* plasmids (Table 1) are pre-grown to log phase in M9 minimal medium, and then diluted back to a low cell density and moderately induced for expression of *hbp* using 40 μM IPTG (Subheading 3.1). The cultures are dispensed into a multi-well plate, grown for 5 min in a plate reader, and then compounds are added at a fixed concentration (usually 10 μM from a DMSO stock solution but not exceeding 1% DMSO in the culture) and incubated for 3 h at 37 °C while shaking. Growth is continued, and the optical cell density at 600 nm (OD_{600}) and the fluorescence intensity of the NG protein (in arbitrary units) are determined at a fixed time point,

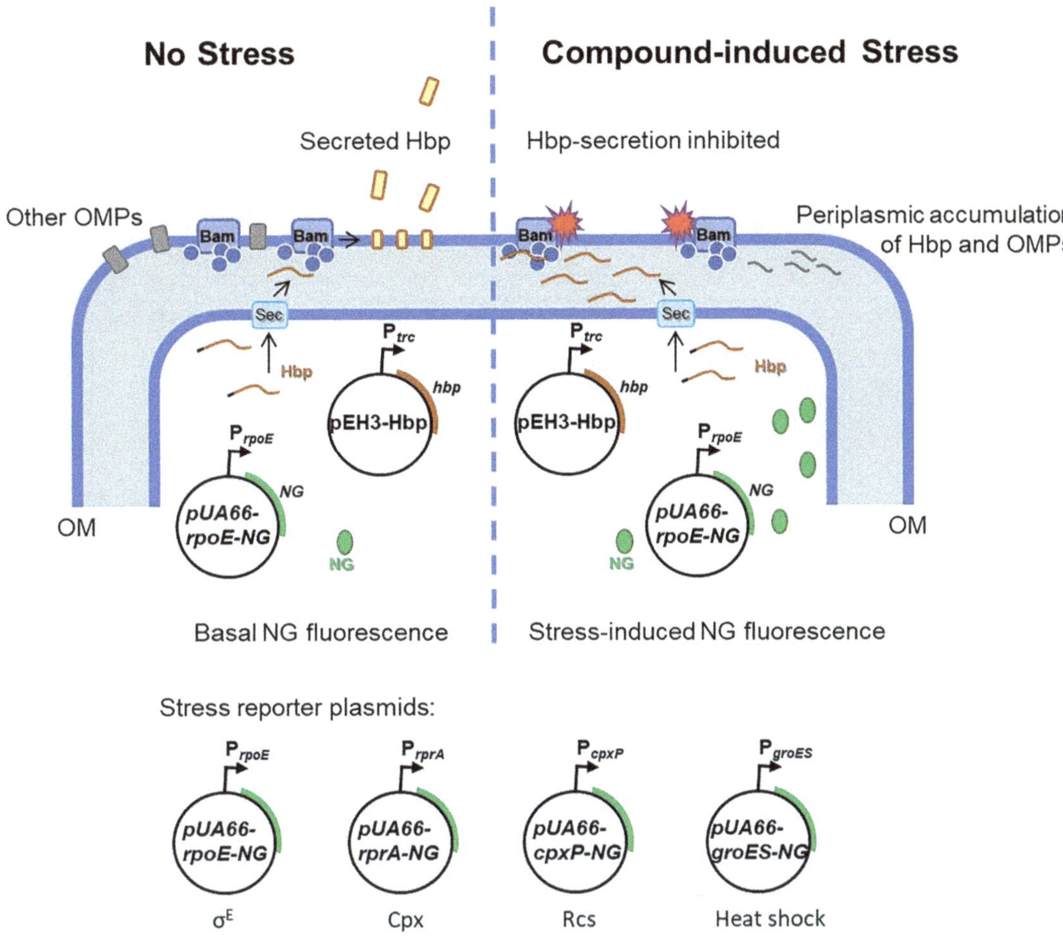

Fig. 1 A two-plasmid system is used to screen compound libraries for cell envelope stress-inducing compounds [10, 12]. *E. coli* Top 10F′ cells carrying plasmids pEH3-Hbp and stress reporter plasmid pUA66-*rpoE-NG* are first induced for *hbp* expression by adding IPTG after which compounds are added. Under normal conditions (left part of the graphic), autotransporter Hbp (pale orange) is secreted across the OM in a step mediated by the Bam complex (dark orange). It results in a large part of Hbp, the so-called passenger, being secreted in the culture medium and an OMP-like β-barrel domain of Hbp being inserted into the OM [14]. Other β-barrel OMPs (gray) also use the Bam complex for OM insertion. A compound that interferes with the translocation of Hbp across the OM (right part of the graphic), or the insertion of OMPs in general, results in the accumulation of precursor molecules of Hbp and OMPs in the periplasm. This accumulation triggers cell envelope stress responses like the σ^E stress response, which is monitored by the increase in fluorescent NG (green)

or can be monitored in real time while growing in a plate reader. This fluorescence is divided by the OD_{600} to compensate for growth effects, and compounds are considered possible hits if they return an increased fluorescence/OD when compared to cells expressing *hbp* incubated with 1% DMSO only (no-compound control). The robustness of the screen can be confirmed by determining the *Z*-factor [15] (*see* **Note 1**). When calculated for the assays described here, the *Z*-factor always exceeded

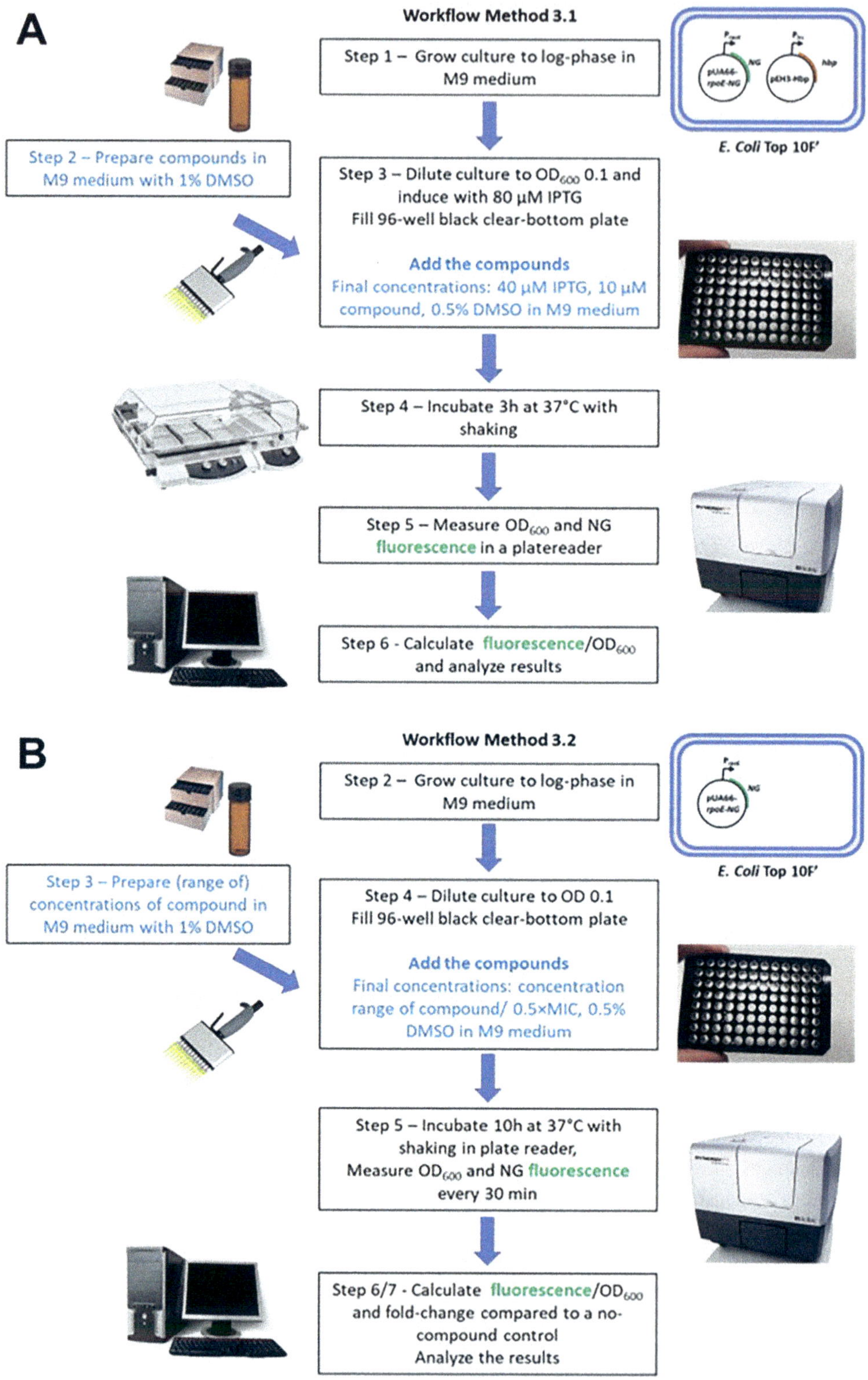

Fig. 2 Graphical representation of the experimental workflow. (**a**) The experimental workflow used to screen compound libraries in Subheading 3.1. (**b**) The experimental workflow used to establish dose-dependent effects of compounds (Subheading 3.2) on the stress systems for which reporter plasmids are available (Table 1)

0.5, which is generally considered the threshold level for trustworthy selection [11]. After this initial screening, subsequent plate reader experiments are conducted using a concentration range of compounds and the other stress reporter plasmids (Subheading 3.2). This analysis is performed without *hbp* expression and serves to determine the dose-response relationship of the compound, filter out false-positive hits, and investigate the effect on other stress systems. Together, this establishes the nature and timing of the stress triggered by a compound. Together, these experiments will yield a list of compounds with interesting features that can be further investigated using more target-focused methods, including those discussed in other chapters of this volume. It is noteworthy that subjecting initial hits to the extra criteria listed above (*see also* **Notes 15** and **16**) led to only a limited number of possible hits that were actually pursued further [10, 12].

2 Materials

Prepare all chemicals using sterile ultrapure water (we use a MilliQ filter system). However, for preparation of the M9 culture media, we use demineralized (demi-) water to include trace elements.

2.1 Plasmids

The plasmids used are listed in Table 1. The vector pUA66, the backbone of the stress reporter plasmids, includes a pSC101 origin of replication rendering it low copy and compatible with other plasmids carrying the colE1 (like pEH3-Hbp used here) or p15A origins (*see* **Note 2**). The coordinates of the promoter regions controlling NG expression are given in Table 1.

2.2 Bacterial Strains

E. coli K-12 strain Top 10F′ (Thermo Fischer Scientific); genotype: F′{*lac*I^{q}Tn10(Tet^R)} *mcr*A Δ(*mrr-hsd*RMS-*mcr*BC) φ80*lac*ZΔM15 Δ*lac*X74 *rec*A1 *ara*D139 Δ(*ara-leu*)7697 *gal*U *gal*K *rps*L *end*A1 *nup*G (*see* **Note 3**).

2.3 M9 Medium

Prepare the following stocks and sterilize these using standard methods. All components can be stored at room temperature, except for the vitamin B1, IPTG, the antibiotic stocks, and the M9 minimal medium mix, which are stored at 4 °C.

1. 10 × M9 salts: 60 g/L Na_2HPO_4, 30 g/L KH_2PO_4, 5 g/L NaCl, 10 g/L NH_4Cl.
2. 20% (w/v) glucose (gently heat the mixture to dissolve glucose, but beware of caramelization)—preferably sterilize using a 0.2 mm filter.
3. $MgSO_4$ stock solution: 100 mM $MgSO_4$.
4. $CaCl_2$ stock solution: 10 mM $CaCl_2$.
5. Cas. Amino Acids stock solution: 10% (w/v) Cas. Amino Acids.

6. 1% (w/v) vitamin B1 (thiamine).
7. Sterile demi-water.
8. 1 mM IPTG in water.
9. 1000× Antibiotic stocks: 30 mg/mL chloramphenicol for selection of pEH3-Hbp containing cells, 50 mg/mL kanamycin for selection of the pUA66 plasmids, and 12.5 mg/mL tetracyclin for selection of the F′ episome in Top 10F′.
10. M9 minimal medium: 1 × M9 Salts, 1 mM $MgSO_4$, 0.1 mM $CaCl_2$, 0.1% Cas. Amino Acids, 0.0001% vitamin B1. Add demi-water to the desired volume, and then add the appropriate antibiotics depending on the strain tested and plasmids present (*see* **Note 4**).

2.4 Compound Stocks

Compounds are normally dissolved to a desired concentration in 100% DMSO. Prior to adding them to the growing cells, dilute them in M9 medium and make sure that the final concentration of DMSO in the wells does not exceed 1%. For screening, a fixed compound concentration is used. We have used 10 μM for a chemical compound library [12] but used a higher concentration of 200 μM when a library of very small compounds (so-called fragments) was tested [10]. These smaller fragments tend to bind their targets with lower affinity and hence require higher concentrations to exert an effect [16].

2.5 Equipment

1. 125 mL Erlenmeyer flasks.
2. 500 mL Erlenmeyer flasks.
3. Black flat- and clear-bottom 96-well plates (*see* **Note 5**).
4. Incubator with rotational shaking, such as the Innova 44 incubator shaker (New Brunswick).
5. Plate incubator such as the Titramax 100 plate incubator (Heidolph).
6. Plate reader, such as the Synergy H1 plate reader or HTX multi-mode reader (Biotek), with filters to enable fluorescence excitation at 485 nm and detection at 535 nm.
7. The Gen5 software (Biotek) supplied with the plate readers (*see* **Note 6**).

3 Methods

3.1 Screening a Compound Library

1. Start an overnight culture of *E. coli* Top10F′ carrying pEH3-Hbp and pUA66-*rpoE-NG* in 5 mL M9 medium in a sterile 25 mL Erlenmeyer flask (*see* **Notes 7–9**). Grow the cultures overnight in an incubator at 37 °C, with 200 rpm shaking. The next day the culture is diluted in 20 mL pre-warmed M9 medium to an OD_{600} of 0.05 in a sterile 100 mL Erlenmeyer flask and grown with shaking to an OD_{600} of 0.3–0.5.

2. For screening of groups/libraries of compounds, dilutions at 20 μM are prepared in M9 medium with 1% DMSO (*see* **Notes 10** and **11**). Pipet 50 μL of these dilutions into the wells of a black clear-bottom 96-well plate (*see* **Note 12**). Also use M9 medium with 1% DMSO as negative control in at least three wells. A positive control serves at least three wells to which 50 μL of 200 μM of compound VUF15259 in M9 with 1% DMSO is added (*see* **Note 13**). Finally, to at least three wells add 100 μL M9 medium with 0.5% DMSO as medium control.
3. The log-phase cells from **step 1** are diluted back to OD_{600} 0.1 using pre-warmed M9 medium, IPTG is added to a concentration of 80 μM, after which the culture is incubated for 5 min in a 37 °C incubator to start expression of *hbp*. Add 50 μL of this culture to the wells (except those of the medium controls). The final concentration now is 40 μM IPTG and 10 μM compound. For the positive control, the final concentration is 100 μM.
4. Incubate the plates for 3 h at 37 °C with 600 rpm orbital shaking on a Titramax 100 plate incubator (*see* **Note 14**). Grow with lid or seal to protect the wells from carryover of cultures.
5. Load the plates into the plate reader and read the OD_{600} and the fluorescence using excitation at 485 nm and emission at 528 nm, using an end-point measurement. *See* **Note 6** for the settings of the reader.
6. Export the data to an Excel file and subtract from each data point the background values for growth and fluorescence, which are the mean of the OD_{600} and fluorescence measurements of the medium controls. Then calibrate the stress response for growth by dividing the fluorescence value by the OD_{600} value and compare this to the no-compound control. Apply appropriate cut-offs to select for positive hits. This may be decided by applying your own criteria (*see* **Notes 15** and **16**). We selected compounds as positive hits when they show an increase of σ^E induction ≥11% compared to the positive control. This percentage is calculated as follows:

$$\%\sigma^E \text{ induction} = ((\sigma^E \text{ value compound} - \text{mean } \sigma^E \text{ value for negative controls})/$$

$$(\text{mean } \sigma^E \text{ value positive.controls} - \text{mean } \sigma^E \text{ for negative controls})) \times 100.$$

7. Selected hits can be re-screened individually, first at 10 μM concentration by repeating **steps 1–6** and by checking for autofluorescence through measuring their fluorescence at 10 μM in M9 medium with 0.5% DMSO with no cells present. A high level of autofluorescence is a counter-selective criterion.

3.2 Dose-Response Kinetics and Orthogonal Stress Responses

1. As a next step, dose-response curves for the selected compounds are tested, as well as the induction of other stress responses. Test a compound for the Rcs and Cpx cell envelope stress responses (*see* Table 2) using Top 10F′, carrying either the pUA66-*rprA*-NG or pUA66-*cpxP-NG* plasmids, to obtain insight into the nature and extent of the cell envelope stress systems induced. To counter-select for compounds that also provoke a cytoplasmic stress response, indicative of a more general effect on the cell metabolism and growth, the compound is tested using Top 10F′ carrying pUA66-*groES-NG*.
2. For this, we use cultures of Top 10F′ carrying only one stress reporter plasmid, that is, either pUA66-*rpoE-NG*, pUA66-*rprA*-NG, pUA66-*cpxP-NG*, or pUA66-*groES-NG*. Plasmid pEH3-Hbp is not present.
3. Start overnight cultures of *E. coli* Top10F′ carrying the stress reporter plasmids in 5 mL M9 medium in a sterile 25 mL Erlenmeyer flask. Grow the cultures in an incubator at 37 °C, with 200 rpm shaking. The next day each culture is diluted in 20 mL pre-warmed M9 medium to an OD_{600} of 0.05 in a sterile 100 mL Erlenmeyer flask and grown with shaking to an OD_{600} of 0.3–0.5.
4. Dilutions of the compounds are prepared in 96-well plates by serial dilution in a 1:1 ratio (*see* **Note 17**). Wells are filled with 60 μL M9 with 0.5% DMSO. Add to the first well of a row or column (depending on the plate layout) 60 μL of the compound in M9 medium at four times the desired start concentration of the range. Then 60 μL are pipetted to the next well in the series, mixed with the medium, and this is repeated for the whole series. Prepare triplicates of the dilution series to obtain technical replicates. Pipet 50 μL of the compound dilutions in the appropriate wells of a black clear-bottom 96-well plate. Add 50 μL of M9 with 1% DMSO to at least three wells for no-compound controls and 100 μL of M9 with 0.5% DMSO to at least three wells for medium controls. Three wells to which 50 μL of 200 μM of compound VUF15259 in M9 with 1% DMSO is added serve as positive controls. The latter control is facultative (*see* **Note 18**).
5. The log-phase cultures from **step 2** are diluted back to OD_{600} 0.1 using pre-warmed M9 medium. Add 50 μL of this culture to the wells of the 96-well plate (except those of the medium controls). The final concentration of the facultative positive control is 100 μM.
6. Plates are incubated in the Biotek plate reader for 10 h at 37 °C, using linear shaking. Every 30 min a measurement of the OD_{600} and the fluorescence using 485 excitation and 535 emission is made. See point 2.7 for the settings of the reader.

7. The data is analyzed by first subtracting the mean values of the medium controls for OD_{600} and fluorescence. Then the mean fluorescence is divided by the mean OD_{600} values of the different concentrations to compensate for the culture growth. This can be plotted over time (*see* **Note 19**). Next, the fold change of the stress induced by the different concentrations over the negative control (no compound added) can be calculated by dividing the Fluorescence/OD_{600} values of the different concentrations by the Fluorescence/OD_{600} value of the no-compound control. This fold change can then be compared to the fold change for the positive control (*see* **Note 20**).
8. These experiments also provide insight into the minimal inhibitory concentration (MIC) of the compound. For this, the relative percentage of inhibition of cell growth after 10 h of incubation is calculated by dividing the mean OD_{600} values of the concentration range by the mean OD_{600} of the negative controls, multiplied by 100. The lowest concentration of compound leading to 10% growth can be considered a MIC value for that specific compound.

4 Notes

1. We have tested the reliability of the stress-induction assays by establishing the *Z*-factor. This factor reports on the dynamic range between the outcome of positive and negative controls and on the data variation [15]. We performed the plate assays according to the protocols and added to the 96-well plates only positive controls (compounds at a concentration of 0.5 × MIC) and negative controls (no-compound added). The fluorescence of the wells was compensated for growth by dividing the outcome by the OD600 measurements. Subsequently, the *Z*-factor was calculated as follows: $Z = 1 - ((3 \times SD_{pos} + 3 \times SD_{neg})/(\mu_{pos} - \mu_{neg}))$, with SD being the standard deviation of the values of the positive and negative controls and μ indicating the mean of the values of the positive and negative control. A score >0.5 for the *Z*-factor indicates an outcome that is reliable and discriminative between positive and negative samples [15].
2. Plasmids can be introduced in a strain using standard methods, for example, electroporation or chemical transformation.
3. We have also introduced the stress reporter plasmids in other *E. coli* K-12 and B strains, either alone or in combination with other plasmids, and they are active in those backgrounds.
4. Use and store prepared M9 medium for about 3 days. If kept longer, salts may precipitate, reducing the quality.

5. We use 96-well black Greiner Cell Culture polystyrene microplates with μClear bottom, including a sterile cover or sterile microtiter seals (e.g., MicroAmp Optical Adhesive Film of Applied Biosystems), but also other suppliers have black plates with clear bottoms in 96-, 384-, and 1536-well formats.
6. The Gen5 software (Biotek) provides the possibility to use standard protocols for the reading of plate assays. For end-point measurements, we use the following protocol: 30 s of 3 mm linear shaking, followed by measuring the OD_{600} (optics set to the bottom), and fluorescence reading by using the 485/535 filter for excitation and emission of the fluorescent NG signal (optics set to the bottom). Indicate that a lid is present and set the gain (for the H1 we use 80, for the HTX 45). The gain differs between plate reader types and may need to be adjusted when the output is too high (overflow), which requires testing. For kinetic readings over time, the protocol is set for 10 h at 37 °C with linear 3 mm shaking. Every 20 min a measurement is taken of the OD_{600} (optics set to the bottom) followed by the fluorescence (optics set to the bottom) using the 485/535 filters, with the gain set at 80. Indicate that a lid is present.
7. *E. coli* Top 10F′ is the strain used in our publications. However, we have introduced the stress reporter plasmids in strains like MC4100 and BL21 DE3, which yielded comparable output. The selection of the F′ episome of Top 10F′ cells with tetracycline is facultative.
8. The M9 culture medium can be supplemented with 1% standard LB (lysogeny broth) to improve the growth of bacteria. Other media could be tried as well. M9 is used as standard culture medium because it is nonfluorescent, whereas broths like LB yield high background levels.
9. The initial screen order presented here is based on our published screens [10, 12]. However, the pUA66 plasmid used for the initial screening can also be adapted by using one of the other stress reporter plasmids (Table 1), either with or without expression of *hbp* from pEH3-Hbp to sensitize the assay.
10. The growing Top 10F′ cells tolerate DMSO concentrations up to 2% without much growth inhibition or stress. Always ensure that the controls without compounds added do include the correct percentage of DMSO.
11. For larger screens, we used libraries of over 300,000 compounds in collaboration with the Centre for Drug Discovery and Design (CD3, KU Leuven, Belgium) [12]. Pipetting of larger compound libraries was facilitated by using the Echo 550 liquid handling robot system (Labcyte). In that case, black 384-well plates with clear bottoms were used in which

30 μL cultures with 40 μM IPTG were added to 180 nL of 1.66 mM compound dilutions, yielding final concentrations of 10 μM of compound.

12. Black multi-well plates are used to prevent carryover of fluorescent signal from neighboring wells.
13. We included compound VUF15259 as positive control in our published screens [10, 12]. However, if a positive control is not available, you could also use the fold change value (*see* Subheading 3.2, **step 6**) and apply a cut-off of how much this fold change should be to warrant further analysis.
14. Incubation on a plate incubator with orbital shaking might result in concentration of cells in the middle of the well. However, the linear shaking step incorporated in the end-point measurement protocol of the plate reader will prevent uneven distribution of cells in the medium. Incubating the plates at 37 °C without shaking can also work though growth kinetics will be different. Always shake linearly prior to measuring the plates, as included in the protocol.
15. Compounds are selected as positive hits in a screen by comparing their effect to that of compound VUF 15259 [10], but these can also be related to the effect of another known Bam-inhibiting compound that is added to the plates as positive control. Examples of such controls are compound 2 [12], MRL-494 [17], or Darobactin [18]. Alternatively, compounds like polymyxin B nonapeptide have been found to induce cell envelope stress and can be used as a control for stress induction [11], although the incubation time of 3 h is less optimal for measuring its response. Alternatively, when a positive control is not available, a cut-off chosen for the fold change with respect to the no-compound control could already signify an interesting compound. In our experiments, positive hits yielded levels of σ^E stress that were about two to three times higher than that of the no-compound control [11, 12]. The fold change observed for σ^E cell envelope stress-inducing compounds, when compared to untreated cells, was generally below five and values found for MRL-494 concurred with those reported by others [17].
16. Exclusion of known drugs is an elimination criterion that can be applied [12]. This ensures a focus on novel compounds, but should, of course, be avoided when re-purposing of known drugs is considered.
17. Alternatively, test a compound at a fixed concentration. We routinely use 0.5 × MIC (as established in Subheading 3.2, **step 6**). Prepare a dilution of a compound to a concentration that is 1 × MIC in M9 medium with 1% DMSO and pipet 50 μL of this dilution to the appropriate well of a black, clear-

bottom 96-well plate. Triplicates are generally tested for each stress system to get technical replicates. A positive control serves at least three wells to which 50 μL of 200 μM Bam-inhibiting compound VUF15259 in M9 with 1% DMSO is added (adding a positive control is facultative; *see* **Note 13**). Finally, to at least three wells add 100 μL M9 medium with 0.5% DMSO as medium control.

18. Having a positive control is facultative, since the fold change of a tested compound compared to nontreated cells can be indicative for a genuine effect. However, the positive control compares such an effect to a known inhibitor of a Bam and can be used to calculate the *Z*-factor (*see* **Note 1**). Alternatively, a compound or condition known to be a trigger for the stress response can also be used (*see* Table 2).
19. The measurements allow for analysis of the fold change over time, providing insight in early and late responses. However, also the induction at a fixed time point can be analyzed, for example, at the 3 h time point.
20. Testing a compound using the stress reporter plasmids works best at concentrations of the compound that lead to mild inhibition of growth but not to a full growth arrest or to cell death. We, therefore, recommend to use concentrations that are below the MIC value. Concentrations at 2 × MIC and higher tend to decrease the observed stress levels again, as the overall metabolism of the cells can be too severely inhibited. Of course, at this stage, concentration ranges can also be tested (apply Subheading 3.2, **step 3**).

Acknowledgments

This work is in part supported by a grant of the Dutch Society for Scientific Research (NWO), Grant no. 19384.

References

1. WHO (2017) Prioritization of pathogens to guide discovery, research and development of new antibiotics for drug-resistant bacterial infections, including tuberculosis. World Health Organization, Geneva. (WHO/EMP/IAU/2017.12)
2. Bos MP, Robert V, Tommassen J (2007) Biogenesis of the Gram-negative bacterial outer membrane. Annu Rev Microbiol 61:191–214. https://doi.org/10.1146/annurev.micro.61.080706.093245. PMID: 17506684
3. Cho SH, Dekoninck K, Collet JF (2023) Envelope-stress sensing mechanism of Rcs and Cpx signaling pathways in Gram-negative bacteria. J Microbiol 61(3):317–329. https://doi.org/10.1007/s12275-023-00030-y. Epub 2023 Mar 9. PMID: 36892778
4. Saha S, Lach SR, Konovalova A (2021) Homeostasis of the Gram-negative cell envelope. Curr Opin Microbiol 61:99–106. https://doi.org/10.1016/j.mib.2021.03.008. Epub 2021 Apr 23. PMID: 33901778; PMCID: PMC8577651
5. Liu X, Xiao J, Wang S, Zhou J, Qin J, Jia Z, Wang Y, Wang Z, Zhang Y, Hao H (2022) Research progress on bacterial membrane

vesicles and antibiotic resistance. Int J Mol Sci 23(19):11553. https://doi.org/10.3390/ijms231911553. PMID: 36232856; PMCID: PMC9569563

6. Konovalova A, Kahne DE, Silhavy TJ (2017) Outer membrane biogenesis. Annu Rev Microbiol 71:539–556. https://doi.org/10.1146/annurev-micro-090816-093754. PMID: 28886680; PMCID: PMC5778897
7. Wu T, Malinverni J, Ruiz N, Kim S, Silhavy TJ, Kahne D (2005) Identification of a multicomponent complex required for outer membrane biogenesis in Escherichia coli. Cell 121(2): 235–245. https://doi.org/10.1016/j.cell.2005.02.015. PMID: 15851030
8. Overly Cottom C, Stephenson R, Wilson L, Noinaj N (2023) Targeting BAM for novel therapeutics against pathogenic Gram-negative bacteria. Antibiotics (Basel) 12(4):679. https://doi.org/10.3390/antibiotics12040679. PMID: 37107041; PMCID: PMC10135246
9. Steenhuis M, van Ulsen P, Martin NI, Luirink J (2021) A ban on BAM: an update on inhibitors of the β-barrel assembly machinery. FEMS Microbiol Lett 368(11):fnab059. https://doi.org/10.1093/femsle/fnab059. PMID: 34048543
10. Steenhuis M, Abdallah AM, de Munnik SM, Kuhne S, Sterk GJ, van den Berg van Saparoea B, Westerhausen S, Wagner S, van der Wel NN, Wijtmans M, van Ulsen P, Jong WSP, Luirink J (2019) Inhibition of autotransporter biogenesis by small molecules. Mol Microbiol 112(1):81–98. https://doi.org/10.1111/mmi.14255. Epub 2019 May 3. PMID: 30983025; PMCID: PMC6850105
11. Steenhuis M, Corona F, Ten Hagen-Jongman CM, Vollmer W, Lambin D, Selhorst P, Klaassen H, Versele M, Chaltin P, Luirink J (2021) Combining cell envelope stress reporter assays in a screening approach to identify BAM complex inhibitors. ACS Infect Dis 7(8):2250–2263. https://doi.org/10.1021/acsinfecdis.0c00728. Epub 2021 Jun 14. PMID: 34125508; PMCID: PMC8369490
12. Steenhuis M, Ten Hagen-Jongman CM, van Ulsen P, Luirink J (2020) Stress-based high-throughput screening assays to identify inhibitors of cell envelope biogenesis. Antibiotics (Basel) 9(11):808. https://doi.org/10.3390/antibiotics9110808. PMID: 33202774; PMCID: PMC7698014
13. Jong WS, ten Hagen-Jongman CM, den Blaauwen T, Slotboom DJ, Tame JR, Wickström D, de Gier JW, Otto BR, Luirink J (2007) Limited tolerance towards folded elements during secretion of the autotransporter Hbp. Mol Microbiol 63(5):1524–1536. https://doi.org/10.1111/j.1365-2958.2007.05605.x. PMID: 17302825
14. van Ulsen P, Zinner KM, Jong WSP, Luirink J (2018) On display: autotransporter secretion and application. FEMS Microbiol Lett 365(18):fny165. https://doi.org/10.1093/femsle/fny165. PMID: 30085010
15. Zhang JH, Chung TD, Oldenburg KR (1999) A simple statistical parameter for use in evaluation and validation of high throughput screening assays. J Biomol Screen 4(2):67–73. https://doi.org/10.1177/108705719900400206. PMID: 10838414
16. Joseph-McCarthy D, Campbell AJ, Kern G, Moustakas D (2014) Fragment-based lead discovery and design. J Chem Inf Model 54(3): 693–704. https://doi.org/10.1021/ci400731w. Epub 2014 Feb 19. PMID: 24490951
17. Hart EM, Mitchell AM, Konovalova A, Grabowicz M, Sheng J, Han X, Rodriguez-Rivera FP, Schwaid AG, Malinverni JC, Balibar CJ, Bodea S, Si Q, Wang H, Homsher MF, Painter RE, Ogawa AK, Sutterlin H, Roemer T, Black TA, Rothman DM, Walker SS, Silhavy TJ (2019) A small-molecule inhibitor of BamA impervious to efflux and the outer membrane permeability barrier. Proc Natl Acad Sci U S A 116(43):21748–21757. https://doi.org/10.1073/pnas.1912345116. Epub 2019 Oct 7. PMID: 31591200; PMCID: PMC6815139
18. Imai Y, Meyer KJ, Iinishi A, Favre-Godal Q, Green R, Manuse S, Caboni M, Mori M, Niles S, Ghiglieri M, Honrao C, Ma X, Guo JJ, Makriyannis A, Linares-Otoya L, Böhringer N, Wuisan ZG, Kaur H, Wu R, Mateus A, Typas A, Savitski MM, Espinoza JL, O'Rourke A, Nelson KE, Hiller S, Noinaj N, Schäberle TF, D'Onofrio A, Lewis K (2019) A new antibiotic selectively kills Gram-negative pathogens. Nature 576(7787): 459–464. https://doi.org/10.1038/s41586-019-1791-1. Epub 2019 Nov 20. Erratum in: Nature. 2020;580(7802):E3. PMID: 31747680; PMCID: PMC7188312

Index

Raffaele Ieva (ed.), *Transmembrane β-Barrel Proteins: Methods and Protocols*, Methods in Molecular Biology, vol. 2778, https://doi.org/10.1007/978-1-0716-3734-0,

O

P

R

S

GPSR Compliance
The European Union's (EU) General Product Safety Regulation (GPSR) is a set of rules that requires consumer products to be safe and our obligations to ensure this.

If you have any concerns about our products, you can contact us on

ProductSafety@springernature.com

In case Publisher is established outside the EU, the EU authorized representative is:

Springer Nature Customer Service Center GmbH
Europaplatz 3
69115 Heidelberg, Germany

www.ingramcontent.com/pod-product-compliance
Ingram Content Group UK Ltd.
Pitfield, Milton Keynes, MK11 3LW, UK
UKHW052001051026
15624UKWH00003B/507
* 9 7 8 1 0 7 1 6 3 7 3 3 3 *